DEFORMATION CHARACTERISTICS OF GEOMATERIALS

RECENT INVESTIGATIONS AND PROSPECTS

Deformation Characteristics of Geomaterials

Recent Investigations and Prospects

Edited by

H. Di Benedetto, T. Doanh, H. Geoffroy & C. Sauzéat
Département Génie Civil et Bâtiment (DGCB, CNRS), Ecole Nationale des Travaux Publics de l'Etat, Vaulx-en-Velin, France

A.A. BALKEMA PUBLISHERS   LEIDEN / LONDON / NEW YORK / PHILADELPHIA / SINGAPORE

Published by: A.A. Balkema Publishers, a member of Taylor & Francis Group plc
www.balkema.nl and www.tandf.co.uk

ISBN (set): 04 1536 701 8
ISBN Book: 04 1537 000 0
ISBN CD-ROM: 04 1537 001 9

Printed in Great Britain

Deformation Characteristics of Geomaterials – Di Benedetto et al (eds)
© 2005 Taylor & Francis Group, London, ISBN 04 1536 701 8

Table of Contents

Foreword

Solutions for soil engineering and soil-structure interaction problems require the use of realistic and pertinent experimental and modelling tools. Geomaterials show very complex behaviour, which may change with the amplitude of the loading domain. Generally, the behaviour is rather linear for very small strain amplitude while strong non-linearities and irreversibilities appear when loading increases. Failure occurs as a localised or diffuse pattern depending on materials and boundary conditions. Anisotropy and time effects may have a very high influence in some practical cases. Considering this complex behaviour, research and developments aiming to improve modelling of soils and soft rocks over a wide range of loading, from very small strains up to beyond failure is still a big challenge with important practical applications. These developments include interpretation of laboratory, in-situ and field observations.

This book on "Deformation Characteristics of Geomaterials" is published in conjunction with the International Symposium, ISLyon 03, held in Lyon, France, in September 2003. It presents some extensive developments and new trends proposed by the Keynote and Invited Lecturers of this Symposium, who are experts in the covered field, which includes:

– experimental investigations into deformation properties – from very small strains to beyond failure;
– laboratory, in-situ and field observation interpretations;
– behaviour characterisation and modelling;
– case histories.

The eight contributions place emphasis on exploring recent investigations into anisotropy and non-linearity, the effects of stress-strain-time history, ageing and time effects, yielding, failure and flow, cyclic and dynamic behaviour. In addition it is aimed to apply advanced geotechnical testing to real engineering problems, and to ways of synthesising information from a range of different sources while engaging in practical site characterisation studies.

In addition to this book, the proceedings of "ISLyon 03", delivered to each participant, include classical conference papers in the proceedings volume I [1] and a special issue of "Soils and Foundations" [2]. I invite you to consult these 2 documents, which include valuable and up to date works on the "Deformation Characteristics of Geomaterials" presented respectively in 200 and 11 papers.

I sincerely hope that these proceedings and this current book, will be considered as an excellent and privileged source of information, data and ideas for researchers and engineers involved in this complex and fascinating topic.

A CD accompanies this book. In addition to the text of the 8 contributions, pictures from working and special events of ISLyon03 are included. The organisers do hope these pictures will correspond to a pleasant souvenir from the Symposium and the Lyon area.

[1] "Deformation Characteristics of Geomaterials", Proceedings of the international Symposium on Deformation Characteristics of Geomaterials, IS LYON 2003, Eds H. Di Benedetto, T. Doanh, H. Geoffroy & C. Sauzéat, A.A. Balkema Publishers, Leiden, 1425 pp., 2003.
[2] "Deformation Characteristics of Geomaterials", Soils & Foundations, Special Issue, Vol. 43, n°4, 243 pp., 2003.

Hervé Di Benedetto

Testing and apparatus
Essais et appareils

Recent developments in deformation and strength testing of geomaterials

S. Shibuya
Graduate School of Engineering, Hokkaido University, Sapporo, Japan

J. Koseki
Institute of Industrial Science, University of Tokyo, Tokyo, Japan

T. Kawaguchi
Hakodate National College of Technology, Hakodate, Japan

ABSTRACT: Recent developments in equipments and techniques regarding deformation and strength testing of geomaterials in the laboratory are reviewed. The scope of strain measurement discussed in this keynote paper is wide, ranging from the elastic behaviour at very small strains to the generation of shear bands at large strains beyond failure. Some problems overlooked in characterising engineering properties of geomaterials in the laboratory are cited, and their countermeasures are discussed. Practical application of the laboratory test results is discussed with a limited number of case histories.

1 INTRODUCTION

In common with preceding symposia, i.e., IS-Hokkaido (Shibuya et al., 1994) and IS-Torino (Jamiolkowski et al., 1999), the deformation behaviour of geomaterials at relatively small strains was a centralized subject. Little attention was paid to the failure and post-failure behaviour. It was an unintentional consequence that reflected a worldwide trend in geotechnical research in the '90s. In cope with a warning to TC29 not to become 'exclusive club' given by Professor Mike Jamiolkowski at the occasion of Géotechnique Symposium in 1997, the term "pre-failure" was removed from the title of this Symposium (IS-Lyon'03). Reflecting this new policy of TC29, the contributed papers to this symposium deal with not only the pre-failure but also the post-failure behaviour of geomaterials.

Figure 1 shows a typical of stress-strain curve of soil element when subjected to shearing in the laboratory. Pulse/vibration tests to obtain elastic or quasi-elastic

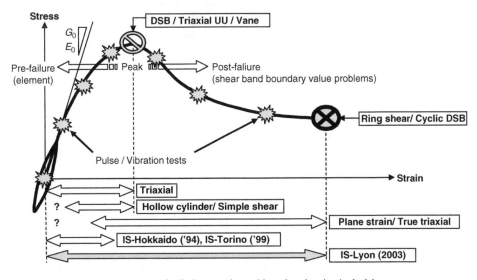

Figure 1. A typical of stress-strain curve of soil element when subjected to shearing in the laboratory.

stiffness, G_0 or E_0, at very small strains can technically be performed at any stages. The interpretation of the result as the element may be appropriate over the pre-failure regime, but it is not so after the peak where strain localization (i.e., the development of shear bands) is usually seen.

In the current world-class standard of geotechnical laboratories, the deformation behaviour over the whole pre-failure regime, including the elastic stiffness at very small strains can successfully be observed in tri-axial test. Conversely, the scope of strain measurement seems inferior in HC/simple shear tests due to stress/strain non-uniformities in the specimens, but these tests enable us to examine more general behaviour such as anisotropy and the effects of principal stress rotation. Plane strain/true triaxial tests both using a rectangular specimen provide an opportunity to examine the behaviour of post-peak strain localization by using, for example, photo-image techniques.

This paper consists of two parts. In part I, recent developments in measurement/control and data acquisition systems are reviewed. The scope of strain measurement discussed is wide, ranging from the elastic behaviour at very small strains to the generation of shear bands at large strains beyond failure. In Part II, practical implications of the laboratory test results are cited with a limited number of case histories.

2 PART I: RECENT DEVELOPMENTS IN MEASUREMENT, CONTROL AND DATA ACQUISITION

2.1 Servo-motor

Recently, the use of servo-motor has greatly enhanced the capability of a variety of equipments in geotechnical engineering laboratory. Figures 2 and 3 show two types of servo-motor equipped for control of axial loading piston in triaxial apparatus. The AC servo-motor (see Fig.2) with a feature of zero-backlash is extensively in use for soil testing at University of Tokyo (Santucci de Magistris et al., 1999). As it can be seen in Fig.2, the loading system consists of a AC servo-motor with two gears. In this system, the motor always drives in one direction. Simultaneously, the upper gear is rotating in one direction, whereas the lower gear is rotating in the opposite direction. The movement of the loading piston can be switched from downwards to upwards without any backlash by using an electric clutch.

The direct-drive motor coupled with nearly zero-backlash reduction unit (see Fig.3) has been developed at Hokkaido University (Shibuya and Mitachi, 1997, Kawaguchi et al., 2002, Kawaguchi, 2002). In this system, 'nearly-zero' backlash on reversal of loading direction can be achieved with the combination of the servo-motor, reduction unit and ballspline screw.

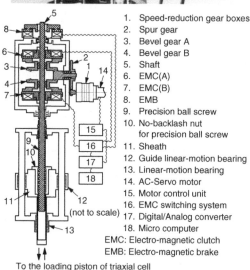

1.	Speed-reduction gear boxes
2.	Spur gear
3.	Bevel gear A
4.	Bevel gear B
5.	Shaft
6.	EMC(A)
7.	EMC(B)
8.	EMB
9.	Precision ball screw
10.	No-backlash nut for precision ball screw
11.	Sheath
12.	Guide linear-motion bearing
13.	Linear-motion bearing
14.	AC-Servo motor
15.	Motor control unit
16.	EMC switching system
17.	Digital/Analog converter
18.	Micro computer

EMC: Electro-magnetic clutch
EMB: Electro-magnetic brake

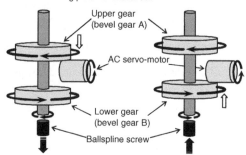

Figure 2. Axial loading system using AC servo-motor (Santucci de Magistris et al., 1999).

The control of axial loading can be fully automated by using a personal computer. Furthermore, the loading system shows a distinct capability of achieving a minimum control for the axial displacement of 0.00015 micrometer spanning over several orders of the rate of axial straining. Note that the resolution of axial displacement is equivalent to 1.5×10^{-9} of strain for a sample with 10 cm high. On reversal of loading direction, the system shows virtually zero backlash, the characteristic of which is vitally important to observe the hysteretic damping ratio of geomaterials under

4

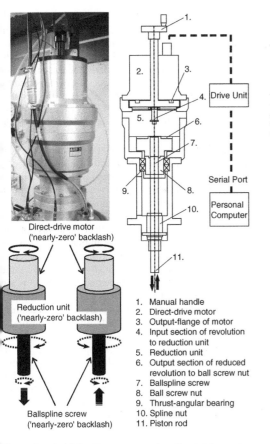

Reduction unit
('nearly-zero' backlash)

Ballspline screw
('nearly-zero' backlash)

Direct-drive motor
('nearly-zero' backlash)

1. Manual handle
2. Direct-drive motor
3. Output-flange of motor
4. Input section of revolution to reduction unit
5. Reduction unit
6. Output section of reduced revolution to ball screw nut
7. Ballspline screw
8. Ball screw nut
9. Thrust-angular bearing
10. Spline nut
11. Piston rod

Figure 3. Axial loading system using direct-drive motor (Shibuya and Mitachi, 1997, Kawaguchi, 2002).

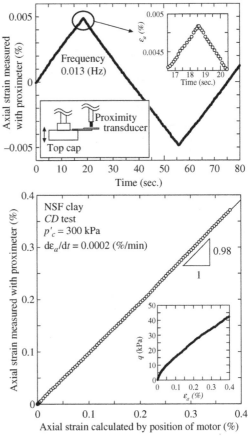

Figure 4. Axial strain from the servo-motor plotted against the comparable measurement using a proximity transducer mounted at the top cap (Kawaguchi et al., 2002).

strictly controlled strain or stress with time (see Tatsuoka and Kohata, 1994). These characteristics when coupled with adequate instrumentation for the axial strain measurement enables examination of geomaterial response under strictly controlled rate of axial straining. The scope of tests that can be carried out in MFT apparatus is wide, including K_0–consolidation, cyclic and monotonic shear with constant as well as altering rate of axial straining.

The direct drive motor has a function to give digital information of the current positioning. Figure 4 shows the axial strain from the servo-motor plotted against the comparable measurement using a proximity transducer mounted at the top cap (Kawaguchi et al., 2002a). As seen in this figure, nearly one-to-one relation is observed between these two measurements. Therefore, an external gage for the axial deformation measurement is no longer needed when this system is employed. Also, it can be seen that the time lag on reversal of the loading direction is virtually nothing.

This feature of nearly zero-backlash enables us to precisely measure, for example, the hysteretic damping of geomaterials showing time-dependent stress-strain characteristics. Figure 5 shows the stress-strain curve of a clay sample subjected to complicated strain path (Kawaguchi, 2002). It is successfully demonstrated that the elastic stiffness can be observed at any stage of shear by using this loading system.

The servo motors with distinctive features of high-precision of displacement control and zero-backlash may be employed for a variety of purposes. Figure 6 shows a large-scale true triaxial apparatus in which the lateral load or lateral displacement, hence σ_y or ε_y can precisely be controlled by using the servo motor system (AnhDan and Koseki, 2003). When the ball-screw driven by the AC servo motor moves up and down, the loading wedges also move vertically, which in turn triggers horizontal movement of the lateral loading plate. Figure 7 shows the results of test on Toyoura

5

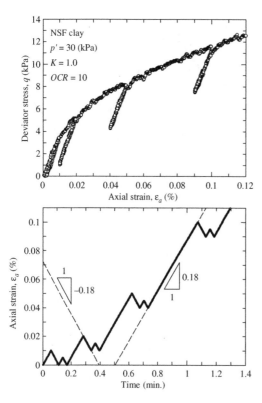

Figure 5. Stress-strain curve of a clay sample subjected to complicated strain path (Kawaguchi, 2002).

sand in the true triaxial apparatus (AnhDan, 2001). It is demonstrated that the elastic Young's modulus in the vertical (z) direction was measured by changing σ_z while the stress in horizontal (y) direction was maintained at a constant value. Similarly, the elastic Young's modulus in y-direction was successfully measured by following similar techniques. It seems this system enables us to measure the small-strain stiffness anisotropy from the direct measurement of stress-strain relationship.

The direct-drive motor is employed in direct shear box (DSB) apparatus, in which the horizontal displacement of the lower box against the fixed upper box can be controlled cyclically without any time lag on reversal of the loading direction. Figure 8 shows the results from a cyclic test (Mitachi et al., 2003). In this test on clay, the horizontal displacement was applied in a cyclic manner with a single amplitude of displacement of 6 mm. As seen in this figure, the shear stress on the horizontal plane gradually decreased as the cumulative displacement increased in value to reach the residual state. The envelopes associated with fully softened state and the residual state were successfully obtained in a series of tests with different consolidation pressures.

Figure 9 shows another example for the use of servo motor. In this torsional shear apparatus, a couple of direct-drive motors are employed for generating the axial load and the torque (Yamashita and Suzuki, 1999a). Figure 10 shows the results of undrained cyclic test performed in this set of apparatus. When the loose sample of Toyoura sand was subjected to cyclic application of shear strain, the ratio of shear stress to

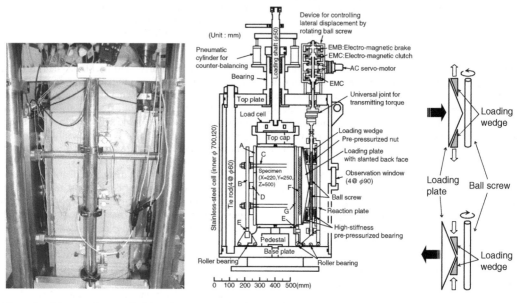

Figure 6. A large-scale true triaxial testing apparatus (AnhDan and Koseki, 2003).

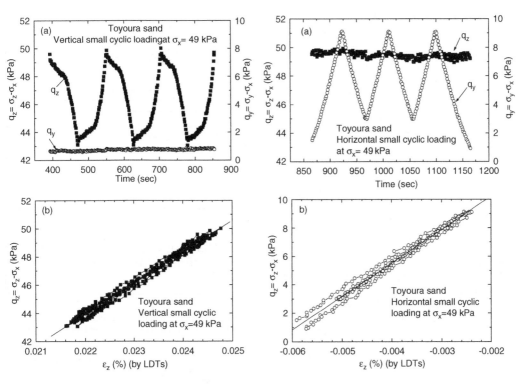

Figure 7. Typical results by applying small amplitude cyclic loading in vertical and horizontal directions (AnhDan, 2001).

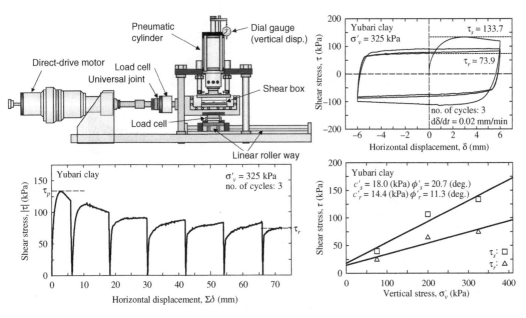

Figure 8. Direct shear box (DSB) apparatus with results from a cyclic test (Mitachi et al., 2003).

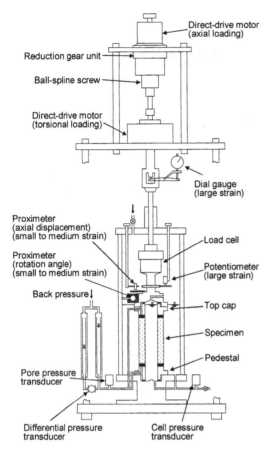

Figure 9. Torsional shear apparatus equipped with a couple of direct-drive motors for generating the axial load and the torque (Yamashita and Suzuki, 1999a).

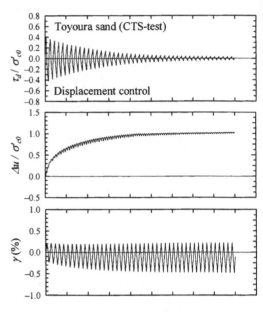

Figure 10. Results of undrained cyclic test performed in the torsional shear apparatus (Yamashita and Suzuki, 1999b).

the isotropic consolidation stress gradually decreased involved with the steady accumulation of excess pore pressure. It seems the value of this system in simulating in-situ shear strain history regarding sand liquefaction.

2.2 Local strain measurement

Figure 11 shows a triaxial specimen whose axial deformation was measured by using a couple of proximity transducers. The metal targets are pinned into the sample so that potential error due to slipping between the target and the rubber membrane and also between the rubber membrane and the soil specimen may be minimized (refer to Lo Presti et al., 2000). We have found that this type of slipping may take place in tests on soft clay and also on stiff geomaterials like soft rocks.

Local deformation transducer (LDT) developed by Goto et al. (1991) is now widely used in geotechnical laboratories. Figure 12 shows a hollow cylinder (HC) specimen on the surface of which three LDTs are

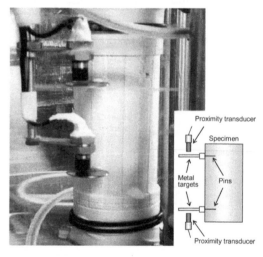

Figure 11. A triaxial specimen whose axial deformation was measured by using a couple of proximity transducers.

mounted to form a triangle in shape. In addition, a horizontal LDT is mounted on the inner wall. In this configuration, four components of strains; i.e., vertical strain (ε_z), circumferential strain (ε_θ), radial strain (ε_r) and shear strain ($\gamma_{z\theta}$) can be measured simultaneously (HongNam and Koseki, 2003). Figure 13 shows the results of such measurements (HongNam, 2004). The shear modulus $G_{z\theta}$ was measured by cyclically

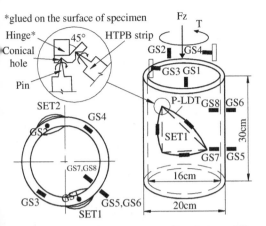

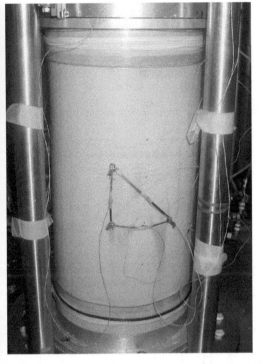

Figure 12. Pin-type triangular LDTs employed for hollow cylinder specimen (HongNam and Koseki, 2003, HongNam, 2004).

applying the shear stress on the horizontal plane. On the other hand, Young's modulus E_z, together with the Poisson's ratio, $\nu_{z\theta}$ was measured through observation of the response of vertical and circumferential strains when σ_z was cyclically changed.

Figure 14 shows a true triaxial apparatus in which 'true' plane strain conditions were achieved by maintaining the normal strain ε_y, when locally measured by LDTs, zero throughout the test (Maqbool and Koseki,

2003). This methodology is termed here as active control in contrast to the conventional passive control, in which the plane strain conditions may not be satisfied owing to the effects of bedding error between the rigid platen and the membrane. In the test performed on coarse-grained gravel, there observed significant difference in the stress-strain behaviour. In the conventional test, the ε_y observed was as much as 1.5%, resulting in the stress-strain response similar to triaxial compression at an early stage of testing. Conversely, the stress-strain relationship in the active test exhibited stiffer and stronger response as compared to the conventional test. As well demonstrated in Fig.14, we should be careful enough about the fact that 'true plane strain' conditions for the small-strain behaviour in particular are difficult to be achieved when a rigid boundary with lubrication is employed. The use of local strain measurement coupled with a deformation control system may be needed for performing this kind of test.

Local strain measurement shows the value in observing strain localization. Strain localization is of interest in two cases; i.e., to give insights into non-uniform strain (or stress) distribution of granular materials and natural cohesive soils and into the development of shear band(s) at large strains. Table 1 shows innovative techniques developed for observing localized or discrete deformation of soil specimen in the laboratory.

Figure 15 shows a high-rigidity plane strain apparatus for testing stiff geomaterials (Salas-Monge et al., 2003). The confining plate is made of well-polished 3 cm-thick transparent Plexiglas. As can be seen, a series of dots are marked on the membrane mounted on the plane of zero strains. Figure 16 shows shear-strain contours of cement-treated sandy soil at four stages of cyclic shear in compression. As it can be seen, the development of strain localization of this stiff geomaterial may be readily visualized in a digital form. The pictures analyzed were taken by a digital camera with a resolution of about 3 megapixels that was available in a market.

2.3 Elastic wave velocity measurement

Elasticity refers to materials' property involved with no energy dissipation for any closed stress cycle. It is irrelevant whether the stress-strain relationship is linear or non-linear. In addition, the elastic properties ought to be time-independent, for example, the stiffness must not be influenced by the rate of stressing or straining.

Quasi-elastic geomaterial properties may be defined for the behaviour at small strains because of the above-mentioned strain (or stress) history with time. For example, the stress-strain response at extremely small strains is quasi-elastic in respect that it shows a hysteretic loop implying a small amount of energy loss, but the stiffness is unaffected by the rate of shearing over

9

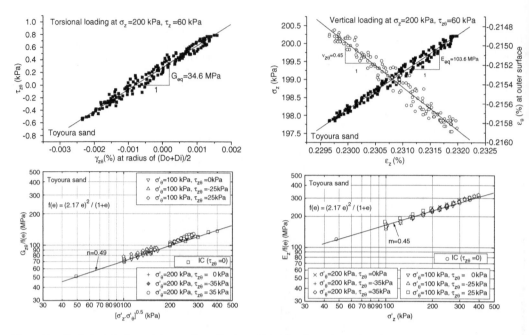

Figure 13. Typical results by applying small amplitude cyclic loading in vertical and torsional directions (HongNam, 2004).

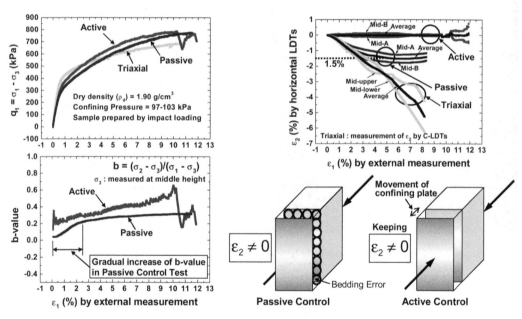

Figure 14. Plane strain compression tests on Chiba gravel under passive and active controls (Maqbool and Koseki, 2003).

normal range encountered in geotechnical engineering (Tatsuoka and Shibuya, 1992, Shibuya et al., 1992).

Seismic tests involved with the measurement of shear body wave velocity, V_s, manifests the quasi-elastic shear modulus, G, the estimate of which is based on;

$$G = \rho_t \cdot V_s^2 \qquad (1)$$

where ρ_t denotes bulk density of soil.

The bender element (BE) that is designated for measuring G, is handy, inexpensive and durable. The

10

Table 1. Innovative techniques for localized deformation.

Image analysis (e.g., PIV)	Frost & Yang, 2003
	Bowman & Soga, 2003
	Kobayashi & Fukagawa, 2003
Laser aided tomography	Matsushima et al., 2003
(LAT)	Konagai et al., 1992
CCD sensor	Kishi & Tani, 2003
Digital microscope	Towhata & Lin, 2003
X-ray CT method	Otani et al., 2000

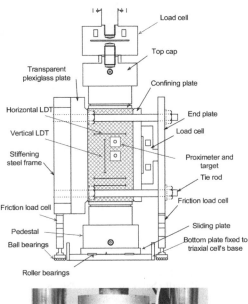

Figure 15. High-rigidity plane strain apparatus for testing stiff geomaterials (Salas-Monge et al., 2003).

instrument suits well for measuring V_s in the laboratory (Shirley and Hampton, 1977, Dyvik and Madshus, 1985). In fact, the BE has been plugged into a few testing devices such as consolidometer (Jamiolkowski

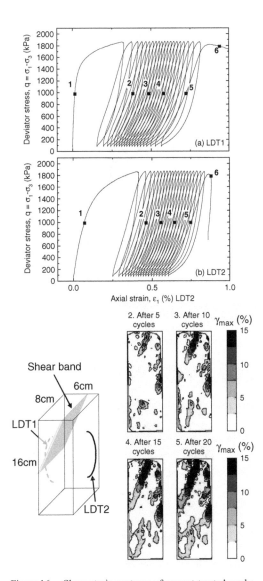

Figure 16. Shear-strain contours of cement-treated sandy soil at four stages of cyclic shear (Salas-Monge et al., 2003).

et al., 1994, Shibuya et al., 1995, Kawaguchi et al., 2001 among others), triaxial apparatus (Tanizawa et al., 1994, Viggiani and Atkinson, 1995, Jovičić et al., 1996, Kuwano and Jardine, 1997 among others), and also applied to in-situ measurement of G_f (Nishio and Katsura, 1994). Strength as well as G can be measured simultaneously when used in shear devices such as direct shear box and triaxial devices.

Figure 17 shows a simple consolidometer equipped with a pair of BEs (Shibuya et al., 1997a). Figures 18 and 19 show similar cases of direct shear

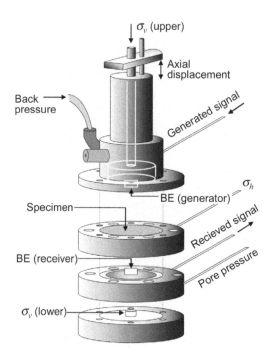

Figure 17. A consolidometer equipped with a pair of BEs (Shibuya et al., 1997a).

Figure 18. Direct shear box apparatus equipped with BEs.

box apparatus and triaxial apparatus (Hwang et al., 1998), respectively.

In the consolidometer developed at Hokkaido University (see Fig.17), a disk-shaped soil specimen of nominal dimensions 60 mm in diameter and 20 mm high undergoes 1D consolidation. It features by extra capabilities of

i) measurement of σ_h as well as σ_v,
ii) back air pressure application,
iii) pore pressure monitoring at the base, and
iv) dual measurement of σ_v at the base and over the apparatus,

The measurement of two principal stresses enables us to determine K_0-value of the clay sample.

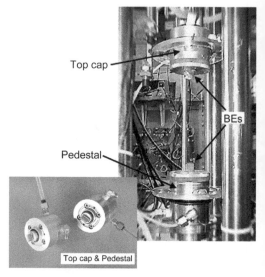

Figure 19. Triaxial apparatus equipped with BEs (Hwang et al., 1998).

Application of back air pressure enhances the degree of saturation in the sample, which in turn improves accuracy of void ratio determination, hence e-logσ_v' curve on reconsolidation.

An unsolved issue regarding the BE test is how to determine the correct travel time of the shear wave. Near field (NF) effects are a real nuisance which makes it difficult (Salinero et al., 1986, Kawaguchi et al., 2001 among others). Yet, no international consensus has been made on this issue. Therefore, TC29 has organized international parallel test on BE test and the latest information regarding BE test is available on the official website of TC29 (http://www.jiban.or.jp/e/tc29/index.htm, 2003).

Anisotropy of elastic stiffness is an interest for practical engineers when they interpret the result of in-situ seismic survey performed using different methods such as cross-hole and down-hole methods (for example, Butcher and Powell, 1995). Anisotropy in shear modulus can also be measured using BEs (refer to Jamiolkowski et al., 1994, Fioravante, 2000 among others). Figure 20 shows a triaxial specimen of Toyoura sand in which the shear wave velocity is measured using three sets of BEs (Yamashita et al., 2003). A set of BEs mounted on top cap and the pedestal enables us to measure shear wave velocity with the wave propagating in the vertical direction involved with the soil grain movement in the horizontal direction, VH. Similarly, the velocities associated with HV and HH shear can be measured with other two sets of BEs. Figure 21 shows the results of such measurements on Toyoura sand and a volcanic ash, suggesting that G_{HH} is slightly larger than G_{VH} ($\cong G_{HV}$).

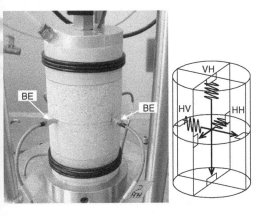

Figure 20. Triaxial specimen of Toyoura sand in which the shear wave velocity is measured using three sets of BEs (Yamashita et al., 2003).

Soil 'structure' may be estimated by knowing the quasi-elastic stiffness. Figure 22 shows the relationship between void ratio, e, and the elastic shear modulus, G when subjected to 1D compression. The triangular symbol refers to the behaviour of reconstituted clay, whereas the comparable results of two natural samples, each recompressed to in-situ geostatic stresses, are shown using squares (Li et al., 2003). As it can be seen in this figure, the 'structure' of the aged natural samples may be evaluated quantitatively by metastability index, $MI(G)$, which refers to the difference of void ratio between the non-structured reconstituted sample and the natural sample at a common G (Shibuya, 2000).

Figure 23 shows the measurement of P-S wave velocities (AnhDan et al., 2002). In this system, a multi-piezo-ceramic unit, which is more powerful than a single bender, is employed as the trigger to generate not only S-wave but also P-wave. The P and S wave velocities are measured using accelerometers mounted on the lateral surface of the specimen. This system seems most suitable for the elastic wave velocity measurement of large size soil specimen.

It is well known that in tests on coarse-grained soil or soft rocks, the elastic wave velocity measurement undergoes the effects of sample heterogeneity (for example, refer to Tatsuoka and Shibuya, 1992). As illustrated in Figure 24, the stiffness from the stress-strain curve in the laboratory refers to the overall deformation behaviour, whereas the elastic wave measurement reflects predominantly on those of the stiff part (Tanaka et al., 1994).

Figure 25 shows the ratio of the equivalent elastic wave velocity from static test to the measured velocity from dynamic test plotted against the ratio of the mean grain diameter to the half of wave length (Tanaka et al., 2000, AnhDan et al., 2002, Maqbool et al., 2004).

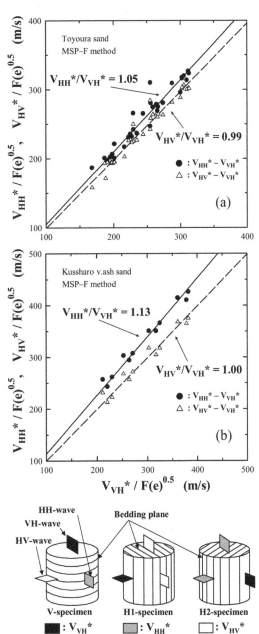

Figure 21. Results of BE tests on Toyoura sand and a volcanic ash (Yamashita et al., 2003).

As it can be seen, the ratio using both S and P waves tends to decrease as the wave length gradually approaches to the grain size. The results strongly suggest that we should use the wave length, say roughly 1,000 times or more, the D_{50} of the soil specimen in order to obtain the overall elastic stiffness on average.

13

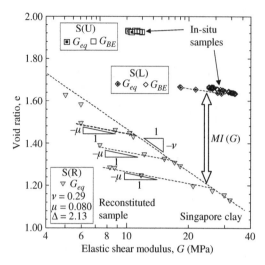

Figure 22. Relationship between void ratio and elastic shear modulus of Singapore clay when subjected to 1D compression (Li et al., 2003).

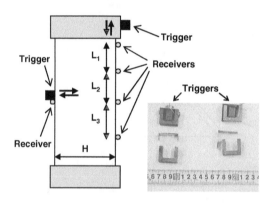

Figure 23. Measurement of *P-S* wave velocities in vertical and horizontal directions (AnhDan et al., 2002).

2.4 *Friction on rigid interface*

The last topic in Part I is the effects of friction on rigid interface in the laboratory test. When soil moves relative to the rigid wall, friction force develops at the interface. Error in the σ_y measurement due to the wall friction has been found quite significant in conventional testing devices in which soil specimen is confined in a rigid box or ring (Shibuya et al., 1997b). Like the consolidometer already shown in Fig.17, the vertical load at the base, W_{lower} as well as the measurement at the top, W_{upper} should therefore be measured by which the effect of wall friction is properly accounted for. Averaged σ_v in consolidometer test may be given by

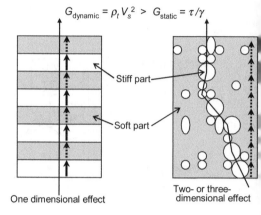

Figure 24. Effects of sample heterogeneity (modified from Tanaka et al., 1994).

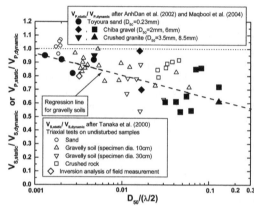

Figure 25. Effects of particle size and wave length on ratio of equivalent elastic wave velocities by static and dynamic measurements (modified from Tanaka et al., 2000, AnhDan et al., 2002, Maqbool et al., 2004).

$$\sigma'_{(oedometer)} = \left(W_{upper} + W_{lower}\right)/2A \qquad (2)$$

where A stands for cross-section area of specimen.

Figure 26 shows the relationship between void ratio e and σ'_v of Osaka Bay clay (Li et al., 2004) The λ value refers to the slope of NC line in terms of e-σ'_v relationship. The effect of interface friction acting on the lateral surface of the specimen was significant for the σ'_v measurement. The conventional measurement at the top grossly overestimated the σ'_v as compared to the average of the upper and lower measurements in compression, and vice versa in swelling. As seen in Figure 27, the conventional σ'_v measurement yielded the well-known design value of the overconsolidation ratio (*OCR*) of 1.2, whereas the *OCR* value from the averaged σ'_v yielded approximately unity. Accordingly, we have chosen the averaged vertical stress in order to

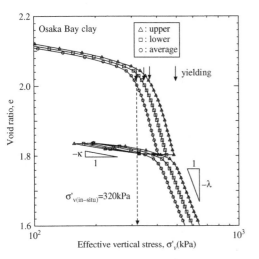

Figure 26. Relationship between e and σ'_v of Osaka Bay clay in CRS test (Li et al., 2004).

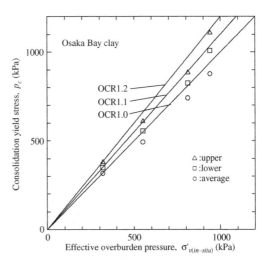

Figure 27. Determination of OCR of Osaka Bay clay in CRS test (Li et al., 2004).

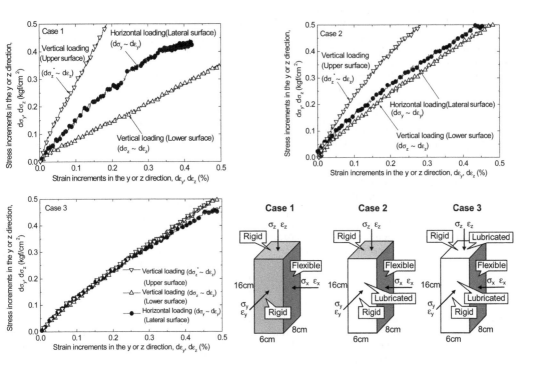

Figure 28. Small-scale true triaxial test on rubber dummy (Hayano, 2001).

determine the yield stress, hence the OCR value. We should be more careful about the effects of interface friction even in conventional oedometer test using a specimen of merely 2 cm high.

The effects of interface friction of a rectangular specimen are demonstrated in Figure 28 by conducting

a small-scale true triaxial test on a rubber dummy (Hayano, 2001). In case where two interfaces are not lubricated, three normal strains exhibited values different to each other when subjected to isotropic stress application. The divergence virtually diminished when the two surfaces are both lubricated. The results

15

strongly suggest the need for lubrication on the rigid soil to metal interface in laboratory element test.

3 PART II: APPLICATION IN ENGINEERING PRACTICE

3.1 Introduction

In Part II, four cases all regarding the application of deformation and strength test results in engineering practice will be described separately. These are;

i) deformation of soft clay ground in deep excavation,
ii) lateral subgrade reaction of large foundation,
iii) earth pressure at high seismic loading, and
iv) seismic stability of fill dam.

Figure 29 shows Geo-cow illustrated by Shibuya et al. (2001). The original concept described by Shibuya et al. (2001) was to distinguish the difference between the reconstituted specimens for hamburger and natural soils for fillet with effects of ageing, anisotropy, and so on. However, in this paper, Geo-cow is meant to distinguish the difference in the extents how we make use of laboratory test results. "Hamburger" stands for the tests that are performed frequently in practice, such as unconfined compression test to evaluate q_u and E_{50} values, which are employed in the conventional design of earth structures and foundations. On the other hand, "fillet" stands for the tests that are not frequently performed in practice, but provide detailed data on the strength and deformation properties.

All of the four case histories described in this part II can be classified as "fillet", since they are based on the results of elaborated laboratory tests. Some of them employ the detailed test results directly in the project, while the others employ the relevant test results in order to develop rational design procedures reflecting the actual behavior of natural soils.

3.2 Deformation of soft clay ground in deep excavation

In the preceding conference in Torino, Simpson (2000) has drawn our attention to a couple of engineering needs. One is the technical need for a good model of soil behavior in the range of strain of engineering significance, generally 0.001% to 0.1%. The other is the pragmatic need for rapid analysis, and especially for more rapid testing methods, notably for clays.

In this section, a case history in Thailand on deep excavation in soft clay with concrete diaphragm wall is described with a particular attention paid to the ground deformation behind the diaphragm wall. As summarized in Table 2, there have been several

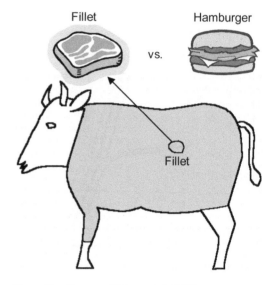

Figure 29. Geo-cow (Shibuya et al., 2001).

Table 2. Elastic deformation modulus employed for FE analysis in the past.

$E = \alpha S_u$	Author	
Embankment loading analysis		
$E_u = 200–500 S_u$	Bjerrum (1964)	
$E' = 70–250 S_{uFVS}$	Balasubramaniam et al. (1981)	
$E_u = 15–40 S_{uFVS}$	Bergado et al. (1990)	(Back-calculation)
Excavation works analysis		
$E_u = 200–500 S_u$	Bowels (1988)	
$E_u = 280–500 S_u$	Hock (1997)	(soft clay)
$E_u = 1200–1600 S_u$	Hock (1997)	(stiff clay)
$E' = E'_{max}/2$	Simpson (2000)	

empirical proposals how we should select the soil stiffness employed for FE analysis. For embankment loading, the use of several hundreds or tens times of undrained shear strength s_u has been proposed for the equivalent Young's modulus. For excavation works, on the other hand, much higher factors have been proposed by Bowels (1988) and Hock (1997) for the conversion from s_u, or the use of half of the small strain stiffness E_{max} obtained by in-situ seismic survey has been proposed by Simpson (2000).

Figure 30 shows the variations of secant Young's modulus normalized using s_u and E_{max} with axial strain, that were obtained in undrained triaxial compression test on normally consolidated clays collected from worldwide (Temma et al., 2000, 2001). In the present case history, the ground strains behind the diaphragm wall was on the order of 0.1%, which is consistent with the proposals by Simpson and Hock. Similar strain level has been also reported by Ou et al. (2000) for deep excavation with diaphragm wall in Taipei, Taiwan. It should be pointed out that the axial strain associated with $E = E_{max}/2$ corresponds to a narrow range from 0.03% to 0.3%. The use of $E = E_{max}/2$, therefore, matches well the ground strains of soft clay behind diaphragm wall induced by deep excavation. Note also that the E_{sec}/s_u value corresponding to the axial strain of 0.1% ranges from 200 to 400.

Figure 31 shows cross-section of deep excavation with diaphragm wall for the construction of subway tunnels in Bangkok by open-cut method. The concrete diaphragm walls were pre-installed to a depth of 39 m, and deep excavation was afterwards carried out in soft and stiff clays down to a depth of 22 m (Tamrakar et al., 2001). Instrumentations were a piezometer, two series of inclinometers set along the wall axis and the borehole located at a horizontal distance of 17 m from the diaphragm wall, and a series of markers for monitoring ground settlement.

Figure 32 shows the stratigraphy with representative profiles of OCR, compression and swell indices, λ and κ, and geostatic stresses and pore pressures.

Figure 33 shows the results of a series of undrained triaxial compression tests on undisturbed clay specimens retrieved from several depths. Although the small strain stiffness values vary with depth, the degradation curves of the normalized stiffness were almost similar to each other.

By considering the proposal by Simpson, half of the small strain stiffness corresponding to the strain level of about 0.1% was employed in the numerical simulation of the full scale behavior using an equivalent-linear elastic approach. In addition, Kovacevic et al. (2003) has performed a non-linear elastic analysis using a model called the small strain stiffness or SSS model. The results of boundary values predicted using these two kinds of analysis are compared in Figures 34 and 35. The horizontal deformation of the diaphragm wall was compared at several excavation stages (see Fig.34). From a practical point of view, the result from equivalent-linear analysis presented using dash lines was not bad, while the deformation profile could be better captured by the non-linear analysis shown in solid lines assuming full moment connection between the roof slab and the diaphragm wall. Similarly, the non-linear analysis using a small-strain stiffness (SSS) model could better capture the ground settlement profile than the linear analysis (see Fig.35), suggesting an importance of using proper soil stiffness considering its strain level and stress state dependencies.

3.3 Lateral subgrade reaction of large foundation

Nowadays, the size of foundations is becoming larger and larger, including those of cast-in-place

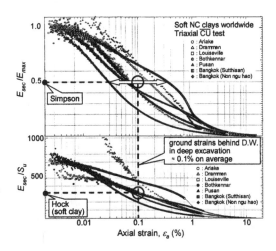

Figure 30. Normalized secant Young's modulus of normally consolidated clays (Temma et al., 2000, 2001).

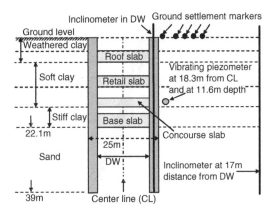

Figure 31. Cross-section of deep excavation with diaphragm wall in Bangkok (Tamrakar et al., 2001).

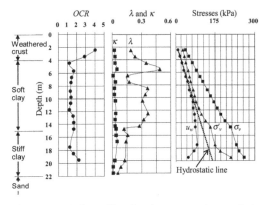

Figure 32. Soil profiles for deep excavation work in Bangkok (Tamrakar et al., 2001).

17

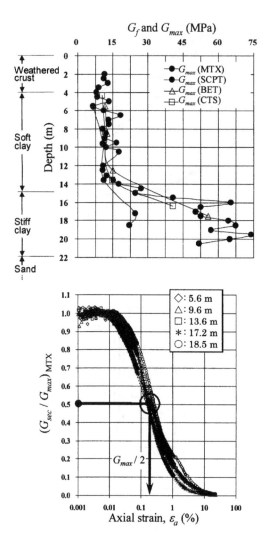

Figure 33. Undrained triaxial compression test results on undisturbed clay specimens (Shibuya et al., 2001).

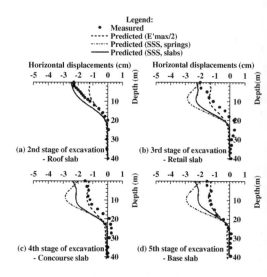

Figure 34. Horizontal deformation of diaphragm wall at four excavation stages (Kovacevic et al., 2003).

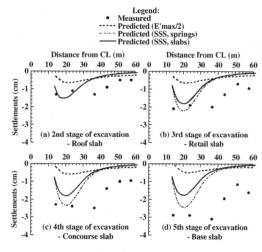

Figure 35. Ground surface settlement at four excavation stages (Kovacevic et al., 2003).

reinforced-concrete piles, diaphragm wall foundations, and well foundations.

When such large foundations are constructed in an idealized linear-elastic ground, the coefficient of subgrade reaction that is used to compute their displacement should be reduced with the increase in the size of foundations. However, the actual ground is not linear elastic. Thus, in order to investigate the scale effects on the horizontal subgrade reactions, a series of in-situ lateral loading tests was conducted by using a part of well foundation having a diameter of 2.5 m constructed in a clayey gravel deposit, while changing the width of the loading plate between 0.5 and 2.0 m, as schematically shown in Figure 36 (Ogata et al., 1999).

Figure 37 shows the result of triaxial compression tests on undisturbed specimens retrieved from the site. Based on this result, the stress-strain relationship was modeled in a non-linear elastic form by considering the dependency of initial stiffness, the peak strength on the confining stress, and the dependency of tangential stiffness on the shear stress level. As shown in Figure 38, a 3-D non-linear elastic analysis with this model could simulate the in-situ lateral loading test result reasonably. It can also be seen that the other predictions largely underestimated the subgrade reaction, which were obtained based on

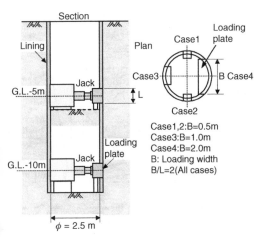

Figure 36. A schematic diagram of in-situ lateral loading tests (Ogata et al., 1999).

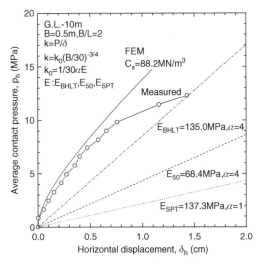

Figure 38. Comparison of in-situ lateral loading test results with its simulations (Ogata et al., 1999).

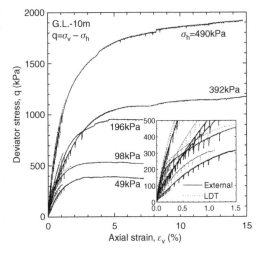

Figure 37. Drained triaxial compression test results on undisturbed gravel specimens (Ogata et al., 1999).

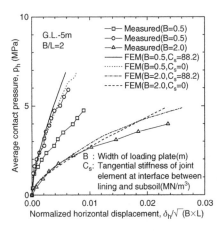

Figure 39. Comparison of in-situ lateral loading test results with results from FE analysis. (Ogata et al., 1999).

conventional designs with linear-elastic approach, in which the stiffness was converted from the results of a bore-hole loading test, the SPT N-value and the secant Young's modulus in a triaxial compression test defined at a shear stress level equal to half of the peak state.

Figure 39 compares the load-displacement curves obtained by using loading plates having a width of 0.5 and 2.0 m, respectively. The 3-D non-linear elastic analysis could capture reasonably the scale effects seen in the observations.

This analysis was applied to predict the behavior of well foundations each having a diameter of 1, 2.5 and 5 m (Koseki et al., 2001). Three different soil properties

were assigned based on relevant triaxial test results. Figure 40 shows the scale effects on the coefficients of horizontal subgrade reaction, defined at a displacement of 1 cm and normalized with the value at a diameter of 1 m. Although such formulation with the power n equal to minus three quarters has been employed in the conventional design of highway bridges in Japan (e.g., JRA, 1994), its correction for large foundations with less scale effects has been introduced by the Japan Highway Public Corporation based on such results considering the dependencies of soil stiffness on strain level and stress states.

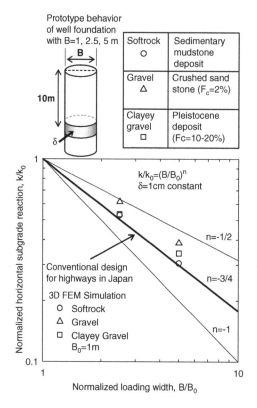

Figure 40. Scale effects on coefficients of horizontal subgrade reaction (Koseki et al., 2001).

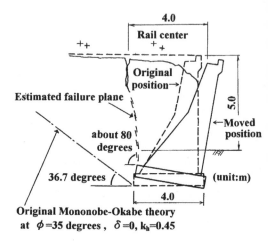

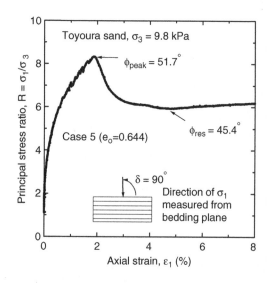

Figure 41. Damage to gravity type retaining wall by the 1995 Hyogoken-nambu earthquake (Tatsuoka et al., 1996).

Figure 42. Drained plane strain compression test results on dense sand (Koseki et al., 2003).

3.4 Earth pressure at high seismic loading

Figure 41 shows a retaining wall structure for railway that was damaged by the 1995 Hyogoken-nanbu earthquake (Tatsuoka et al., 1996). Before this earthquake, relatively low soil strength values, such as an internal friction angle of 30 degrees, have been employed on the assumption of relatively low design loads in the seismic design of retaining walls in Japan (e.g., JRA, 1987). However, if a well-compacted backfill material undergoes shearing under plane strain conditions, which is typical boundary conditions for retaining walls with a sufficient length in the longitudinal direction, it can exhibit a peak strength that may approach to as large as 50 degrees.

After the Hyogoken-nanbu earthquake, it was attempted in Japan to establish a rational design procedure to evaluate stability of retaining walls at high seismic loading, typically more than 0.5 g. Under such a high design seismic load, a rational design was not possible without taking advantage of the high peak strength of the backfill. At the same time, it was also required to consider the effects of strain softening

from peak to residual strengths as typically shown in Figure 42 (Koseki et al., 2003).

Figure 43 shows a force equilibrium assumed in the original Mononobe-Okabe method (Mononobe and Matsuo, 1929; Okabe, 1924), which has been employed in many of the conventional seismic designs in which seismic earth pressures are estimated for design. This method is an extended version of the Coulomb's earth pressure theory, where the effects of pseudo-static seismic forces in both the horizontal and vertical directions were considered.

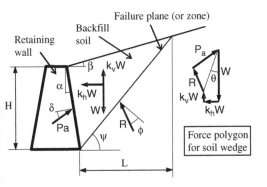

Figure 43. Force equilibrium assumed in Mononobe-Okabe method (modified from Koseki et al., 1998).

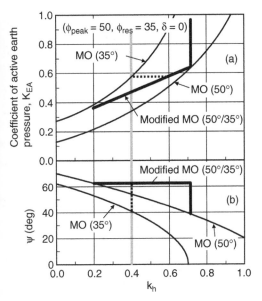

Figure 44. Comparison of original and modified Mononobe-Okabe methods (RTRI, 1999).

It should be noted that, in this method, the mobilized strength of backfill along the potential failure plane (or shear band) is assumed to be constant and distributed uniformly.

Figure 44a shows a relationship between the coefficient of active earth pressure and the horizontal seismic coefficient. When the original Mononobe-Okabe method is applied with a relatively low ϕ value, the coefficient of active earth pressure becomes extremely large. As shown in Figure 44b, it also induces unrealistically small value of ψ, defined as the angle between the shear band and the horizontal direction.

Table 3. Peak and residual strengths assigned for modified Mononobe-Okabe method (RTRI, 1999).

	ϕ_{peak}	ϕ_{res}
Well-graded sand	55°	40°
Sand/Gravel	50°	35°
Poorly-graded sand	45°	30°
Clay	45°	45°

On the other hand, when considering the strain softening behavior of well-compacted backfill as mentioned before, the use of the original Mononobe-Okabe method with a high ϕ value corresponding to the peak strength is obviously less conservative. Therefore, a modified version of the Mononobe-Okabe method has been developed and adopted in some design codes in Japan. In this modified version, the location of shear band in the backfill is determined by using the peak strength. After the formation of the shear band, the strength mobilized on the shear band will drop from the peak to the residual state, while the other region will remain to potentially mobilize the peak strength (Koseki et al., 1998).

The modified version can yield reasonable values of the coefficient of active earth pressure and the angle ψ. A typical example is shown in this figure with the peak and residual ϕ values set equal to 50 and 35 degrees, respectively. Referring to the relevant laboratory test results, the peak and residual ϕ values to be used in the design of railway retaining walls in Japan (RTRI, 1997 and 1999) are assigned as shown in Table 3. Thus, proper use of soil strength considering strain softening behavior from peak to residual states and associated shear band formation has been implemented in practice.

Since the modified version yields a step-wise change in the coefficient of active earth pressure, it was further simplified by approximating it with a linear function as shown in Figure 45 and adopted in the design code of highway bridge abutments in Japan (JRA, 2002).

3.5 Seismic stability of fill dam

As is the case with retaining walls, a relatively low soil strength has been used in the conventional seismic design of fill dams in Japan. In order to urge dam construction projects in a cost-effective manner, it is required to establish a rational design procedure with which the earthquake-induced displacement of fill dams may be properly evaluated. In this section, a modified Newmark's sliding block method considering the strain-softening properties of coarse materials along shear band is introduced.

The rock material used for constructing rockfill dam is coarse. In order to investigate the shear banding

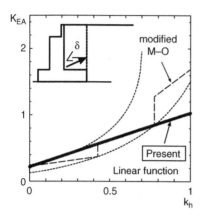

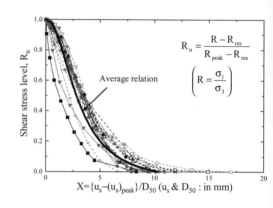

$$R_n = \frac{R - R_{res}}{R_{peak} - R_{res}}$$

$$\left(R = \frac{\sigma_1}{\sigma_3} \right)$$

Average relation

$X = \{u_s - (u_s)_{peak}\}/D_{50}$ (u_s & D_{50} : in mm)

Figure 45. Simplification of modified Mononobe-Okabe method (JRA, 2002).

Figure 47. A summary of strain softening properties of well-graded materials (Okuyama et al., 2003).

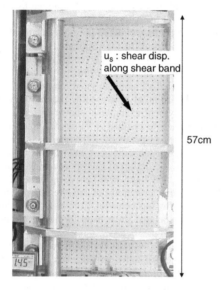

u_s : shear disp. along shear band

57cm

Figure 46. Shear band observed in large-scale plane strain compression test (Okuyama et al., 2003).

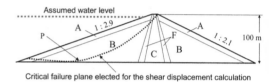

Assumed water level

Critical failure plane elected for the shear displacement calculation

Figure 48. Dam model (Okuyama et al., 2003)

properties of such coarse geomaterials, a series of large scale plane compression tests was conducted by Okuyama et al. (2003) on a variety of samples having different grading curves. Figure 46 shows a typical formation of shear band, obtained for an Andesite-origin sample with a mean diameter of 2.5 mm. By analyzing such image data, the shear displacement along shear band was evaluated. In this case, the thickness of the shear band was about 15 mm, and a clear strain-softening behavior was observed as shown in this stress-strain relationship.

Figure 47 shows a summary for the tested samples on the normalized relationship between the post-peak drop of the principal stress ratio and the increment of shear displacement along the shear band from the peak state. At the peak and residual states, the normalized principal stress ratios are defined to be unity and zero, respectively. By normalizing the increment of shear displacement along the shear band with the mean diameter powered by 0.66, rather a unique relationship could be obtained among different test results. A trial computation of earthquake-induced displacement of a rockfill dam was conducted using this normalized relationship.

Figure 48 shows a cross-section of a rockfill dam having a height of 100 m that was analyzed by Okuyama et al. (2003). An input earthquake motion having a maximum amplitude of about 0.5 g was employed for its earthquake response analysis. A pseudo-static stability analysis was conducted using the dynamic response. Since the computed factor of safety was always larger than unity, the dynamic response was increased by a factor of 1.7. As a result, when the largest peak strengths of each material are used in the stability analysis, the critical failure plane to yield a factor of safety equal to unity was obtained as shown in Fig.48.

By assuming that strain softening takes place along the critical failure plane, cumulative shear displacements under four different peak strength conditions corresponding to different degrees of compaction were computed based on the Newmark's sliding block method, which was modified to accommodate the

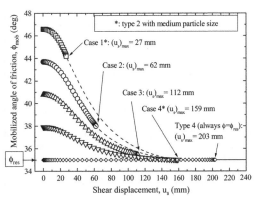

Figure 49. Computed shear displacement along critical failure plane (Okuyama et al., 2003)

strain softening behavior. It was also assumed that the residual strength of each material is constant irrespective of the different peak strengths, and the strain softening property was assumed to follow the normalized relationship as mentioned before.

As shown in Figure 49, when the largest peak strengths corresponding to a case with heavy compaction were used, the computed shear displacement along the failure plane was 27 mm, and the mobilized shear resistance did not drop down to the residual state. On the other hand, when the mobilized shear resistance was assumed to be equal to the residual state from the beginning, as is the case with the conventional approach, the shear displacement accumulated up to 203 mm. Such a difference suggests an importance of considering the strain softening behavior that depends on the degree of compaction and the particle size.

4 CONCLUDING REMARKS

The contents of Part I: Recent developments in control/measurement/data acquisition can be summarized as follows;

Servo-motors for control

i) The use of servo-motor featured with 'zero' backlash and variable speed enables us to achieve fullautomation and high-precision of testing, and also enhances the scope of testing to be performed (e.g., monotonic/cyclic loadings, elastic stiffness at small strains to residual strength at large strains, strain rate/acceleration effects, etc.).

ii) Such servo-motors may be facilitated in various laboratory machines with various purposes (e.g., the control of cell pressure, loading plate, etc.)

Local strain measurement (LSM)

i) Dual-measurement with two independent proximity transducers is compatible to mono-gauge measurement such as LDT.

ii) The hinge (or target) of clip gauge like LDT (or of proximity transducer) should be glued on the membrane (or pinned into the sample) in tests on stiff (or soft) geomaterials.

iii) 'True plane-strain' test can exclusively be carried out by maintaining zero strain with LSM.

iv) Stiffness anisotropy at small strains can be investigated in true triaxial apparatus with LSM.

v) The LSM should be employed in hollow cylinder testing, since the effects of end restraint are significant.

vi) Innovative techniques (e.g., LAT, PIV) are now available so as to observe strain localization of not only a rectangular specimen but also a cylindrical specimen in shape.

Elastic wave velocity measurement

i) Bender element when designated to generate S-wave is suitable for measuring quasi-elastic shear modulus of relatively small sample of soft soils.

ii) Determination of correct travel time is still an unsolved issue involved in bender test. An international round-robin test as to the measurement of elastic shear wave by benders is now underway to provide an insight into it.

iii) An piezo-ceramic vibrator may be mounted on top cap to generate P-S waves, and the velocities may be measured on the lateral surface of the specimen by using accelerometers. This system is suitable for large sample of geomaterials.

iv) Some of important characteristics such as structure, anisotropy, sample disturbance, etc. may be understood by measuring quasi-elastic soil properties.

Friction on soil/metal interface

i) The effects of interface friction are significant even in oedometer test on soft clay using a standard specimen of 2 cm in height. The conventional measurement of vertical load grossly overestimates the stress at yielding, which in turn brings about a considerable overestimate of OCR value.

ii) Lubricated ends are needed to achieve reasonable stress/strain uniformities in true triaxial test using a rectangular specimen.

The contents of Part II: application in engineering practice may suggest the following;

Engineering needs for deformation/strength testing of geomaterials

There are a variety of engineering needs for application of deformation and strength test results, such as excavation works in urban areas, design of large-scale structures or foundations, and seismic design against large earthquakes. In order to meet these needs, conventional approaches based on stability analysis using conservative soil strength, or deformation analysis using conservative soil stiffness may not work.

Use of proper soil stiffness

The use of proper soil stiffness considering its strain level dependency and stress state dependency in the deformation analysis is vitally important. If the strain level can be predicted in advance, a linear elastic analysis using a well-chosen stiffness would be of practical value. If not, however, non-linear elastic or elasto-plastic approaches considering these factors would be more appropriate.

Use of proper soil strength

In addition, use of proper soil strength considering strain softening from peak to residual states, accompanied by shear banding or strain localization, is required in establishing rational design procedures against large earthquake loads.

It should be noted that, relevant laboratory tests in combination with proper in-situ tests can reveal all these important soil properties mentioned in the above.

ACKNOWLEDGEMENTS

A lot of people as shown below kindly supported us in preparing this paper; Prof. Tatsuoka, F. (Univ. of Tokyo), Prof. Mitachi, T. (Hokkaido Univ.), Prof. Miura, S. (Hokkaido Univ.), Dr. Yamashita, S. (Kitami Institute of Technology), Mr. Sato, T. (IIS, Univ. of Tokyo), Dr. Hayano, K. (Port and Airport Research Institute), Dr. Fukuda, F. (Hokkaido Univ.), Dr. Li, D.J. (Hokkaido Univ.), Dr. AnhDan, L.Q. (Univ. of Notre Dam) and Dr. HongNam, N. and Mr. Maqbool, S. (IIS, Univ. of Tokyo).

REFERENCES

AnhDan, L.Q. 2001. Study on small strain behaviour and time effects on deformation characteristics of dense gravel by triaxial and true triaxial tests, *PhD thesis*, The University of Tokyo.

AnhDan, L.Q., Koseki, J. and Sato, T. 2002. Comparison of Young's moduli of dense sand and gravel measured by dynamic and static methods, *Geotechnical Testing Journal* 25(4), ASTM: 349–368.

AnhDan, L.Q. and Koseki, J. 2003. Anisotropic deformation properties of dense granular soils by large-scale true triaxial tests, *Deformation Characteristics of Geomaterials* 1, IS-Lyon, Balkema: 305–312.

Balasubramaniam, A.S. and Brenner, R.P. 1981. Consolidation and settlement of soft clay, *Soft clay engineering*, Brand, E.W. and Brenner, R.P., eds.: Elsevier Scientific Publishing Co.

Bergado, D.T., Ahmeed, S., Sampaco, C.L. and Balasubramaniam, A.S. 1990. Settlement of Bangna-Bangapakong Highway on soft Bangkok clay, *Geotechnical engineering* 116(1): 136–155.

Bjerrum, L. 1964 Observed versus computed settlements of structures on clay and sand, *Lecture presented at Massachusetts Institute of Technology*, Cambridge, Massachusetts.

Bowels, J.E. 1988. Foundation analysis and Design, *4th Edition, McGraw-Hill.*

Bowman, E.T. and Soga, K. 2003. Creep, ageing and microstructural change in dense granular materials, *Soils and Foundations* 43(4), Special Issue Deformation Characteristics of Geomaterials, IS-Lyon: 107–117.

Butcher, A.P. and Powell, J.J.M. 1995. The effects of geological history on the dynamic stiffness in soils, *Proc. of the 11th European Conference on SMFE* 1, Copenhagen: 27–36.

Dyvik, R. and Madshus, C. 1985. Laboratory measurement of G_{max} using bender elements, *Proceeding of American Society of Civil Engineering Annual Convention*: 186–196.

Fioravante V. 2000. Anisotropy of small strain stiffness of Ticino and Kenya sands from seismic wave propagation measured in triaxial testing, *Soils and Foundations* 40(4): 129–142.

Frost, J.D. and Yang, C.T. 2003. Effect of end platens on microstructure evolution in dilatant specimens, *Soils and Foundations* 43(4), Special Issue Deformation Characteristics of Geomaterials, IS-Lyon: 107–117.

Goto, S., Tatsuoka, F., Shibuya, S., Kim, Y.S. and Sato, T. 1991. A simple gauge for local small strain measurements in the laboratory, *Soils and Foundations* 31(1): 169–180.

Hayano, K. 2001. Pre-failure deformation characteristics of sedimentary soft rocks, *PhD thesis*, The University of Tokyo. (in Japanese)

Hock, G.C. 1997. Review and analysis of ground movements of braced excavation in Bangkok subsoil using diaphragm walls, *M.Eng.Thesis*, Bangkok Thailand.

HongNam, N. and Koseki, J. 2003. Modeling quasi-elastic deformation properties of sand, *Deformation Characteristics of Geomaterials* 1, IS-Lyon: 275–283.

HongNam, N. 2004. Locally measured deformation properties of Toyoura sand in cyclic triaxial and torsional loadings and their modelling, *PhD thesis*, The University of Tokyo.

Hwang, S.C., Mitachi, T., Shibuya, S. and Tateichi, K. 1998. Stress-strain characteristics in the wide strain rate from small strain to failure state and undrained shear strength of natural clays, *Journal of Japan Society of Civil Engineering* 589(III-42): 305–319 (in Japanese).

Jamiolkowski, M., Lancellotta, R. and Lo Presti, D.C.F. 1994. Remarks on the stiffness at small strains of six Italian clays, *Pre-failure Deformation of Geomaterials* 2, IS-Hokkaido, Balkema: 817–836.

Jamiolkowski, M., Lancellotta, R. and Lo Presti, D.C.F. eds. 1999. *Pre-failure Deformation of Geomaterials* 1&2, IS-Torino: Balkema.

Japan Road Association 1987. Earthworks manual – retaining walls, culverts and temporary structures: 25: (in Japanese)

Japan Road Association 1994. Specifications for highway bridges, part IV – foundations: 201–203 (in Japanese).

Japan Road Association 2002: Specifications for highway bridges, part V – seismic design: 65–68 (in Japanese).

Jovičić, V., Coop, M. R. and Simi, M. 1996. Objective criteria for determining G_{max} from bender element tests, *Géotechnique* 46(2): 357–362.

Kawaguchi, T., Mitachi, T. and Shibuya, S. 2001. Evaluation of shear wave travel time in laboratory bender element test *Proc. of the 15th ICSMGE* 1: 155–158. Istanbul: Balkema.

Kawaguchi, T., Mitachi, T., Shibuya, S. and Sano, Y. 2002. Development of an elaborate triaxial testing system for deformation of clay, *Journal of Japan Society of Civil Engineering* 708(III-59): 175–186. (in Japanese)

Kawaguchi, T. 2002. Study on measurement and evaluation of elastic modulus in clays, *PhD thesis*, Hokkaido University. (in Japanese)

Kishi, M. and Tani, K. 2003. Development of measuring method for axial and lateral strain distribution using CCD sensor in triaxial test, *Deformation Characteristics of Geomaterials* 1, IS-Lyon, Balkema: 31–36.

Kobayashi, T. and Fukagawa, R. 2003. Characterization of deformation process of CPT using X-ray TV imaging technique, *Deformation Characteristics of Geomaterials* 1, IS-Lyon, Balkema: 43–47.

Konagai, K., Tamura, C., Rangelow, P. and Matsushima, T. 1992. Laser-Aided tomography: A tool for visualization of changes in the fabric of granular assemblage, Structural Engrg./Earthquake Engrg., JSCE 9(3): 193–201.

Koseki, J., Tatsuoka, F., Munaf, Y., Tateyama, M. and Kojima, K. 1998. A modified procedure to evaluate active earth pressure at high seismic loads, *Soils and Foundations*, Special Issue on Geotechnical Aspects of the January 17 1995 Hyogoken-Nambu Earthquake 2: 209–216.

Koseki, J., Kurachi, Y. and Ogata, T. 2001. Dependency of horizontal and vertical subgrade reaction coefficients on loading width, *Advanced Laboratory Stress-Strain Testing of Geomaterials* (Tatsuoka et al. eds.), Balkema: 259–264.

Koseki, J., Tatsuoka, F., Watanabe, K., Tateyama, M., Kojima, K. and Munaf, Y. 2003. Model tests on seismic stability of several types of soil retaining walls, *Reinforced Soil Engineering*, Ling, Leshchinsky and Tatsuoka (eds.), Dekker: 317–358.

Kovacevic, N., Hight, D.W. and Potts, D.M. 2003. A comparison between observed and predicted behaviour of a deep excavation in soft Bangkok clay, *Deformation Characteristics of Geomaterials* 1, IS-Lyon, Balkema: 983–989.

Kuwano, R. and Jardine, R.J. 1997. Stiffness measurements in a stress-path cell, Pre-failure Deformation Behaviour of Geomaterials, *Géotechnique Symposium In Print*: 391–394.

Li, D.J., Shibuya, S., Mitachi, T. and Kawaguchi, T. 2003. Judging fabric bonding of natural sedimentary clay, *Deformation Characteristics of Geomaterials* 1, IS-Lyon, Balkema: 203–209.

Li, D.J., Shibuya, S. and Mitachi, T. 2004. Engineering properties of Osaka bay clay, *Proc. of the International Symposium on Engineering Practice of Soft Deposits*, IS-Osaka. (in print)

Lo Presti, D.C.F., Puci, I., Pallara, O., Maniscalco, R. and Pedroni, S. 2000. Experimental Laboratory determination of the steady state of sands, *Soils and Foundations* 40(1): 113–122.

Maqbool, S. and Koseki, J. 2003. Comparison of plane strain compression tests using active and passive controls with triaxial compression tests on gravel, *Proc. of 5th International Summer Symposium*, International Activities Committee, JSCE: 229–232.

Maqbool, S., Koseki, J. and Sato, T. 2004. Effects of compaction on small strain Young's moduli by dynamic and static measurements, *Bulletin of ERS* 37, Institute of Industrial Science, University of Tokyo. (in print)

Matsushima, T., Saomoto, H., Tsubokawa, Y. and Yamada, Y. 2003. Grain rotation versus continuum rotation during shear deformation of granular assembly, *Soils and Foundations* 43(4), Special Issue Deformation Characteristics of Geomaterials, IS-Lyon: 95–105.

Mitachi, T., Kuda, T., Okawara, M. and Ishibashi, M. 2003. Determination of strength parameters for landslide slope stability analysis by laboratory test and inverse calculation engagement, *Journal of the Japan Landslide Society* 40(2): 105–116.

Mononobe, N. and Matsuo, H. 1929. On determination of earth pressure during earthquake, *Proc. World Engineering Congress* 9, Tokyo: 177–185.

Nishio, S. and Katsura, Y. 1994. Shear wave anisotropy in Edogawa Pleistocene deposit, *Pre-failure Deformation of Geomaterials* 1, IS-Hokkaido, Balkema: 169–174.

Ogata, T., Kurachi, Y., Oishi, M., Ouchi, M., Maeda, Y. and Koseki, J. 1999. In-situ horizontal loading tests on gravelly subsoil and their numerical simulations, *Prefailure Deformation Characteristics of Geomaterials* 1, IS-Trino, Balkema: 379–386.

Otani, J. Mukunoki, T. and Obara, Y. 2000. Application of X-ray CT method for characterization of failure soils, *Soils and Foundations* 40(2): 111–118.

Okabe, S 1924. General theory on earth pressure and seismic stability of retaining wall and dam, *Journal of Japan Society of Civil Engineers* 10(6): 1277–1323.

Okuyama Y, Yoshida, T., Tatsuoka, F., Koseki, J., Uchimura, T., Sato, N. and Oie, M. 2003. Shear banding characteristics of granular materials and particle size effects on the seismic stability of earth structures, *Deformation Characteristics of Geomaterials* 1, IS-Lyon, Balkema: 607–616.

Ou, C.Y., Liao, J.T. and Cheng, W.L. 2000. Building response and ground movements induced by a deep excavation, *Géotechnique* 30(3): 209–220.

Railway Technical Research Institute 1997. Railway structure design standard – foundations/soil retaining structures, Maruzen: 364 and 371 (in Japanese).

Railway Technical Research Institute 1999. Railway structure design standard – seismic design, Maruzen: 89–90. (in Japanese).

Salas-Monge, R., Koseki, J. and Sato, T. 2003. Cyclic plane strain compression tests on cement treated sand, *Bulletin of ERS* 36, Institute of Industrial Science, University of Tokyo: 131–141.

Salinero, I.S., Roesset, J.M. and Stokoe, K.H. 1986. Analytical studies of body wave propagation and attenuation, *Report GR86-15*, University of Texas at Austin.

Santucci de Magistris, F., Koseki, J., Amaya, M., Hamaya, S., Sato, T. and Tatsuoka, F. 1999. A triaxial testing system to evaluate stress-strain behaviour of soils for wide range of strain and strain rate, *Geotechnical Testing Journal* 22(1), ASTM: 44–60.

Shibuya, S., Tatsuoka, F., Teachavorasinskun, S., Kong, X.J., Abe, F., Kim, Y.S. and Park, C.S. 1992. Elastic deformation properties of geomaterials, *Soils and Foundations* 32(3): 26–46.

Shibuya, S., Mitachi, T. and Miura, S. eds. 1994. *Pre-failure Deformation of Geomaterials* 1&2, IS-Hokkaido: Balkema.

Shibuya, S., Mitachi, T., Fukuda, F. and Degoshi, T. 1995. Strain rate effects on shear modulus and damping of normally consolidated clays, *Geotechnical Testing Journal* 18(3): 365–375.

Shibuya, S. and Mitachi, T. 1997. Development of a fully digitized triaxial apparatus for testing soils and soft rocks, *Geotechnical Engineering Journal* 28(2): 183–207.

Shibuya, S., Hwang, S.C. and Mitachi, T. 1997a. Elastic shear modulus of soft clays from shear wave velocity measurement, *Géotechnique* 47(3): 593–601.

Shibuya, S., Mitachi, T. and Tamate, S. 1997b. Interpretation of direct shear box testing of sands as quasi-simple shear, *Géotechnique* 47(4): 769–790.

Shibuya, S. 2000. Assessing structure of aged natural sedimentary clays, *Soils and Foundations* 40(3): 1–16.

Shibuya, S., Mitachi, T., Tanaka, H., Kawaguchi, T. & Lee, I-M. 2001. Measurement and application of quasi-elastic properties in geotechnical site characterization, Theme Lecture for Plenary Session 1, *Proc. of 11th Asian Regional Conference on Soil Mechanics and Geotechnical Engineering* 2: 639–710: Balkema.

Shirley, D.J. and Hampton, L.D. 1977. Shear-wave measurements in laboratory sediments, *Journal of the Acoustical Society of America* 63(2): 607–613.

Simpson, B. 2000. Engineering needs. *Prefailure Deformatioms Characteristics of Geomaterials.* 2, Theme and Keynote Lectures, IS-Torino, Balkema: 1011–1026.

Tamrakar, S.B., Shibuya, S. and Mitachi, T. 2001. A practical FE analysis for predicting deformation of soft clay subjected to deep excavation, *Proc. of 3rd International Conference on Soft Soil Engineering* 1, Hong Kong: 377-382.

Tanaka, Y., Kudo, K., Nishi, K. and Okamoto, T. 1994. Shear modulus and damping ratio of gravelly soils measured by several methods, *Pre-failure Deformation of Geomaterials* 1, IS-Hokkaido, Balkema: 47–53.

Tanaka, Y., Kudo, K., Nishi, K., Okamoto, T., Kataoka, T. and Ueshima, T. 2000. Small strain characteristics of soils in Hualien, Taiwan. *Soils and Foundations* 40(3): 111–125.

Tanizawa, F., Teachavorasinskun, S. Yamaguchi, J., Sueoka, T. and Goto, S. 1994. Measurement of shear wave velocity of sand before liquefaction and during cyclic mobility, *Pre-failure Deformation of Geomaterials* 1, IS-Hokkaido, Balkema: 63–68.

Tatsuoka, F. and Shibuya, S. 1992. Deformation characteristics of soils and rocks from field and laboratory tests, Keynote Paper, *Proc. of 9th Asian Regional Conf. on SMFE* 2: 101–170.

Tatsuoka, F. and Kohata, Y. 1994. Stiffness of hard soils and soft rocks in engineering applications, *Pre-failure Deformation of Geomaterials* 2, IS-Hokkaido, Balkema: 947–1063.

Tatsuoka, F., Tateyama, M. and Koseki, J. 1996. Performance of Soil Retaining Walls for Railway Embankments, *Soils and Foundations*, Special Issue of Soils and Foundations on Geotechnical Aspects of the January 17 1995 Hyogoken-Nanbu Earthquake: 311–324.

Temma, M., Shibuya, S. and Mitachi, T. 2000. Evaluating ageing effects of undrained shear strength of soft clays, *Proc. of International Symposium on Coastal Geotechnical Engineering in Practice* 1, Yokohama: 173–179.

Temma, M., Shibuya, S., Mitachi, T. and Yamamoto, N. 2001. Interlink between metastability index, MI(G) and undrained shear strength in aged holocene clay deposit, *Soils and Foundations* 41(2): 133–142. (in Japanese)

Towhata, I. and Lin, C.E. 2003. Microscopic observation of shear behavior of granular material, *Deformation Characteristics of Geomaterials* 1, IS-Lyon, Balkema: 113–118.

Viggiani, G. and Atkinson, J. H. 1995. Interpretation of bender element tests, *Géotechnique* 45(1): 149–154.

Yamashita, S. and Suzuki, T. 1999a. Stress and strain controlled cyclic undrained tests on soils, *Proc. of the 11th Asian Regional Conference on Soil Mechanics and Geotechnical Engineering.* 1: 533–536.

Yamashita, S. and Suzuki, T. 1999b. Young's and shear moduli under different principal stress directions of sand, *Prefailure Deformation Characteristics of Geomaterials* 1, IS-Trino, Balkema: 149–158.

Yamashita, S., Hori, T. and Suzuki, T. 2003. Effects of fabric anisotropy and stress condition on small strain stiffness of sands, *Deformation Characteristics of Geomaterials* 1, IS-Lyon, Balkema: 187–194.

Characterisation
Charactérisation

Deformation Characteristics of Geomaterials – Di Benedetto et al (eds)
© 2005 Taylor & Francis Group, London, ISBN 04 1536 701 8

On the mechanics of reconstituted and natural sands

M.R. Coop
Imperial College, London, UK

ABSTRACT: A framework for the mechanics of sands is presented, based on the concepts of Critical State Soil Mechanics, which has been used to analyze the behaviour of sands of a variety of mineralogies from small to large strains. The paper emphasizes the differences in the underlying micro-mechanics of sands and clays, highlighting the breakage of sand particles in both compression and shear and the differences that therefore arise between the behaviour of sands and clays. The framework developed for the behaviour of reconstituted sands is then used as a basis for the interpretation of the behaviour of natural sands, highlighting the separate effects of bonding and fabric and emphasizing the importance of when, within the geological history of the sand, the structure was created. Limitations of the framework at very large strains are identified, as are difficulties in finding an appropriate critical state type of framework for transitional soils between sands and clays.

1 INTRODUCTION

The behaviour of sands at any particular strain level; small, intermediate or large, cannot easily be described in isolation, and is better understood within a general framework that encompasses all strain levels. This paper will therefore endeavour to present a useful framework, based on the concepts of Critical State Soil Mechanics (Schofield & Wroth, 1968) but identifying how particle fracture both influences the behaviour within the framework and leads to its breaking down at very large strains. The approach used throughout is to work from high pressures towards low pressures since the behaviour of sands at high pressures is generally simpler to interpret.

Having developed a framework for the *intrinsic* behaviour of sands based on data from reconstituted samples, the behaviour of natural sands and sandstones is then examined, using the intrinsic behaviour as a reference to identify the effects of the natural structure on the mechanics.

2 RECONSTITUTED SANDS

2.1 *Intermediate-large strains*

Typical isotropic compression data for reconstituted sands are shown in Fig. 1. The sands are of very diverse geological origin and hence grading, mineralogy and particle type. Dog's Bay sand is a biogenic carbonate

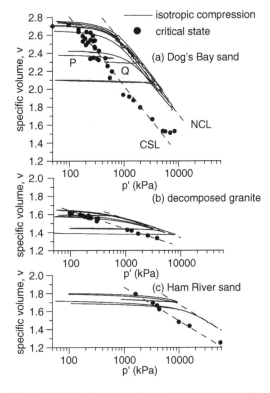

Figure 1. Isotropic compression and critical state data for three sands (adapted from Coop & Lee, 1993).

sand with delicate, angular and often hollow particles formed from shell fragments while the Ham River sand is a simple quartzitic sand. The decomposed granite has a complex mineralogy and particle structure, described in detail by Lee & Coop (1995). It is the only one of the three sands to be well-graded.

For each soil, samples had been created at a variety of densities, but the majority were moderately loose. Apart from the technique of creating the samples, the initial specific volume, v (=1 + e), is controlled by the nature of the particles, being highest for the open, angular particles of the carbonate sand and lowest for the decomposed granite because it is well-graded.

Despite the variety of geological origins, the patterns of behaviour for the three soils on Fig. 1 are the same. At higher mean normal effective stresses, p', the virgin compression curves for each density converge towards a unique isotropic normal compression line (NCL). The plastic volumetric strain that occurs as the soil is compressed down the NCL is associated with particle breakage, as can be seen from the data for Dog's Bay sand in Fig. 2 where there is a unique relationship between p' and the amount of particle breakage for states on the NCL. Here particle breakage is quantified using the relative breakage, B_r, of Hardin (1985), defined from the change in the particle size distribution curve, as indicated in Fig. 3. McDowell & Bolton (1998) have shown that isotropic normal compression is associated with the sand tending towards a fractal grading, with a fractal dimension of 2.5.

Yield in isotropic compression is then associated with the onset of particle breakage, which is gradual so that the first yield is indistinct. The yield stress in compression will then be related to the location of the NCL, which must be controlled by the strength of the particles, and also to the initial density, which will determine the number of inter-particle contacts and hence the contact stresses. Kwag et al. (1999) have related the yield stress in compression to the strength of individual particles, but as Fig. 1 demonstrates it is as much the high initial specific volume of the carbonate sands as the low strength of the particles and hence low location of the NCL in the v:ln p' plane that determines the relatively low yield stress in compression by comparison with the quartzitic sand. In unload/reload the behaviour is very much stiffer than in first loading with a much better defined yield point, which occurs at the previous maximum stress. One of the key differences between sands and clays is that the yield during first loading occurs at very much higher stresses, so that within the engineering stress range most sands will not reach their NCL and so the in situ density, and hence behaviour, will depend critically on the depositional density.

Critical states can be difficult to define for sands as large strains have to be reached during shearing, typically over 30% shear strain. Provided they are correctly determined the critical state line (CSL) is

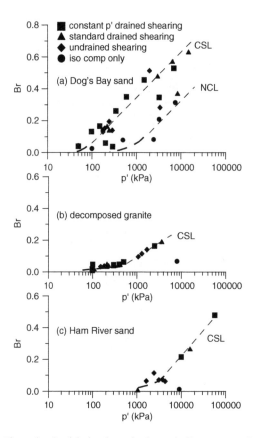

Figure 2. Particle breakage for isotropically compressed and sheared samples (after Coop & Lee, 1993).

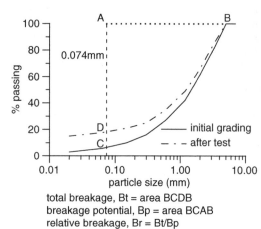

total breakage, Bt = area BCDB
breakage potential, Bp = area BCAB
relative breakage, Br = Bt/Bp

Figure 3. Definition of relative breakage, Br (Hardin, 1985).

generally found to be parallel to the NCL in the v:ln p' plane (Fig. 1) in the higher stress region where the behaviour is dominated by particle breakage. The amount of particle breakage at the critical state is then independent of the path taken to reach it (Fig. 2) so that a CSL can also be defined in the B_r:ln p' plane.

The CSL is generally believed to curve towards the horizontal at lower stress levels, so that it is represented as being curved (e.g. Verdugo & Ishihara, 1996) or bi-linear (e.g. Konrad, 1998). The increase in gradient of the line at higher values of p' is associated with the onset of particle breakage during shearing, even if Luzzani & Coop (2002) have shown that even for quartz sands sheared at low stress levels, there is a still small amount of particle breakage. One of the key problems for the mechanics of sands, as will be discussed later, is that the location of the critical state line at lower stresses is very difficult to determine, leading to problems in the application of state parameter methods with the original definition of state parameter as used by Been & Jefferies (1985).

In the q:p' plane, providing the critical states have been correctly defined, the CSL is typically found to be a straight line, so that the CSL gradient, M, does not change with particle breakage at higher stresses. The locations and gradients of the CSL and NCL are related to the grading of the soil, well-graded soils tending to have lower locations in the v:ln p' plane and lower gradients, λ, as can be seen in Fig. 4 where the grading of the carbonate sand has been varied by adding 24% fines. This is confirmed by a comparison of the NCL/CSL locations in Fig.1 for the poorly-graded quartz and carbonate sands with that of the well-graded decomposed granite. The critical state angle of shearing resistance, ϕ'_{cs} (or M) is however much less sensitive to the grading (Coop & Atkinson, 1993). Since the locations of the NCL/CSL in the v:ln p' plane are sensitive to the grading, as Pan (1999) has shown, if at any point during loading the soil is recovered and reconstituted, then on reloading the NCL/CSL locations will have changed. The behaviour of the soil is therefore controlled by its initial grading not its current grading. Care therefore needs to be exercised in interpreting the behaviour of a sand in situ from tests on reconstituted samples if the sand has experienced any particle breakage during its geological history. Quite apart from the loss of any effects of natural structure on reconstitution the difference in grading between the current soil and that which was originally deposited would alter the behaviour.

Figure 5 shows the locations of the NCL/CSLs in the higher pressure range of a variety of quartzitic and carbonate sands from the literature. The carbonate sands show a much greater variation, partly as a result of a slightly wider range of gradings (Coop & Cuccovillo, 1999) but mostly because of the much greater range of geological origins and hence particle

types, from delicate shell fragments to solid ooliths. In comparison the quartzitic sands have a much narrower range of behaviour perhaps as a result of greater similarity in their origins, particle types and gradings.

Typical stress:dilatancy data for a sand are shown in Fig. 6. At larger strains the data tend towards a unique straight line, but at smaller strains there is a curved approach path. By normalizing the shearing data, state boundary surfaces (SBS) can be identified (Fig. 7), which however show some differences with those seen for clays. The definition of the normalising equivalent pressure p'_p is given in Fig. 8. In particular the CSL is not at the apex of the surface but is to the left, so that the separation of the CSL and isotropic

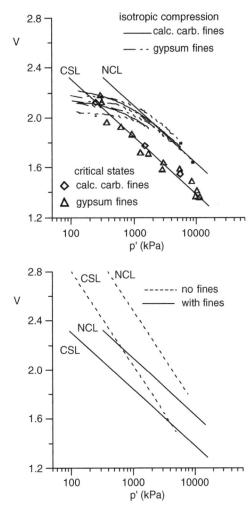

Figure 4. Isotropic compression data and critical states for Dog's Bay sand with added fines (adapted from Coop & Atkinson, 1993).

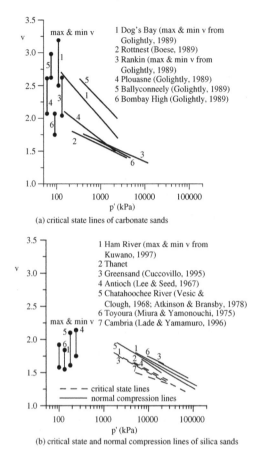

(a) critical state lines of carbonate sands

1 Dog's Bay (max & min v from Golightly, 1989)
2 Rottnest (Boese, 1989)
3 Rankin (max & min v from Golightly, 1989)
4 Plouasne (Golightly, 1989)
5 Ballyconneely (Golightly, 1989)
6 Bombay High (Golightly, 1989)

1 Ham River (max & min v from Kuwano, 1997)
2 Thanet
3 Greensand (Cuccovillo, 1995)
4 Antioch (Lee & Seed, 1967)
5 Chatahoochee River (Vesic & Clough, 1968; Atkinson & Bransby, 1978)
6 Toyoura (Miura & Yamonouchi, 1975)
7 Cambria (Lade & Yamamuro, 1996)

(b) critical state and normal compression lines of silica sands

Figure 5. Critical state and normal compression lines for various sands (after Coop & Cuccovillo, 1999).

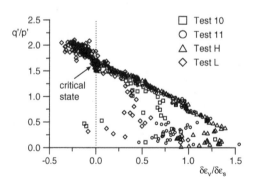

Figure 6. Stress-dilatancy data for Dog's Bay sand (after Coop, 1990).

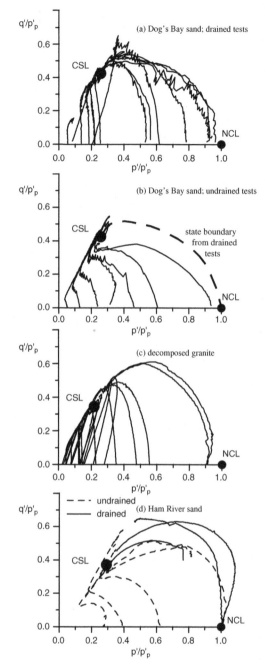

Figure 7. Normalised shearing data (after Coop & Lee, 1993).

NCL is greater and normality cannot apply to the plastic strain increment vectors. The other key difference is that undrained tests follow paths under the SBS identified from the drained.

This is most noticeable on the wet side of the CSL but also on the dry side there is a slight tendency for the paths to lie a little under the SBS. The success of the normalisation in producing unique surfaces

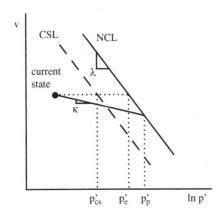

Figure 8. Definition of the normalising parameters.

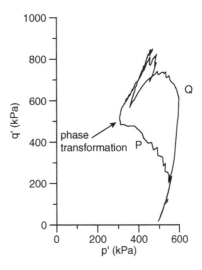

Figure 9. Comparison of undrained tests on compacted (P) and overconsolidated (Q) samples of Dog's Bay sand (after Coop, 1990).

indicates that it is the distance of the current state from the locations of the CSL/NCL that controls the behaviour of the soil, for example the peak angle of shearing resistance, ϕ'_p, for states on the dry side of the CSL. To quantify this distance a ratio of stresses is preferred to a difference in volume such as the state parameter, ψ, (Been & Jefferies, 1985), as will be discussed later.

Another key difference between sands and clays is that sands can reach a given initial state by a variety of paths. For example, on Fig.1 samples P and Q have reached the same initial v and p', but whereas Q has achieved its initial state by means of overconsolidation from the NCL, P has undergone first loading only after an initial compaction. The difference in particle breakage that they will have undergone then has an impact on their subsequent shearing behaviour, as illustrated in Fig. 9. In undrained loading, the compacted sample has the typical S-shaped path often seen for tests on reconstituted sands, with a point of phase transformation at intermediate strains. However the overconsolidated sample does not show a phase transformation and is considerably stiffer. Again care would be needed in interpreting the in situ behaviour of a sand from tests on reconstituted samples. If it had been overconsolidated geologically then compacting it back to the same density in the laboratory would not give the same behaviour.

2.2 Small strains

Figure 10 presents elastic shear moduli, G_0, during isotropic first loading and unload-reload for the Dog's Bay sand for which compression data were given in Fig. 1(a). Here the term "compacted" will be used to refer simply to samples that are under first loading only in compression, whether they are loose or dense. The measurements were made with platen mounted bender elements in a high pressure triaxial apparatus,

and therefore correspond to G_{vh}, with vertical propagation and horizontal polarization. The initial values of G_0 depend on the initial density of the sample, but as the confining stress increases and the state reaches the NCL in the v:ln p' plane, the values of G_0 also converge towards a unique line that corresponds to the NCL, denoted $G_{0(NCL)}$. This line may be represented by the same equation that Viggiani & Atkinson (1995) used for normally consolidated reconstituted clays:

$$\left(\frac{G_{0(nc)}}{p_r}\right) = A_0 \left(\frac{p'}{p_r}\right)^{n_0} \tag{1}$$

where $G_{0(nc)}$ is the value of G_0 on the $G_{0(NCL)}$, p_r is a reference pressure and A_0 and n_0 are constants.

On Fig. 11 the locations of a number of $G_{0(NCL)}$ for sands in the literature have been identified. There is very much less variation in the location of the $G_{0(NCL)}$ for sands of different mineralogies than for the NCL/CSL in the v:ln p' plane (Fig. 5).

On unload/reload from states on the NCL, the stiffnesses are significantly higher than for first loading (Fig. 10), but the unload and reload data are indistinguishable and there is no evidence of the small hysterisis loop that can be seen in the v:ln p' plane.

In order to make a correct comparison between the data for first loading and the overconsolidated samples, the data have first to be normalised for the effects of their different densities. This has been done in Fig. 12 using p'_e for volume and the normalising parameter $G_{0(nc)}$ for stiffness, which is the value of G_0 on the $G_{0(NCL)}$ at the current value of p', as indicated in

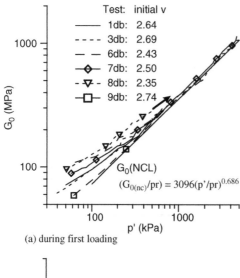

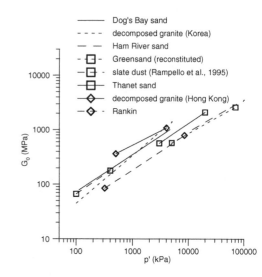

Figure 11. G_0 on the NCL for various sands. (after Coop & Jovicic, 1999).

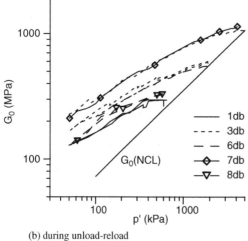

(a) during first loading

(b) during unload-reload

Figure 10. Measurements of G_0 for Dog's Bay sand (after Jovicic & Coop, 1997).

Fig. 13. Using this normalisation the $G_{0(NCL)}$ plots as a single point at the coordinates 1,1. The resulting plots illustrate some important features of the small strain behaviour of sands. The first is that once normalised in this way the data for first loading (compacted) for all initial densities lie on a unique line, so that as for the larger strain behaviour, it is clear that it is the state relative to the NCL that controls stiffness. Similarly, the data for all overconsolidated samples plot on a unique line, but a different line to that for the compacted soil, again emphasising the effect that different means of reaching the same state have on the behaviour.

Using the logarithmic axes chosen, then the two unique lines may be represented, if imperfectly, by two straight lines, in which case the same equation as was used by Rampello et al. (1995) for reconstituted clays may be used:

$$G_0 = G_{0(nc)} \left(\frac{p'_e}{p'} \right)^c \qquad (2$$

The only difference is that two values of c are required one for the compacted samples and one for the overconsolidated. This again emphasises dangers in interpreting the in situ stiffness of a geologically over-consolidated sand from tests on reconstituted samples even if they are compacted to the same density.

The traditional approach in analysing the stiffness of sands is to define it in terms of two separate functions, one for p' and one for voids ratio, and there is much debate in the literature as to suitable functions Such an approach could never reproduce accurately the patterns of behaviour seen here firstly because it fails to recognise that p' and v are not independent, whereas p' and state (i.e. p'/p'_e) are, and secondly because it could not capture the difference between compacted and overconsolidated samples. The gradient c represents the effect of state, or distance from the NCL on the stiffness, and it can be seen that this is much smaller for quartzitic sands than for the carbonate sands. The difference between the overconsolidated and compacted samples is also least for the quartzitic sand and so failure to account properly for these effects will be least serious in these sands.

As illustrated in Fig. 14, it is not only possible to reach a given state by overconsolidation from the NCL

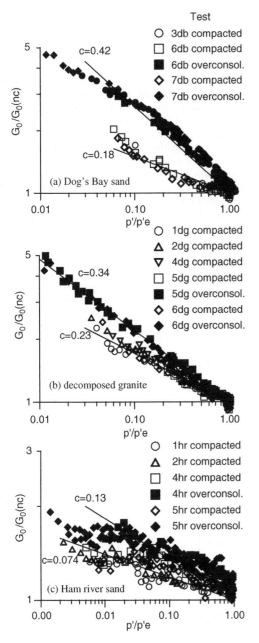

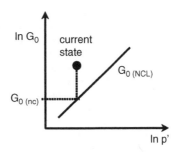

Figure 13. Definition of $G_{0(nc)}$.

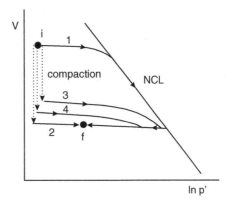

Figure 14. Combinations of compaction and overconsolidation to reach a given final state (f).

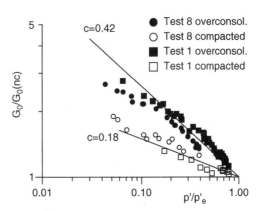

Figure 12. Variation of G_0 with normalised volumetric state (after Coop & Jovicic, 1999).

Figure 15. Variation of G_0 with normalised volumetric state for Dog's Bay sand unloaded prior to reaching the NCL (adapted from Jovicic & Coop, 1997).

(Path 1) or compaction followed by first loading (Path 2), but also by a reduced degree of compaction followed by unloading from a state prior to reaching the NCL (Paths 3 and 4), so that a given state may be reached by an infinite number of paths. As shown in Fig. 15 the consequence of unloading prior to reaching

the NCL is that the stiffnesses plot below the overconsolidated line (e.g. Test 8). It can therefore be seen that the two unique lines representing compacted and overconsolidated samples actually define the boundaries of a zone of possible stiffnessess that can be reached by varying degrees of compaction and overconsolidation.

Fig. 16 shows tangent stiffnesses for larger strain levels from samples sheared undrained from states on the NCL. The stiffnesses were measured using a system of LVDTs attached directly to the sample (Cuccovillo & Coop, 1997a). The good agreement between the tangent stiffnesses (G_u) at the smallest strain levels for which they could be defined (0.0001%) and the bender element stiffnesses (G_{vh}) indicate that in this case for states on the NCL there is no significant anisotropy of stiffness. For each strain level a straight line is obtained, the gradient of which tends to increase with strain level while the intercept reduces, so that they tend to converge as p' increases. These may again be represented by Equation 1 but using variable values of A and n rather than the fixed values A_0 and n_0 for the elastic stiffness. Figure 17 shows the variation of A and n with strain level, illustrating how as strain increases A tends to zero as n seems to tend towards unity. Again a very similar pattern of behaviour was seen by Viggiani & Atkinson (1995) for reconstituted clays.

For clays, the stiffness at intermediate strain levels has been found to depend strongly on the previous loading path (Atkinson et al., 1990). For consistency with the definitions used for clays, the effects of recent stress history on sands should be investigated for truly overconsolidated states only, i.e. for samples that have all been unloaded from the NCL. This has been done for most of the tests on Fig. 18. The different approach paths are illustrated schematically in Fig. 18a. These followed an initial isotropic compression to the isotropic NCL reaching isotropic stresses of 800 kPa for the tests on Dog's Bay sand and 1200 kPa for those on decomposed granite, while the

initial p' for shearing was 300 kPa for the Dog's Bay sand and 600 kPa for the decomposed granite. The shearing was undrained for the Dog's Bay sand but followed a constant p' drained path for the decomposed granite. The other key difference between the two sets of data is that while for the tests on Dog's Bay sand no waiting time was allowed between the approach and shearing paths, for those on the decomposed granite 24 hours waiting time was used.

While there is a clear effect of recent stress history for zero waiting time, 24 hours allowed for creep appears to erase this effect in the case of the decomposed granite. Figure 18(c) also illustrates the importance of investigating the effect of recent stress history only on truly overconsolidated samples, as the stiffness of a compacted sample that has reached the initial state by first loading only is significantly lower those of the other tests.

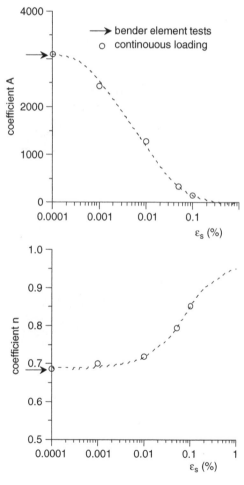

Figure 17. Variation of parameters A and n with strain for Dog's Bay sand (after Jovicic & Coop, 1997).

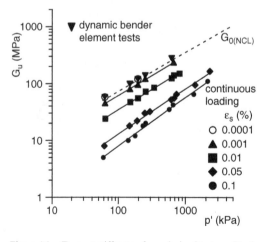

Figure 16. Tangent stiffnesses for undrained tests on Dog's Bay sand at states on or near the NCL (after Jovicic & Coop, 1997).

2.3 Quantifying state

As for the larger strain behaviour, the normalisation for state in the v:ln p' plane has been in terms of a ratio

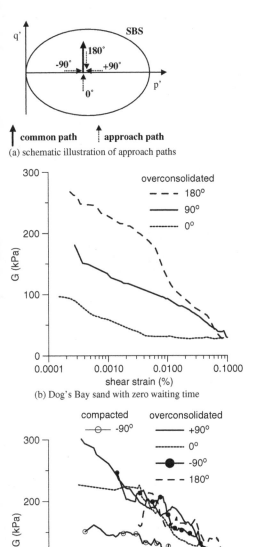

(a) schematic illustration of approach paths

(b) Dog's Bay sand with zero waiting time

(c) decomposed granite with 24 hours waiting time

Figure 18. Recent stress history effects in sand (after Amoroso, 1997).

of stresses, thereby quantifying the horizontal distance of the state from the NCL or CSL rather than using the difference in volume as the vertical distance, such as the state parameter Ψ. If the CSL were straight over the complete range of interest, then the two methods should be interchangeable. However, as Fig. 19 illustrates schematically, for curved CSLs there will be significant differences. In particular, in the typical range of engineering stresses, for a given density ψ will vary much less with stress level than the ratio of stresses. It is also much more important for ψ than for a stress ratio to establish exactly the location of the CSL for a particular sand at lower stress levels, but as Klotz & Coop (2002) among others have highlighted the identification of the CSL in this region is far from straight forward.

A series of tests was conducted using both local instrumentation and global measurements of volumetric strain. The local instrumentation consisted of a radial strain belt mounted at the sample mid-height and two local axial strain transducers. Figure 20

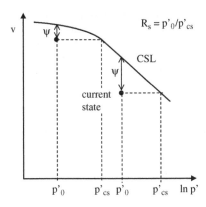

Figure 19. Definition of state parameters.

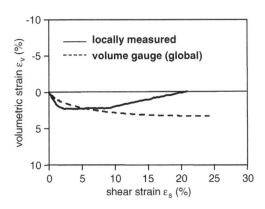

Figure 20. Comparison of local and global measurements of volumetric strain (adapted from Klotz & Coop, 2002).

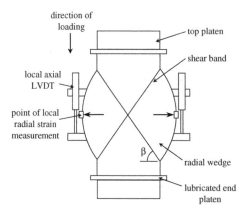

Figure 21. Schematic illustration of effect of localization of strains on radial strain measurements (after Klotz & Coop, 2002).

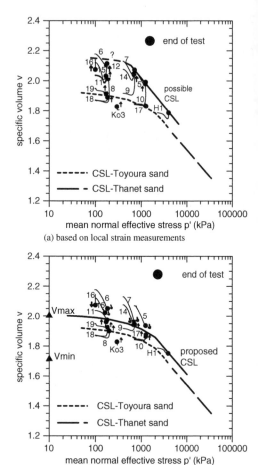

(a) based on local strain measurements

(b) based on global strain measurements

Figure 22. The identification of the critical state line for Leighton Buzzard sand at lower stress levels (after Klotz & Coop, 2002).

shows a comparison between the local and global measurements of volumetric strain for a drained test at 100 kPa effective cell pressure on a medium dense sample of a quartzitic sand (Leighton Buzzard sand). The local measurement, calculated from $\varepsilon_a + 2\varepsilon_r$, is significantly different to that from the volume gauge, with a tendency towards continued dilation at larger strains rather than reaching an apparent critical state like the global strains. This was found to be partly because the sample was found to barrel slightly at larger strains, despite the use of lubricated end platens, but also partly because of the localization of strains, as illustrated in Fig. 21, so that continued axial displacement gave rise to a continued increase of the local measurement of radial strain through a rigid body mechanism.

The choice then of whether to follow the local or global measurements of volumetric strain will determine the location of the CSL in the v:ln p' plane as illustrated in Fig. 22. If the local strain data are used (Fig. 22a) then the paths followed during the test indicate a large dilation for all initial densities, which generally does not cease, as indicated by the arrows. If the local strain data are corrected to agree with the global measurements of ε_v (Fig. 22b) then the looser samples compress and the denser ones dilate. This allows a CSL to be identified that then passes through the minimum density at low stress levels, following an exponential path as proposed by Gudehus (1996), while being straight at higher pressures. This CSL, based on global measurements of volumetric strain is consistent with those proposed by others that have used the state parameter method (e.g. Konrad, 1998) as well as other authors who have investigated in detail the location of the CSL at lower stresses, for example that for the Toyoura sand shown on the figure, reported by Verdugo & Ishihara (1996).

The location of the CSL for Ham River sand has been reexamined at lower stresses in Fig. 23, using additional data from Skinner (1975) and Ovando-Shelley (1986) and choosing again an exponential function that passes through the minimum density at low stresses. The G_0 data from Fig. 12 have then been reanalysed in terms of state parameter, ψ, as shown in Fig. 24. The method is much less successful in determining any consistent pattern of behaviour. There is a much greater scatter of data than on Fig. 12 and the method fails to distinguish between the compacted and overconsolidated samples. This calls into question whether the CSL identified is indeed correct and perhaps the usefulness of trying to identify a CSL in the region where the data are so badly affected by strain localization, particularly when the stress ratio type of state parameter does seem to work well.

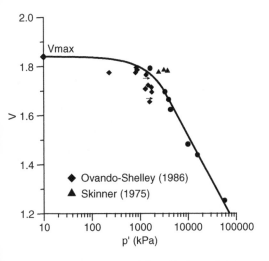

Figure 23. Reinterpretation of the critical state line location for Ham River sand at lower stress levels.

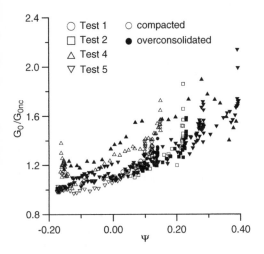

Figure 24. Analysis of G_0 data for Ham River sand in terms of state parameter.

2.4 Creep

Creep has a significant effect on the stiffness of sands, as shown in Fig. 25. In each case the current stiffness is normalised with respect to that measured when a given stress was first reached ($G_{0(t=0)}$), having applied the initial stress quickly. The effect it has depends on the state of the soil, with a much greater effect at a given stress in first loading than for unloading or reloading (Fig. 25(a)). Perhaps surprisingly the effect of creep for a given state reduces as stress level increases as shown by the data for first loading states in Fig. 25(b).

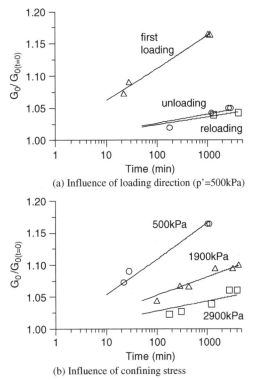

(a) Influence of loading direction (p'=500kPa)

(b) Influence of confining stress

Figure 25. Effect of creep on G_0 for Dog's Bay sand (after Jovicic & Coop, 1997).

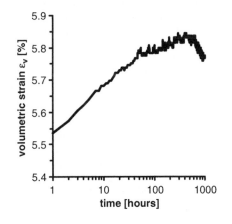

Figure 26. Volumetric creep data for Dog's Bay sand (after Gasparre et al., 2003).

Creep also gives rise to a significant volumetric strain as can be seen in Fig. 26. Having applied an isotropic effective stress to sample of Dog's Bay sand of 500 kPa in about 1 hour, giving an initial volumetric strain of 5.54%, over a period of about a month the

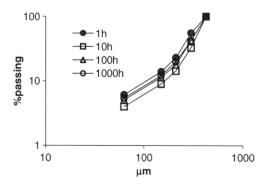

Figure 27. Particle size distributions for samples of Dog's Bay sand under an isotropic stress of 500 kPa (after Gasparre et al., 2003).

volumetric strain increased by around 5% of this initial value to about 5.8% where it eventually appears to stabilise. The volumetric strains in Fig. 26 were calculated from measurements of local axial and radial strains made with LVDTs (Klotz & Coop, 2002) so that there would be no errors in the data that might become evident over such a long period from small deficiencies in saturation or leaks within the drainage system.

While the effects of creep on the volumetric strain and stiffness are clear, the effect on particle breakage is not. Figure 27 shows particle size distribution data not from creep tests but from samples isotropically compressed to the same final p' of 500 kPa over different periods of time from 1 to 1000 hours. The pattern of particle breakage is uncertain and it seems that there is even greater breakage for 1 hour than for the slower rates. However it is clear that the increases of volumetric strain and stiffness that occur during creep are not simply linked to an increase of particle breakage. This may be because the stresses within sands are not evenly distributed and instead are transmitted through strong force columns in which the particle contacts are highly stressed, while other particles between the columns remain lightly loaded. It is perhaps the case that only very small amounts of breakage, which are undetectable by crude sieving techniques, are required at the particle contacts for the columns to realign in a more stable configuration, allowing a small amount of particle rearrangement and there to be a volumetric strain and an increase of stiffness.

2.5 Dry sands

Some sands have particles that are stronger when they are dry than when they are saturated. An example is decomposed granite for which typical stress:strain data are given in Fig. 28. For the same initial density the dry soil has a higher peak strength, corresponding to greater dilation. Whenever the soil is flooded it

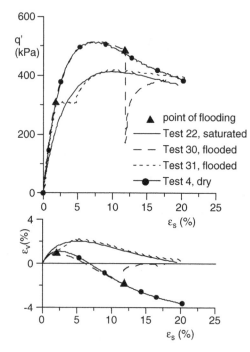

Figure 28. Stress:strain behaviour for dry, saturated and flooded samples of decomposed granite (after Lee & Coop, 1995).

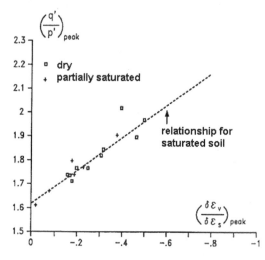

Figure 29. Stress: dilatancy relationship for dry and partially saturated samples (after Lee & Coop, 1995).

quickly reverts to the behaviour of the saturated soil. In Fig. 29 the relationship between the rates of dilation at peak and the peak stress ratio, q/p' can be seen to be the same as that for the saturated soil, and the

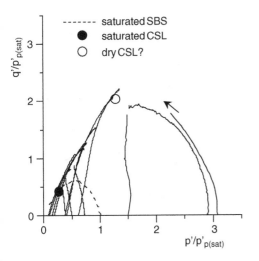

Figure 30. Normalised shearing data for dry decomposed granite (after Lee & Coop, 1995).

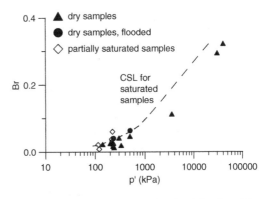

Figure 31. Particle breakage for dry, flooded and partially saturated samples (after Lee & Coop, 1995).

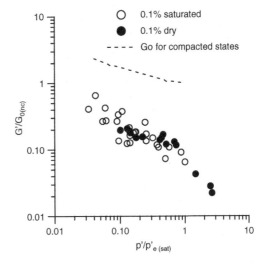

Figure 32. Comparison of tangent shear stiffnesses, G' measured for drained tests on dry and saturated samples of decomposed granite (after Jovicic & Coop, 1997).

key difference between them is therefore only that the dry soil has a greater tendency to dilation, with its CSL lying above that of the saturated soil in the v:ln p' plane. Normalising the shearing data for the dry samples with respect to the NCL of the saturated soil ($p'_{p(sat)}$), shows that the SBS of the dry soil is considerably larger than that of the saturated (Fig. 30). This difference of behaviour in the dry state was found to be due to the particles being stronger, as shown by lower amount of particle breakage that they undergo in Fig. 31. Coop & Lee (1995) found that the carbonate Dog's Bay sand was similarly affected, but that the quartzitic Ham River sand was not.

Small strain data for the decomposed granite are given in Fig. 32, where again the state has been normalised with respect to the location of the saturated NCL, so that some of the dry samples plot at values of p'/p'_p greater than unity. There is no significant difference between the stiffnesses of the dry and saturated soil, which demonstrates that the particle strength and particle breakage affect only the large strain shear behaviour and not that at small strains. This may explain why on Fig. 11 sands of very different particle types and strengths had fairly similar elastic stiffnesses.

3 TRANSITIONAL SOILS

The framework for reconstituted sands described above has much in common with the behaviour of clays, even if the mechanisms of plastic strain are very different, with the behaviour of sands being dominated by particle breakage at higher stress levels. The question then arises as to what the behaviour would be of transitional soils, intermediate to sands and clays and clays. An example of the compression behaviour of reconstituted samples of a well-graded residual soil is given in Fig. 33; the gradings curve is shown in Fig. 34. The compression behaviour is very sensitive to the initial density, and there is little convergence of the data for different densities and there is no evidence of a NCL.

For sands the initial compression curves for different densities are relatively flat, so when a sufficiently high stress is reached, particle breakage commences and the compression paths converge on a NCL. In contrast the compression paths for different initial densities of the transitional soil are relatively steep, so

41

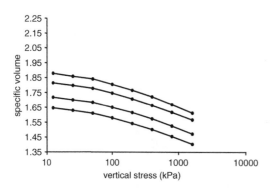

Figure 33. Compression data for reconstituted Botucatu sandstone (adapted from Burmeister-Martins et al., 2002).

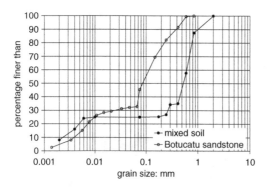

Figure 34. Gradings of the two transitional soils (after Burmeister-Martins et al., 2002).

that it is unlikely that convergence would be achieved no matter what stress level was applied, as the paths approach zero voids ratio more quickly than any possible convergence between them.

It was believed by Burmeister-Martins et al. (2002) that the reason for this unusual behaviour might have been the slight gap-grading of the Botucatu residual sandstone and a second series of tests was carried out by them on a model soil created from a quartzitic sand and kaolin. This "mixed" soil, the gradings curve for which is also given in Fig. 34, emphasised the gap-grading, and again gave a non-convergent compression behaviour (Fig. 35). However, more recently work by Colleselli et al. (2003) has identified similar behaviour for well-graded silts, for which the gradings curves are continuous. It may therefore be a more widespread feature of the behaviour of transitional soils and it is important that future research should identify which gradings and particle mineralogies are affected as many natural sands have some fines content. Soils of similar grading are also often used in research on partially saturated soils as they combine

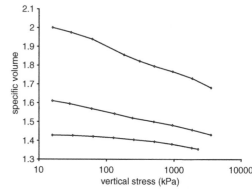

Figure 35. Compression data for quartz sand and kaolin mixed soil (adapted from Burmeister-Martins et al., 2002).

the relatively high permeabilities required to minimize test durations with a reasonably plastic behaviour, so that it may not be wise to use constitutive models for clays as the base for the saturated behaviour to which the effect of suction is then added.

4   NATURAL SANDS

4.1   *Introduction*

For clays, structure has been defined as the combination of fabric and bonding (Mitchell, 1976) and which may often be more easily identified through their effects on the mechanical behaviour than through direct observation, for example with an electron microscope. Comparisons between the soil in its natural state with the same soil reconstituted in a standard way can easily define the contributions to strength and stiffness of the natural structure (e.g. Burland, 1990), although separating the effects of fabric and bonding is more problematic (e.g. Baudet & Stallebrass, 2004).

The effects of structure have been identified in general terms for a wider range of materials, including residual soils and weak rocks (e.g. Leroueil & Vaughan, 1990), although for other materials the intrinsic behaviour may be more difficult to determine as it can be difficult to reconstitute the soil. Many early studies of the behaviour of weak rocks (e.g. Coop & Atkinson, 1993) used artificially cemented materials, so that the intrinsic behaviour could be determined from a soil comprising the same constituent materials but left uncemented. More recently Cuccovillo & Coop (1997b) have managed to reconstitute weak sandstones, breaking the cement without significant damage to the particles.

For sands bonding typically takes the form of interparticle cementing, and is usually recognised more easily than for clays, generally being identifiable through optical microscopy. The concept of fabric in sands has received less attention, although the effects of locked fabrics have been investigated by Dusseault & Morgenstern (1979) amongst others.

For sands some of the features of the behaviour of the reconstituted soils, identified above, suggest that the identification of the intrinsic behaviour from reconstituted samples may be more problematic than for clays. In particular, if a sand has undergone significant particle breakage due to geological loading in situ, then the same soil reconstituted would give an incorrect intrinsic behaviour as it would start with a different initial grading to the natural soil. For any natural sand that has been geologically overconsolidated then simple compaction back to the in situ density in the laboratory will also not be sufficient to identify the correct in situ behaviour. For transitional soils, the problems in identifying the intrinsic behaviour are even more acute, because the intrinsic behaviour is so strongly dependent on the initial density.

4.2 Inter-particle cementing

The simplest form of structure in a natural sand is inter-particle cementing; Fig. 36 shows two examples. The first is Rankin calcarenite (Fig. 36a), which is of early Pleistocene age (1.6–1 m years BP). The thin section is shown under cross-polarized light so that the void spaces are black. The particles are shell fragments, similar to the Dog's Bay sand, and the calcium carbonate cement can be seen as a white fringe that surrounds each particle so forming a bond between them. The cement would have been deposited soon after the deposition of the particles, after which the soil underwent continuous burial to its current depth, so that the cement preserved the open fabric of the soil at deposition with high specific volumes (1.63–2.05). The second example is the Lower Greensand (Fig. 36b), which is a quartzitic sandstone. This soil was deposited in the Lower Cretaceous (146–97 m years BP) and was buried deeply to a maximum past vertical effective stress of around 9 MPa before being unloaded back to the ground surface through erosion. In the Scanning Electron Micrograph the cement shows as small patches of white, some of which are at particle contacts and so provide an inter-particle bonding. This cement is iron oxide, which was deposited from water flowing through the soil relatively recently and after geological overconsolidation, when the soil was already near ground level.

Isotropic compression data for the two sandstones are shown in Fig. 37. The samples of Rankin calcarenite were from cores over a range of depths (127–142 m) and since the density and degree of cementing are spatially variable, there is a range of initial specific

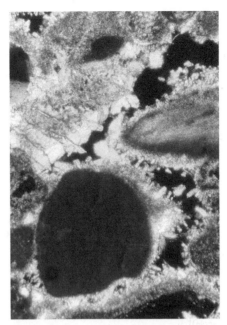

(a) Thin section of a Rankin calcarenite (width = 1mm)

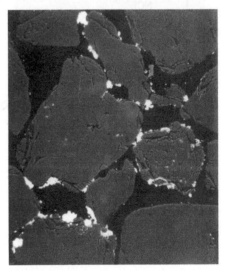

(b) SEM image of Lower Greensand (width = 0.5mm)

Figure 36.   Inter-particle cementing of sands.

volumes. In contrast the Greensand samples were taken from a single block, so that even if there is again spatial variability in this soil, all of the samples tested were identical. The key difference between the two sandstones is that while the calcarenite reaches states outside the intrinsic isotropic compression line defined by tests on the reconstituted soil, the Greensand does not.

43

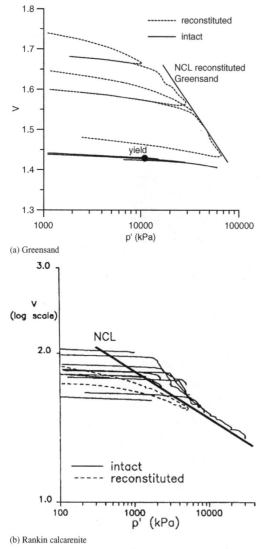

(a) Greensand

(b) Rankin calcarenite

Figure 37. Isotropic compression data for two sandstones (after Cuccovillo & Coop, 1999).

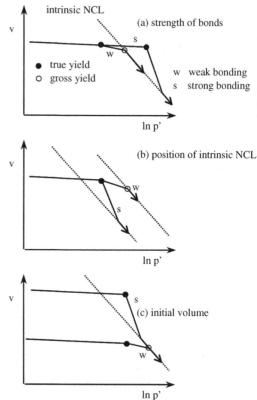

Figure 38. Schematic representation of the factors that might influence the effect that inter-particle cementing has on the compression behaviour of a sand (after Coop & Willson, 2003).

The yield of the Greensand is indistinct but closer examination of the bulk modulus reveals a perfectly elastic initial behaviour followed by a well-defined yield at about 11 MPa.

When yield occurs outside the intrinsic NCL it is defined as "strong" bonding, and "weak" when it occurs inside as illustrated in Fig.38a. In the case of strong bonding the behaviour is elastic up to a yield that occurs at higher stresses than would be required to initiate particle breakage in the uncemented soil, so that the cement is carrying some of the confining stress and preventing breakage. Degradation of the cement bonds then must coincide with the onset of particle breakage at the yield point. However, in the case of weak bonding, the yield of the cement occurs before the onset of the particle breakage, so that the cement yield is less distinct and a second yield point or "gross yield" will be associated with the start of particle breakage.

From the comparison of these two soils it may be concluded that whether the bonding is weak or strong is primarily related to the amount and strength of the cement deposited, with a smaller amount clearly being present for the Greensand. However, it may also be deduced that the position of the intrinsic NCL must be a factor (Fig. 38b) as the Greensand, with stronger quartz particles has its intrinsic NCL at a higher location in the v:ln p' plane than the calcarenite with its weak particles composed of shell fragments. It is therefore the relative strength of the particles and bonding material that is critical.

The initial density is also important, as illustrated in Fig. 38c. The Greensand is much denser than the calcarenite because the latter has angular open particles and cementing preserved its high depositional specific volume while the former has solid quartz particles and was subjected to high compressive stresses for many millions of years prior to cementing. The initial density and location of the intrinsic NCL are of such importance that even if the Greensand has far the higher yield stress in compression, it is the one with the weak mode of behaviour.

For the Rankin calcarenite samples of different initial densities it can be seen that the paths followed after yield coincide, and that the yield points occur slightly below this post-yield compression path. Both yield points and the post-yield compression path converge with the intrinsic NCL at higher pressures. With a natural soil, systematic investigation is hampered by natural variability and many researchers have tested artificially cemented materials. An example is shown in Fig. 39 where Rotta et al. (2003) have tested a sand that has been artificially cemented with Portland cement, curing taking place while the soil was at a variety of confining stresses on the intrinsic NCL, thus replicating cementing at depth. This is in contrast to the calcarenite, which was cemented prior to burial, but the patterns of behaviour are similar, a schematic of which is given in Fig. 40. Examples of two cement strengths are given, the stronger having its yield locus and post-yield compression line further outside the intrinsic NCL. There is a clearer difference between the yield locus and post-yield compression line than for the calcarenite, but again both converge with the intrinsic NCL at higher stresses.

In shearing the behaviour will depend on the state relative to the yield locus of the cement. The patterns of behaviour seen are illustrated schematically in Fig. 41 and in Fig. 42 stress:strain data are given for a calcarenite artificially cemented with gypsum. At low stresses the behaviour is elastic up to a yield that almost coincides with a peak strength that is cohesive in nature and considerably higher than the intrinsic strength of the uncemented soil at similar densities and confining stresses. At intermediate stresses the yield occurs during shearing, but before reaching the critical state strength, so that yield is followed by plastic straining up to the critical state, and the cement therefore only affects the initial stiffness and not the strength.

Normalising the shearing data (Figs 43 & 44) with respect to an equivalent pressure p'_p taken on the intrinsic NCL, the cemented soil reaches states outside the intrinsic state boundary surface on both dry and wet sides, so that the yield surface is larger than the intrinsic SBS. For states beyond yield in isotropic compression the normalised shearing path collapses back towards the intrinsic SBS.

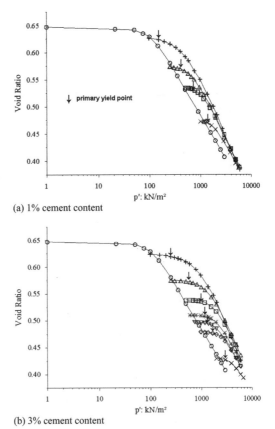

(a) 1% cement content

(b) 3% cement content

Figure 39. Isotropic compression data for sand samples of various densities cemented on their NCL (after Rotta et al., 2003).

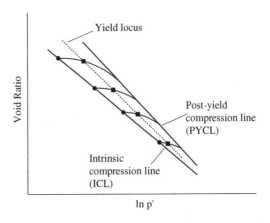

Figure 40. Schematic diagram illustrating patterns of behaviour for cemented sands (after Rotta et al., 2003).

45

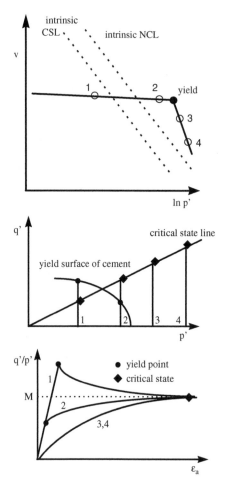

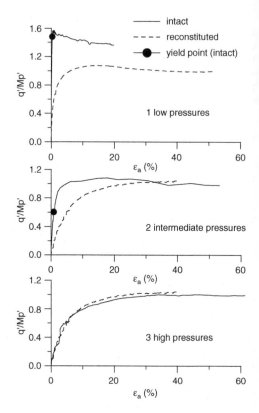

Figure 42. Stress:strain behaviour of the "strongly" bonded artificially cemented calcarenite (adapted from Cuccovillo & Coop, 1993).

Figure 41. Schematic illustration of typical shearing behaviour for strong bonding.

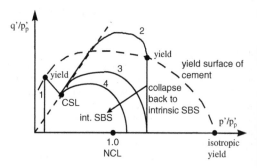

Figure 43. Characteristic normalised shearing behaviour of a "strongly" bonded sand.

The behaviour of the weakly bonded sandstone is illustrated in Figs 45 & 46. The key difference is that at intermediate stresses beyond the yield of the cement, the sandstone still has a peak strength resulting from dilation on shearing, so that the peak strength envelope in not restricted to the yield locus of the cement. Normalising the shearing data (Figs 47 & 48), the yield locus of the cement occupies only a small part of the intrinsic SBS and states outside the intrinsic SBS are only reached on the dry side of critical, at low confining stresses. In Fig. 48 the normalisation is with respect to an equivalent pressure on the CSL rather than the NCL (Fig. 8), as the NCL could not be reached within the pressure capacity of the apparatus. It was also found that the CSL of the cemented soil had a volumetric offset in the v:ln p' plane, lying higher than that of the uncemented soil and so the

shearing behaviour of each soil has been normalised with respect to its respective CSL. It is not known to what extent the different CSLs result in some way from the presence of the cement, and to what extent they are caused by a change of grading that the soil may have experienced in situ due to particle breakage prior to cementing.

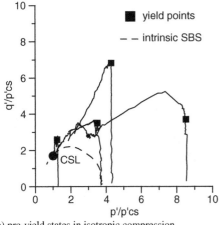

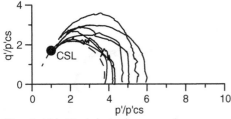

(a) pre-yield states in isotropic compression

(b) post-yield states in isotropic compression

Figure 44. Normalised shearing behaviour for Rankin calcarenite (after Cuccovillo & Coop, 1993).

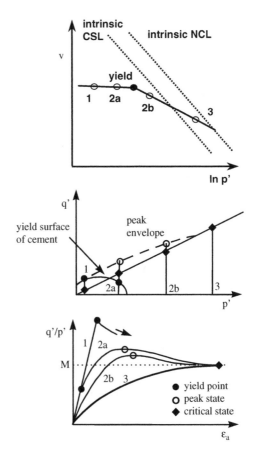

Figure 45. Idealised stress:strain behaviour for weak bonding.

Under monotonic shearing, a cemented sand often shows an initially linear elastic behaviour, with a well-defined yield, as shown in the example in Fig. 49. The initial stiffness, G_0, is insensitive to confining stress for both the calcarenite and the Greensand (Fig. 50), but in both cases is very much higher than for the reconstituted sands. When the "weakly bonded" Greensand yields in isotropic compression, the value of G_0 starts to increase slowly, but for the "strongly bonded" calcarenite, yield in compression is accompanied by a small decrease as the contribution to the stiffness from the bonding is gradually lost.

For small numbers of cycles, cyclic loading below the cement yield locus does not change the stiffness (Fig. 51a). Cycles that cross the yield locus lead firstly to a decrease in the yield stress, indicating that the cement yield locus shrinks, and once the yield stress has reduced to zero, there is then a reduction in G_0, as shown in the example in Fig. 51b.

The linear elasticity seen for the Calcarenite and Greensand is in sharp contrast to the type of behaviour seen by petroleum engineers for hydrocarbon bearing weak sandstones. An example of the type of behaviour seen for these materials is given in Fig. 52, for Castlegate sandstone. This is a chlorite-cemented

quartzitic sandstone with an unconfined compressive strength of around 5 MPa. At all but the highest stress levels, there is a well-defined and almost constant initial stiffness, with again a clear yield as for the Rankin calcarenite and Greensand. However, the strain at yield is an order of magnitude higher and the initial stiffness is both dependent on the confining stress and much lower than for the other cemented materials. Both the pressure dependency and actual values of the initial stiffness are similar to those for reconstituted sands (Fig. 53). This type of behaviour has led petroleum engineers to use a pressure dependent form of elasticity (e.g. Zimmerman, 1991).

Closer examination of the initial part of the stress:strain curve reveals that the although the initial stress:strain behaviour is linear, it is not elastic and so the initial stiffness cannot be G_0. This behaviour has been attributed in the petroleum industry to the presence of micro-cracks within the cement (Plona & Cook, 1995), created by the stress release on sampling

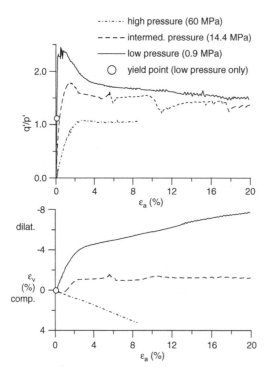

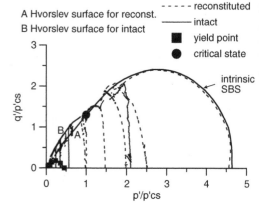

Figure 48. Normalised shearing behaviour of the Greensand (after Cuccovillo & Coop, 1999).

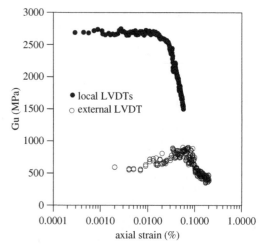

Figure 46. Stress:strain behaviour of the Greensand (adapted from Cuccovillo & Coop, 1999).

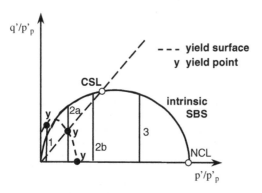

Figure 47. Characteristic normalised behaviour for weak bonding.

Figure 49. Tangent stiffnesses for an undrained shearing probe on Greensand (after Cuccovillo & Coop, 1997a).

from great depth. The effect of unloading on the cement contribution to stiffness has been demonstrated experimentally by Fernandez & Santamarina (2001). Although its behaviour is very similar to that of hydrocarbon bearing sandstones so that it is used as an analogue material, the Castlegate sandstone is not actually one, and it outcrops in Colorado. The unloading that has caused the bonds to be damaged in

this case was geological erosion, but again occurred after deposition of the cement.

The linear elastic behaviour seen for the Rankin calcarenite and Greensand can therefore be seen to be typical of sandstones that have not experienced a large change of stress since the deposition of the cement, so that the bonds remain intact. The calcium carbonate cement of the calcarenite and the iron oxide of the quartzitic Greensand are both typically deposited when the soils are near the surface. In the case of the Greensand there was little change of stress after deposition, and in the case of the Rankin calcarenite the depth of burial of around 127–142 m of the samples tested had been insufficient to yield the bonding. However, other forms of cement are deposited during

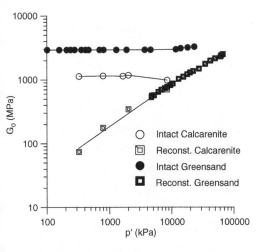

Figure 50. Comparison of the elastic shear moduli for two cemented sands and the equivalent reconstituted soils (adapted from Cuccovillo & Coop, 1997b and Jovicic et al, 1997).

burial. The chlorite bonding of the Castlegate sandstone, which is also typical of many petroleum reservoir sandstones is deposited during burial but at a fairly early stage. The other principal form of cement found in reservoir sandstones, quartz, requires deeper burial to be deposited. Large changes to the stresses applied to a sandstone that has been cemented at depth, either by further burial, geological unloading through erosion, or simply the stress release due to sampling are likely to have a significant effect on the behaviour observed through micro-cracking of the cement, as seen for the Castlegate sandstone.

4.3 Fabric in sands

Figure 54 shows a thin section of the Greensand, with the void spaces stained so that they appear dark. It has a well-developed locked fabric (Dusseault & Morgenstern, 1979), created by pressure solution at particle contacts. Instead of having point contacts as would be expected for rounded particles the contacts are extended and the particles tend to fit one into another. This has been created by extended application of high stresses, the Greensand being Lower Cretaceous and having experienced a past maximum vertical in situ stress of around 9 MPa. The lowest specific volume that could be achieved by compaction of the Greensand in the laboratory was 1.66, and as can be seen in Fig. 37, even after compression to 70 MPa, a far higher stress than the soil has experienced in the ground, the in situ specific volume could not be achieved, emphasising that when the locked particle contacts are disrupted, they cannot be recreated through simple compaction or compression.

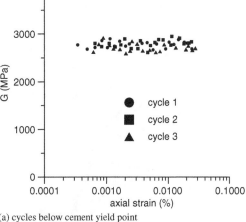

(a) cycles below cement yield point

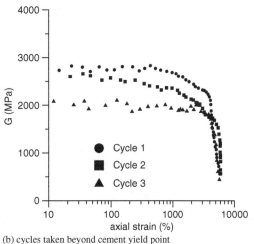

(b) cycles taken beyond cement yield point

Figure 51. Tangent shear moduli for drained probes on Greensand at an initial p' = 1760 kPa (after Cuccovillo & Coop, 1997b).

Although the Greensand is lightly cemented, it is the fabric that in fact dominates its behaviour, particularly its strength. Figure 55 shows stress:dilatancy paths followed during shearing for the Greensand. While the reconstituted sample follows a straight path similar to those of other reconstituted sands (Fig. 6), the intact samples initially reach much higher stress ratios, with little dilation. The dilation seems to be delayed by the cement and locked fabric, so when the cement breaks and the fabric unlocks, there is then a very rapid dilation and correspondingly high peak strength. The relationship between the rate of dilation and peak strength is otherwise the same for cemented and reconstituted soils and the key difference is the much higher rates of dilation for the cemented. Figure 56 shows the maximum rate of dilation at peak plotted

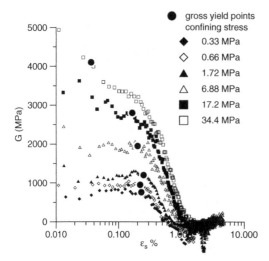

Figure 52. Tangent stiffnesses for drained shearing of Castlegate sandstone (after Coop & Willson, 2003).

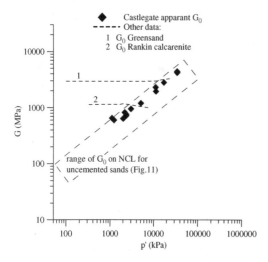

Figure 53. Comparison of stiffnesses of Castlegate sandstone with database for uncemented and cemented sands (adapted from Coop & Willson, 2003).

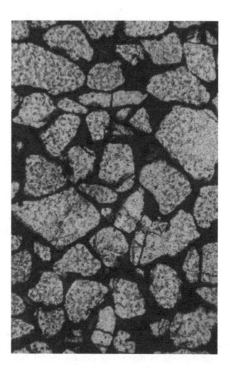

Figure 54. Thin section of Lower Greensand (width = 2 mm).

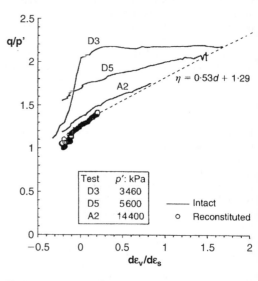

Figure 55. Stress: dilatancy relationship for Greensand (after Cuccovillo & Coop, 1999).

against the state in the v:ln p' plane, quantified using an equivalent pressure on the CSL (p'_{cs}). For a given state the rates of dilation experienced by the intact soil are very much higher than those of the reconstituted. The intercept of the graph, i.e. the value of p'/p'_{cs} at which the rate of dilation becomes zero, is also significantly higher for the cemented soil, indicating that the CSL has a higher location in the v:ln p' plane, as discussed previously.

A similar delayed dilation with increased peak stress ratio can be seen for sands with cementing only and no locked fabric, such as the Castlegate sandstone, but for the Greensand the effect extends to pressures

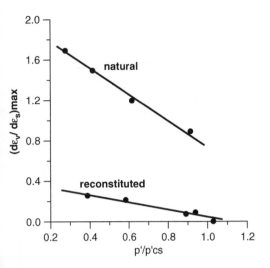

Figure 56. Relationship between rate of dilation at peak and state for Greensand (after Cuccovillo & Coop, 1999).

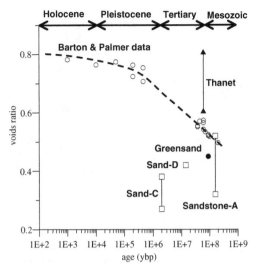

Figure 57. Relationship between voids ratio and age of sands and sandstones (data from Barton & Palmer, 1989).

well beyond the yield of the cement, and this must be the result of the fabric. This can be seen in the normalised shearing data in Fig. 48, where the effect of the increased dilation of the intact soil is to raise the Hvorslev surface above that of the reconstituted soil, even well beyond the yield envelope of the cement.

Barton & Palmer (1989) have examined the in situ dry densities of a range of British sands, correlating the relative densities with their geological age. Their data are reported on Fig. 57, but in terms of voids ratio, having assumed a value of specific gravity of 2.65 for all of them. The sands they considered were all uncemented, fine or fine-medium sands, poorly graded with a low fines content, and of quartzitic mineralogy. Barton & Palmer identified a trend of increasing relative density with age, here reinterpreted as a decrease of voids ratio. This increase of density was explained as being due to the development of a locked fabric, although they emphasized that age could only be an approximate indicator of the development of a locked fabric, since it would actually depend on factors more difficult to determine, such as its depth of burial and the chemical environment. Barton & Palmer recognised that most of the Tertiary and Mesozoic sands had relative densities greater than 100%, reaching values up to around 120%, so that, like the Lower Greensand, it would not be possible to compact a reconstituted sample to so dense a state.

Also shown on Fig. 57 is a point for the Lower Greensand, a range of values for the Thanet sand from recent tests at Imperial College, together with data for hydrocarbon reservoir sands and sandstones of quartzitic mineralogy (Sandstone-A and Sands C & D; Coop & Willson, 2003). The precise locations of

the sites for the hydrocarbon reservoirs cannot be revealed for commercial reasons. The reservoir sands and sandstones had been rotary cored and then frozen to avoid disturbance.

The Lower Greensand is British and plots close to the trend. It has a slightly lower voids ratios than the Cretaceous and Jurassic sands reported by Barton & Palmer, perhaps because it is slightly cemented, although, as was seen in Fig. 36b the amount of cement is small. In contrast, the Thanet sand voids ratios, taken from rotary cored samples from London, plot significantly higher than the trend, and higher than the value that Barton & Palmer report for a block sample of the same soil. The high voids ratio may relate to the fact that the sand has only slight evidence of a locked fabric, as can be seen in the thin section in Fig. 58.

The reservoir sands are interesting in that the voids ratios are generally very low indeed, even if the fabrics are again only slightly locked. Sands C and D are from the Gulf of Mexico and Sandstone-A from the North Sea, but they are all poorly graded and quartzitic. The key difference may be in the geological history, in that all the sands sampled by Barton & Palmer and also the Thanet sand and Greensand are currently relatively close to the surface, even if they may have been deeply buried at some time in their past. In contrast, all the reservoir sands (Sandstone A and Sands C & D) have all remained at their maximum depths of burial, which are all between 3–4.7 km.

The compression data for Sand-C are revealing (Fig. 59). Only one isotropic compression test reached yield and its NCL. The one-dimensional NCL would lie slightly below the isotropic NCL and

Figure 58. Thin section of Thanet sand.

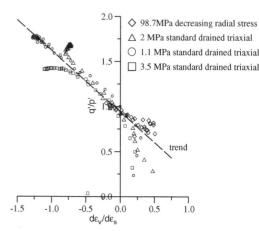

Figure 60. Stress:dilatancy data for reservoir Sand-C (after Coop & Willson, 2003).

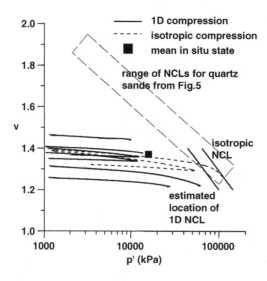

Figure 59. Compression data for reservoir Sand-C (adapted from Coop & Willson, 2003).

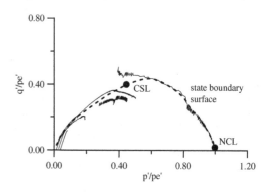

Figure 61. Normalised shearing data for reservoir Sand-C (after Coop & Willson, 2003).

its position has been estimated. The samples were from a range of depths and had a range of initial specific volumes, but the in situ state is represented by an average value. The in situ stresses were measured for this site and are believed to be the maximum stresses the soil has experienced. A key question is then how the in situ volumes were achieved. The densities are much greater than could reasonably be achieved by deposition or compaction of a poorly graded quartzitic sand, but the current state plots far below the NCL and cannot be explained solely by stress history, and yet the fabric is not significantly locked.

Even if Sand-C is very dense, the lack of a significant locked fabric is confirmed by the behaviour in triaxial tests, although unfortunately there are only data for the intact sand and no tests on reconstituted

samples for comparison. The stress:dilatancy data (Fig. 60) are similar to those of reconstituted sands and do not show the exaggerated dilation of the locked sands. Similarly, the normalised shearing data in Fig. 61 identify a state boundary surface very similar to those of reconstituted sands.

In Fig. 62 the database of G_0 for various sands from Fig. 11 is compared to G_0 data from the start of monotonic shearing stages for Sand-C. The states of the Sand-C samples were not on the NCL, which would cause the values of G_0 to plot slightly higher, than if they were, but there is clearly not a large difference between the stiffnesses of this natural sand and those of reconstituted sands. Of the database, the line reported for Thanet sand in Fig. 11 is also for intact samples, but there is again little sign of any significant influence of structure, although tests on reconstituted samples would be required to confirm this.

Many natural sands therefore have a well-defined fabric that must be different to those of reconstituted

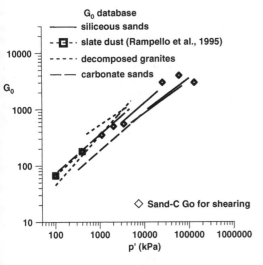

Figure 62. G_0 data for reservoir Sand-C (adapted from Coop & Willson, 2003).

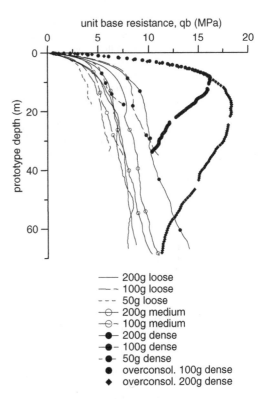

Figure 63. Unit end bearing during installation of a model pile in Dog's Bay sand (after Klotz & Coop, 2001).

sands since they correspond to much higher densities than can be achieved by compaction or compression. In some cases the high densities are indicative of locked fabrics, which may be easily identified through thin sections, and has a clear effect on the mechanical behaviour of the soil. But in other cases the density has not been achieved through pressure solution and a locked fabric, and the mechanical behaviour appears to be similar to reconstituted sands. In these cases the only affect of the fabric has on the behaviour is through the very high density that has been achieved. For example, the very high density of Sand-C means that it would not reach its one-dimensional NCL and yield through particle break-age until the in situ stress had more than doubled, which is key importance in the prediction of the behaviour and production of the reservoir.

The correlation between density and age is poor and since the in situ density seems often to be unre-lated to stress history, even a knowledge of the maxi-mum stress may be insufficient as a predictor of the in situ density. Future research needs therefore to iden-tify how natural sands achieve their in situ densities and the roles that stress history, pressure solution, tec-tonic loading, seismic densification and the physico-chemical environment play in this.

5 APPLICATIONS

To examine how the concepts outlined above might be applied to design of geotechnical structures in sand, Klotz & Coop (2001) carried out a series of cen-trifuge tests on piles jacked into both the Dog's Bay

carbonate sand and a quartzitic sand (Leighton Buzzard sand). The model pile was solid of 16 mm diameter and was jacked up to 375 mm into the soil. It was instrumented to measure the radial stress, shaft friction and end bearing. Tests were carried out at accelerations of 50 g, 100 g and 200 g, so that at the highest acceleration the pile that the model repre-sented would have been of 3.2 m diameter and 75 m in length.

Figure 63 gives some typical data, in this case for the unit end bearing, q_b against prototype penetration, during installation in the Dog's Bay sand. There is quite a wide variation of resistance with density. Two samples were overconsolidated by applying a static pressure of 3 MPa prior to conducting the test. As for the behaviour seen in laboratory element tests, it is clear that overconsolidation has a significant effect on behaviour.

On Fig. 64 the unit end bearing has been nor-malised with respect to the initial in situ vertical effective stress, σ'_v, to give values of N_q, which are then plotted against the relative density D_r, showing no correlation at all. Plotting N_q against a state parameter, ψ, derived from a curved CSL passing through the minimum density at low pressures, similar to Fig. 22b, also gives no useful correlation. As for the stiffness behaviour, a correlation is only obtained if the state parameter is taken in the horizontal rather than vertical direction, as a stress ratio (Fig. 65). Even after the normalisation for state, the overconsolidated samples still show higher capacities, similar to the effect of overconsolidation on stiffnesses on Fig. 12.

Similar data were obtained for the shaft friction and for the tests on the quartzitic sand. Field pile tests in sands rarely include sufficient information about the behaviour and in situ state of the sands to be able to identify both the in situ density and the CSL location and hence quantify the in situ state. For the data available, values of unit shaft friction normalised with respect to the mean initial vertical effective stress over the pile depth to give values of β, are shown in Fig. 66. The data are very scattered because of the variety of sand mineralogies (Table 1) and of pile types and sizes, from model driven piles to full sized grouted sections. However, there is still a clear trend when state is quantified as a stress ratio.

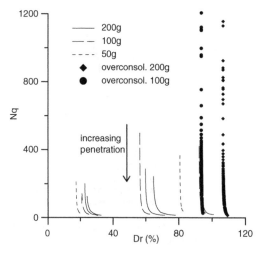

Figure 64. Relationship between N_q and relative density during model pile installation in Dog's Bay sand (after Klotz & Coop, 2001).

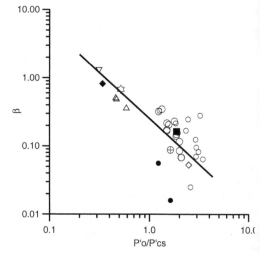

Figure 66. Shaft friction data for field pile tests (adapted from Klotz & Coop, 2001).

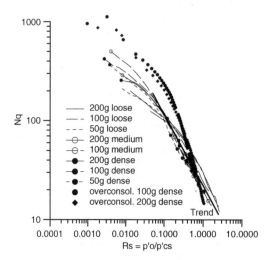

Figure 65. Normalised end bearing during installation of model pile in Dog's Bay sand (after Klotz & Coop, 2001).

Table 1. Legend to Figure 66.

	Reference	Location/Type	Soil
◇	Renfrey et al. (1988)	Rankin driven section	uncem/weak calcarenite
■	Renfrey et al. (1988)	Rankin grouted sections	uncem/weak calcarenite
○	Khorshid et al. (1988)	Rankin steel model pile (SFT)	uncem/weak calcarenite
●	Ripley et al. (1988)	Rankin conductor load test	uncem/weak calcarenite
⊕	Dolwin et al. (1988)	Rankin redrive of foundation pile	uncem/weak calcarenite
◆	Nauroy & Le Tirant (1985)	Plouasne, grouted section	uncem/weakly cement carbonate
▽	Cotecchia et al. (1998)	Naples, grouted pile	Volcanic sand
△	Lehane et al. (1993)	Labenne, jacked steel pile	silica sand
☆	Yasufuku et al. (1997)	Japan, grouted pile	"Shirasu" volcanic sanda

6 LIMITATIONS

Typical stress:strain data for triaxial tests on the Dog's Bay sand are shown in Fig. 67. Although the stresses typically reach well-defined critical state values, there is a tendency, particularly at high stresses, for the volumetric strain never quite to stabilise. This was investigated further by Coop et al. (2004) by

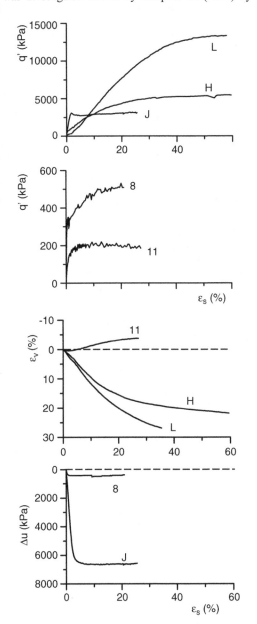

Figure 67. Typical stress:strain data for triaxial tests on Dog's Bay sand (after Coop, 1990).

means of ring shear tests, so that there was no limit to the strain that could be reached.

In Fig. 68 the volumetric strain, ε_v, and relative breakage, B_r, are plotted against the shear strain, γ, for tests conducted with vertical effective stresses in the range 650–930 kPa. Because of strain non-uniformities within the ring shear apparatus the shear and volumetric strains are purely nominal. Similarly, the values of B_r were found to vary through the sample thickness and the values reported in Fig. 68 are only those in the central zone, where the strains are greatest. The data clearly show that volumetric compression and particle breakage continue long after the critical state identified for triaxial tests, although ϕ'_{mob} was found to be relatively unaffected by the continued breakage. The volumetric compression does eventually cease, when the soil reaches a stable grading. Although the data in Fig. 68 are for a carbonate sand, with fragile particles, Luzzani & Coop (2002) identified that small amounts of breakage continued to very large strains also for quartzitic sands, even when sheared at low stress levels.

A clear limitation of the critical state framework presented in this paper, is therefore that for sands the critical state, as identified from single element tests at a few

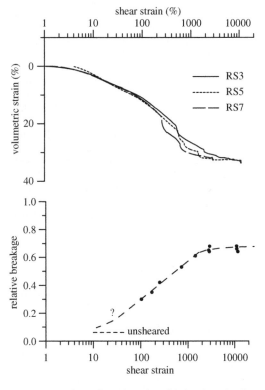

Figure 68. Volumetric strain and particle breakage for ring shear tests on Dog's Bay sand under normal stress = 650–930 kPa (adapted from Coop et al., 2004).

55

tens of percent shear strain, is not a rigorous concept. Although this does not invalidate the usefulness of the framework, it does highlight the need for vigilance in applications where there might be large concentrated shear strains in sands, for example along driven pile shafts. While the ϕ'_{mob} may be insensitive to the continued shearing, the volumetric strain and particle breakage will continue to very large displacements and may greatly exceed those measured in usual laboratory tests.

7 CONCLUSIONS

A critical state type of framework has been shown to be helpful in understanding the behaviour of reconstituted sands at all strain levels, although there are significant differences with the type of framework that might be applied to clays, largely arising from the particle breakage that occurs for sands. This particle breakage also limits the applicability of such a framework at very large strains.

The framework is both useful for understanding the behaviour of soil elements and that of geotechnical structures, but of key importance for both is that the in situ state has to be quantified using the horizontal distance from the critical state line in the v:ln p' plane rather than the more usual vertical offset.

The framework for reconstituted sands provides the intrinsic behaviour with which to identify the effects of bonding and fabric for natural sands. The processes that lead to a sand being cemented are fairly well understood, although stress changes experienced subsequent to the deposition of the cement may lead to significant changes in behaviour if they are sufficient to damage the bonds. These effects still require further investigation, but it is clearly important to understand when in the geological history of the sediment the cement had been deposited. The factors that control the in situ density of natural sands and whether or not a locked fabric is created are less well understood and should also be the focus of future research.

REFERENCES

Amoroso, C.G. (1997) L'effetto della storia tensionale recente sulla rigidezza delle sabbie. MEng Dissertation, University of Trento, Italy.

Atkinson, J.H. & Bransby, P.L. (1978) *The Mechanics of Soils*. London: McGraw-Hill.

Atkinson, J.H., Richardson, D. & Stallebrass, S.E. (1990) Effect of recent stress history on the stiffness of overconsolidated soil. *Géotechnique*, 40, No.4, 531–540.

Barton, M.E. & Palmer, S.N. (1989) The relative density of geologically aged, British fine and fine-medium sands. *Quarterly J. Engng. Geol.*, Vol. 22, 49–58.

Baudet, B.A. & Stallebrass, S.E. (2004) A constitutive model for structured clays. Submitted to *Géotechnique*.

Been, K. & Jefferies, M.G. (1985) A state parameter for sands. *Géotechnique*, 35, No. 2, 99–112.

Boese, R.J. (1989) The study of the mechanical behaviour of uncemented sand. BSc dissertation, City University, London.

Burland, J.B. (1990) On the compressibility and shear strength of natural clays. Thirtieth Rankine Lecture. *Géotechnique*, 40, No. 3, 329–378.

Burmeister-Martins, F., Bressani, L.A., Coop, M.R. & Bica, V.D. (2002) Some aspects of the compressibility behaviour of a clayey sand. *Canadian Geotech. J.*, 38, No. 6, 1177–1186.

Colleselli, F., Coop, M.R. & Nocilla, A. (2003) La meccanica del limo visto come transizione fra il comportamento delle sabbie e delle argille. Incontro Annuale dei Ricercatori di Geotecnica, Potenza, Italy, IARG. In press.

Coop, M.R. (1990). The mechanics of uncemented carbonate sands. *Geotechnique*, 40(4), 607–626.

Coop, M.R. & Atkinson, J.H. (1993) The Mechanics of Cemented Carbonate Sands. *Géotechnique*, 43, No. 1, 53–67.

Coop, M.R. & Cuccovillo, T. (1999) The influence of geological origin on the behaviour of carbonate sands. *Problematic Soils*, Yanagisawa, W., Moroto, N. & Mitachi, T. (eds) (Balkema, Rotterdam), 607–610.

Coop, M.R. & Jovicic, V. (1999) The influence of state on the very small strain stiffness of sands. *2nd Intl. Symp. Pre-failure Deformation of Geomaterials, IS-Torino 99*, Turin, Italy, pp. 175–181.

Coop, M.R. & Lee, I.K. (1993) The behaviour of granular soils at elevated stresses. *Predictive Soil Mechanics, Proc. C.P.Wroth Mem Symp.*, Thomas Telford, London, 186–198.

Coop, M.R. & Lee, I.K. (1995) The Influence of Pore Water on the Mechanics of Granular Soils. *Proc. XI European Conf. Soil Mechs Foundn Engng, Copenhagen*, Danish Geotechnical Society, 1.63–1.72.

Coop, M.R., Sorensen, K.K., Bodas Freitas, T. & Georgoutsos, G. (2004) Particle Breakage During Shearing of a Carbonate Sand. In press for *Géotechnique*.

Coop, M.R. & Willson, S.M. (2003) On the Behavior of Hydrocarbon Reservoir Sands and Sandstones. J. *Geotech.l Engng, ASCE*, Vol. 129, No. 11, 1010–1019.

Cotecchia, F., Santaloia, F., Lagioia, R. & Coop, M.R. (1998) An investigation of the behaviour of bored piles in pyroclastic sands. *Int. Symp. on Problematic Soils, Yanagisawa, Japan, Moroto & Mitachi (eds)*, Balkema, Rotterdam, 715–719.

Cuccovillo, T. (1995) The shear behaviour and stiffness of naturally cemented sands, PhD Thesis, City University, London.

Cuccovillo, T. & Coop, M.R. (1993) The influence of bond strength on the mechanics of carbonate soft rocks. *Proc. Int. Symp. Geotechnical Engng Hard Soils – Soft Rocks, Athens*, Balkema, Rotterdam, 447–455.

Cuccovillo, T. & Coop, M.R. (1997a) The Measurement of Local Axial Strains in Triaxial Tests Using LVDTs. *Géotechnique*, 47, No. 1, 167–171.

Cuccovillo, T. & Coop, M.R. (1997b) Yielding and prefailure deformation of structured sands. *Géotechnique*, 47(3), 491–508.

Cuccovillo, T. and Coop, M.R. (1999) On the mechanics of structured sands, *Géotechnique*, 49, No. 6, 741–760.

Dolwin, J., Khorshid, M.S. & van Goudoever, P. (1988) Evaluation of driven pile capacity – Methods and results. *Proc. Int. Conf. Calcareous Sediments, Perth*, (1), 429–438.

Dusseault, M.B. & Morgenstern, N.R. (1979) Locked sands. *J.Engng Geol.*, (12), 117–131.

Fernandez, A.L. & Santamarina, J.C. (2001) Effect of cementation on the small-strain parameters of sand. *Can. Geotech. J.*, 38, pp. 191–199.

Gasparre, A., Coop, M.R. and Cotecchia, F. (2003) A Laboratory Investigation of a Crushable Sand", *3rd International Symposium on Deformation Characteristics of Geomaterials, Lyon*, Di Benedetto, H., Doanh, T., Geoffroy, H. & Sauzeat, C. eds., Balkema, Rotterdam, Vol. 1, 773–778.

Golightly, C.R. (1989) Engineering properties of carbonate sands. PhD Thesis, University of Bradford.

Gudehus, G. (1996) A comprehensive constitutive equation for granular materials. *Soils and Foundations*, 36, No.1, pp. 1–12.

Hardin, B.O. (1985) Crushing of soil particles. *Journal of Geotechnical Engineering, Proc. ASCE*, 111(10), 1177–1192.

Jovicic, V. & Coop, M.R. (1997). Stiffness of coarse grained soils at small strains. *Geotechnique*, 47(3), 545–561.

Jovicic, V., Coop, M.R. & Atkinson, J.H. (1997) Laboratory measurements of small strain stiffness of a soft rock. *Proc.XIV ICSMFE, Hamburg*, 323–326.

Khorshid, M.S., Haggerty, B.C. & Male, R. (1988) Development of geotechnical aspects of the investigation programme. *Proc. Int. Conf. Calcareous Sediments, Perth*, (2), 377–386.

Klotz, E.U. & Coop, M.R. (2001) An investigation of the effect of soil state on the capacity of driven piles in sands. *Géotechnique*, 51, No. 9, 733–751.

Klotz, E.U. & Coop, M.R. (2002) On the Identification of Critical State Lines for Sands. *Am Soc. Test. Materials, Geotechnical Testing Journal*, 25, No. 3, 289–302.

Konrad, J.M. (1998) Sand state from cone penetrometer tests: a framework considering grain crushing stress. *Géotechnique*, 48, No. 2, 201–215.

Kuwano, R. (1997). Personal communication.

Kwag, J.M., Ochiai, H. & Yasafuku, N. (1999) Yielding stress characteristics of carbonate sand in relation to individual particle fragmentation strength. *Engineering for Calcareous Sediments, Bahrain, Al-Shafei (ed.)*, Balkema, Rotterdam, (1), 79–86.

Lade, P.V. & Yamamuro, J.A. (1996) Undrained sand behavior in axisymmetric tests at high pressures. *Proc. ASCE*, 122(2), 120–129.

Lee, I.K. & Coop, M.R. (1995) The Intrinsic Behaviour of Decomposed Granite Soil. *Géotechnique*, 45, No. 1, 117–130.

Lee, K.L & Seed, H.B (1967) Drained strength characteristics of sands. *Proc. ASCE, Journal Soils Mechs Foundn Engng*, SM6, 117–141.

Lehane, B.M., Jardine, R.J., Bond, A.J. & Frank, R. (1993) Mechanisms of shaft friction in sand from instrumented pile tests. *Proc.ASCE, Journal Geotech. Engng*, 119(GT1), 19–35.

Leroueil, S. & Vaughan, P.R. (1990) The general and congruent effects of structure in natural soils and weak rocks. *Géotechnique*, 40, No. 3, 467–488.

Luzzani, L. & Coop, M.R. (2002) On the relationship between particle breakage and the critical state of sands. *Soils & Foundations*, 42, No. 2, 71–82.

McDowell, G.R. & Bolton, M.D. (1998) On the micro mechanics of crushable aggregates. *Géotechnique*, 48, No. 5, 667–679.

Mitchell, J.K. (1976) *Fundamentals of Soil Behaviour.* New York: Wiley.

Miura, N. & Yamonouchi, T. (1975) Effect of water on the behaviour of quartz-rich sand under high stresses. *Soils & Foundations*, 15(4), 23–34.

Nauroy, J.F. & LeTirant, P. (1985) Driven and drilled and grouted piles in calcareous sands. *Offshore Tech. Conf.*, Paper No.OTC 4850, 83–91.

Ovando-Shelly, E. (1986) Stress–strain behaviour of granular soils tested in the triaxial cell. PhD Thesis, Imperial College, University of London.

Pan, J. (1999) The behaviour of shallow foundations on calcareous soil subjected to inclined load. PhD thesis, University of Sydney.

Plona, T.J. & Cook, J.M. (1995) Effects of stress cycles on static and dynamic Young's moduli in Castlegate sandstone. *Rock Mechanics, 35th Rock Mechanics Symposium*, U. Nevada, Reno, June 1995, Daemen & Schultz eds, Balkema.

Rampello, S., Viggiani, G. & Silvestri, F. (1995) Panellist discussion: The dependence of G_0 on stress state and history in cohesive soils. *Pre-failure Deformation of Geomaterials*, Balkema, Rotterdam, 1155–1160.

Renfrey, G.E., Waterton, C.A. & van Goudoever, P. (1988) Geotechnical data used for the design of the North Rankin 'A' platform. *Engineering for calcareous sediments*, R.J. Jewell & M.S. Khorshid eds., Balkema, Rotterdam, Vol. 2, 343–355.

Ripley, I., Keulers, A.J.C. & Creed, S.G. (1988) Conductor load tests. *Proc. Int. Conf. Calcareous Sediments, Perth*, (2), 429–438.

Rotta, G.V., Consoli, N.C., Prietto, P.D.M., Coop, M.R. & Graham, J. (2003) Isotropic Yielding in an Artificially Cemented Soil Cured Under Stress, *Géotechnique*, No. 5, 493–502.

Schofield, A.N. & Wroth, C.P. (1968) *Critical State Soil Mechanics.* London: McGraw-Hill.

Skinner, A.E. (1975) The effect of high pore water pressures on the mechanical behaviour of sediments. PhD Thesis, Imperial College, University of London.

Verdugo, R. & Ishihara, K. (1996) The steady state of sandy soils. *Soils and Foundations*, 36, No. 2, 81–91.

Vesic, A.S. & Clough, E.W. (1968) Behaviour of granular materials under high stresses. *Proc. ASCE*, 94(SM3), 661–688.

Viggiani, G. & Atkinson, J.H. (1995) Stiffness of fine grained soil at very small strains. *Géotechnique*, 45(2), 249–265.

Yasufuku, N., Ochiai, H. & Maeda, Y. (1997) Geotechnical analysis of skin friction of cast-in-place piles. *Proc. 14th ICSMFE*, (2), 921–924.

Zimmerman, R.W. (1991) *Compressibility of Sandstones*, Elsevier Science.

NOMENCLATURE

Br	relative breakage (Hardin, 1985)
CSL	Critical State Line
Dr	relative density
e	void ratio
G_u	undrained shear modulus
G_0	elastic shear stiffness
$G_{0(NCL)}$	NCL in terms of elastic stiffness
$G_{0(nc)}$	value of G_0 on the $G_{0(NCL)}$ at current p'
G_{vh}	G_0 measured with vertically propagating shear wave and horizontal polarisation
M	gradient of CSL in q':p' plane
N_q	bearing capacity factor ($=q_b/\sigma'_v$)
NCL	Normal Compression Line
p'	mean normal effective stress $[=(\sigma'_a + 2\sigma'_r)/3$ or $(\sigma'_v + 2\sigma'_h)/3]$
p'_0	in situ p'
p'_{cs}	equivalent pressure taken on CSL
p'_e	equivalent pressure taken on NCL
p'_p	equivalent preconsolidation pressure
p_r	reference (atmospheric) pressure
q'	deviatoric stress ($=\sigma'_a - \sigma'_r$ or $\sigma'_v - \sigma'_h$)
q_b	unit end bearing
q_s	unit shaft friction
R_s	normalising parameter for state ($=p'_0/p'_{cs}$)
v	specific volume ($=1 + e$)
Δu	change in pore pressure
β	shaft friction coefficient ($=q_s/\sigma'_v$)
γ	engineers shear strain
a	axial strain
r	radial strain
v	volumetric strain
s	shear strain $[= {}^2/_3(\varepsilon_a - \varepsilon_r)]$
ϕ'_{cs}	critical state angle of shearing resistance
ϕ'_p	peak angle of shearing resistance
ϕ'_{mob}	mobilised angle of shearing resistance
λ	gradient of CSL & NCL in v: ln p' plane
κ	gradient of swelling line in v: ln p' plane
ψ	state parameter
σ'_a	axial effective stress
σ'_h	horizontal effective stress
σ'_r	radial effective stress
σ'_v	vertical effective stress

Deformation Characteristics of Geomaterials – Di Benedetto et al (eds)
© 2005 Taylor & Francis Group, London, ISBN 04 1536 701 8

Time effects on the behaviour of geomaterials

H. Di Benedetto
Département Génie Civil et Bâtiment, Ecole Nationale des Travaux Publics de l'Etat, France

F. Tatsuoka
Department of Civil Engineering, University of Tokyo, Japan

D. Lo Presti
Politecnico di Torino, Italy

C. Sauzéat & H. Geoffroy
Département Génie Civil et Bâtiment, Ecole Nationale des Travaux Publics de l'Etat, France

ABSTRACT: This paper presents an overview of the time effects including i) viscous or loading rate effects and, ii) ageing effects, on the behaviour of geomaterials. The experimental results on a wide range of geomaterials, including sands, reconstituted and undisturbed clays, gravels, natural soils, crushed concretes, soil-cement mixtures, bituminous materials …, are analysed from extensive experimental campaigns performed on specifically designed devices at University of Tokyo and ENTPE. It is observed that viscous effects do exist for all the considered geomaterials. Modelling is proposed in the framework of the three-component model, which is shown to be very versatile and powerful. A new type of behaviour called "viscous evanescent" is identified and modelled by two developed formalisms: the VE and TESRA models. Presented results from physical and full scale models confirm the importance of viscous effects in practical cases. Some preliminary results of FEM calculation show that implementation of the developed laws, in existing elasto-plastic codes, could be made smoothly.

1 INTRODUCTION

1.1 General consideration and definition

The stress-strain response of a geomaterial may change with the time chosen for the beginning of loading and changes in the loading rate. This phenomenon, depending on the type of geomaterial, moist conditions, dry density, temperature, stress and strain conditions, loading history and so on is called **the time effects** in general terms. To develop a relevant constitutive model that can simulate "the time effect" realistically under as possible as different conditions, it is necessary to properly define the time effect including "**ageing effects**" and "**viscous or loading rate effects**". These two different aspects of the time effects should be linked to different basic mechanisms behind, such as "changes with time in the intrinsic material properties due to changes in interface and/or internal particle properties caused by a physico-chemical process" (ageing effects), "viscous sliding at inter-particle contact points" (viscous effects), "time dependent property of the particles", among others.

In this paper only the time effect on the skeleton behaviour are considered. This means, in particular, that the effects created by pore water movement are not analysed. For example, seepage and consolidation aspects are out of the scope.

Following the same consideration, the time dependency created by inertia effects (wave propagation …), which also is a non rheological phenomenon, is not taken into account in the presented developments.

The loading history of any material is described by, i) the path followed in a six dimensional loading space (l_{ij}), corresponding to imposed stress or strain in "direction ij" (ie: $l_{ij} = \varepsilon_{ij}$ or σ_{ij}, for a given "ij" among the six possible choices) and, ii) the chronology (i.e. the time history) at which this path is traced. It has to be noted that, in the general case, loading space can be composed of a combination of stress and strain applied in different directions (ij), that may change with time. The response to this loading history is given by a curve, in the six dimensional response space (r_{ij}) (i.e., for a fixed "ij", $r_{ij} = \sigma_{ij}$ if $\varepsilon_{ij} = l_{ij}$ [belonging to the loading space] or $r_{ij} = \varepsilon_{ij}$ if $\sigma_{ij} = l_{ij}$ [belonging to the loading space]), which is described in function of time. In this presentation, as schematised in Figure 1-1, the space and time aspects are separated even though both are needed and always connected to each other in practical cases.

A general definition of materials, which are sensitive to time (or by opposition which are not sensitive to time effects), can be obtained from the consideration described above. *If in a given domain of loading the response curve is independent of the chronology of loading (or loading rate) the material is not sensitive to time effects.* As a consequence, every plot in the stress-strain axes remains the same whatever the loading rate. The notion of domain of loading is important to take into consideration, as it is well known that geomaterials can exhibit very different behaviours, depending on both considered time scale and level of loading. For example, when considering the human time scale, time effects on hard rocks can generally be neglected while they appear as very time-sensitive during a long geological period.

The above mentioned definition is a generalisation of the traditional viscous criterion based on the existence of strain evolution during creep tests (i.e., at constant stress) or of stress evolution for relaxation tests (i.e., at constant strain).

Geomaterial always exhibits a more-or-less complicated stress-strain-time behaviour including instantaneous non-linearity and viscous effects, which may or may not be simplified in a considered range of loading (considered loading domain). The simplified behaviour, which can only be deduced from experimental observation, is an approximation of the much more complicated real one. All our further developments should then be considered as the first approximation. The pertinence and validation of this approximation is finally linked to respective specific engineering practice aspect. The following two extreme cases can then arise:

– the phenomenon (e.g., structural displacements, in particular residual ones due to the viscous deformations of geomaterial) is under-estimated by the approximated behaviour and the risk is high for the construction. For example Figure 1-2 shows the time history of settlement measured at a representative point at the Kansai International Airport site. It may be seen that the actual settlement rate is recently becoming larger than the value predicted by the conventional consolidation theory that does not take into account the viscous property of clay (i.e., the stress-strain properties of clay is assumed to be elasto-plastic). The difference tends to increase with time. Leroueil and Hight (2003) argued that this settlement difference, due to the viscous deformation of clay, is becoming more important with time.

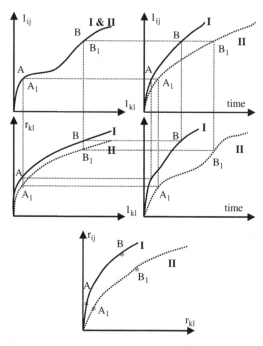

Figure 1-1. Loading and response: different plots and time effect sensitivity (schematic). If curves I and II are always superimposed in the axes $l_{ij} - r_{ij}$ (which correspond to stress-strain curve) and in the axes $r_{ij} - r_{kl}$, the material is not sensitive to time effects (non viscous and non sensitive to ageing) in the considered loading domain.

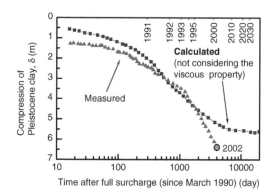

Figure 1-2. Time history of settlement at a representative point at the Kansai International Airport site (Leroueil & Hight, 2003).

– the phenomenon (e.g., structural displacements) is over-estimated. Then, the design of the construction is not optimised, which induces an increase in the construction cost. For example, Figure 1-3 shows the time histories of average contact pressure and

60

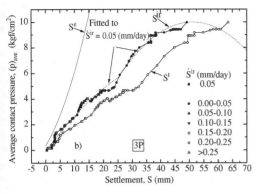

Figure 1-3. Time histories of average contact pressure and settlement of Pier 3 of Akashi Strait Bridge: S^t is the measured settlement, S^e is the elastic part estimated from wave propagation measurements and $S^{ir} = S^t - S^e$ (Tatsuoka *et al.*, 2001a).

settlement of Pier 3, which was constructed on Tertiary sedimentary soft rock, of Akashi Strait Bridge. Tatsuoka *et al.* (2001a) reported that the instant settlement was significantly over-estimated by the prediction based on the stiffness evaluated by the pressure-meter tests converted to the value by plate loading tests. The reason for this deviation is deemed to that the strain operated in these field loading tests was much larger than the one operated in the ground, along the central axis of the pier, during the construction, which was about 0.5% or less (in terms of vertical normal strain). Moreover, the residual compression taking place after nearly full load had been applied, which was due mostly to the viscous properties of the sedimentary soft rock, was over-estimated at the design stage. It is likely that the viscous properties were not properly evaluated, perhaps the effects of sample disturbance, improper axial strain measurements by external measurements and the initial strain rate at the start of creep loading in triaxial compression test was much higher than the field value.

1.2 *Time effects: viscous and ageing*

The two time effect phenomena discussed in this paper are the viscous (or loading rate) effects and the ageing sensitivity.

Non-ageing geomaterials exhibit the properties that *give the same response for two identical loading histories started at different time values (t1 and t2) on the material previously at rest* (stress and strain rate that are equal to zero). A definition taking into account previous loadings, which is more convenient for experimental use, is given in appendix 1.

This definition has a very important consequence on the relevant selection of time variable for constitutive modelling. For non-ageing materials (which can be non-viscous or viscous), the time origin can be chosen arbitrarily and is not linked to the material evolution (ageing). In any case, the time "t" should not appear in the constitutive equation.

For ageing materials, a specific material time "t_c", which characterizes the ageing of the material, has to be added to the common time "t". The two times have the same evolution ($dt = dt_c$) but the origin of the time "t_c" should be connected with an identified date in the life of the material, which should be independent of the loading history. It can be compared to a kind of "birth". For example the starting time of setting is a natural choice for materials that are treated with hydraulic binders.

Ageing can have a positive or negative impact on the respective material. Typical positive ageing effect is associated with the development with time of bonding or cementation at inter-particle contact points, and typical negative one is explained by the deterioration by weathering with time of bonding or cementation at inter-particle contact points. For ageing materials, the intrinsic properties, which are directly related to the particle and inter-particle properties, change with time (t_c) even if no loading is applied.

The following four typical main types of behaviour with respect to time effects can then be considered:

1 Time-independent; whether the stress-strain behaviour is linear or not, reversible or not and path history-dependent or not. The time can be omitted in the description of stress-strain response as only the trajectory of loading in the loading space is needed (cf Figure 1-1).
2 Viscous and non-ageing; In this case, time effects on the stress-strain behaviour are only given by loading rate-dependency, and the time "t" must not appear explicitly in the constitutive equation.
3 Ageing and non-viscous; This case can only be conceived if ageing is negligible when stress-strain response is considered. This means that the variation of "t_c" for the considered loading period (Δt) has no effect on the intrinsic properties (such as modulus, failure surface, viscosity…). But these intrinsic properties may become very different for another age (different "t_c") of the material.
4 General-time dependent; In this case, the variable "t_c" should be added to the formalism. For some specific materials, the ageing and viscous effects may appear as independent phenomena, but in the general case these two effects are coupled.

Figure 1-4 presents a schematic explanation for the types of behaviour considered in this paper.

A wide variety of stress-strain-time behaviour observed by the authors in a great amount of homogeneous element tests and scaled model tests using different kinds of geomaterial that were performed

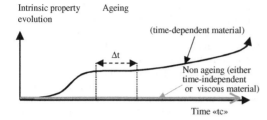

Figure 1-4. Intrinsic property evolution (schematic), no evolution (i.e. non ageing) during Δt means either time independent or viscous behaviour.

following various loading domains, is presented in the second paragraph. These experimental results show that the four typical types of behaviour listed above, can effectively be observed for respective specific conditions existing in different geotechnical constructions. In addition specific and peculiar stress-strain responses for time-dependent materials were observed. A correct modelling of these specific behaviours corresponds to a challenge, which will help in improving construction design.

1.3 Modelling and framework of the considered models

A number of different constitutive models have been proposed to simulate the viscous or loading rate effects observed on geomaterials (e.g., Perzyna, 1963; Suklje, 1969; Akai et al., 1975; Sekiguchi and Ohta, 1977; Adachi and Oka, 1982, 1995; Kaliakin, 1988; Yin and Graham, 1994; Adachi et al., 1996; Di Prisco and Imposimato, 1996; Fodil et al., 1997; Modaressi and Laloui, 1997; Cristescu and Hunsche, 1998; Nawrocki and Mroz, 1998; Vermeer and Neher, 1999; Schanz et al., 1999; Hashiguchi and Okayasu, 2000; Namikawa, 2001; among many others).

A first and essential question when considering time effects, and in a more general way description of irreversibilities, is the type of decomposition chosen for the strain increment ($d\varepsilon$) or strain rate ($\dot{\varepsilon} = d\varepsilon/dt$) (where the symbol "d" denotes an objective increment and "t" is the time). A common decomposition is given by:

$$\dot{\varepsilon} = \dot{\varepsilon}^e + \dot{\varepsilon}^p + \dot{\varepsilon}^{purelyviscous}$$

$$\text{or } \dot{\varepsilon} = \dot{\varepsilon}^{ve} + \dot{\varepsilon}^p + \dot{\varepsilon}^{purelyviscous} \tag{1-1}$$

where $\dot{\varepsilon}^e$ (respectively $\dot{\varepsilon}^{ve}$) is an elastic (respectively viscoelastic) component, $\dot{\varepsilon}^p$ a non-viscous irreversible (plastic) component. As the component $\dot{\varepsilon}^{purelyviscous}$ has the property of presenting creep strains that always increase with time, this decomposition results in an infinite creep strain, which is however not relevant to most types of geomaterial. It is worth noting that any

model including explicitly the time "t" as variable is not objective therefore not relevant as explained in the previous paragraph.

In an attempt to develop a more geomaterial-like constitutive model that can simulate realistically the viscous properties in particular, the authors also proposed several different types of constitutive models based on a great amount of experimental data obtained from triaxial tests (two and three dimensional), plane strain compression tests, one-dimensional loading tests and torsional tests on a wide variety of geomaterials (see for example: Di Benedetto et al., 2001a, 2001b, 2001c, 2002; Hayano et al., 2001; Tatsuoka et al., 2001a; and papers from the authors published in volume 1 of the symposium ISLyon03, among others). Based on the experimental observations, it is assumed that the strain is the sum of a non-viscous (or instantaneous) part $\dot{\varepsilon}^{nv}$ and a viscous (or deferred) part $\dot{\varepsilon}^{vp}$ (Di Benedetto, 1987; Di Benedetto and Tatsuoka, 1997; Di Benedetto et al., 2001b):

$$\dot{\varepsilon} = \dot{\varepsilon}^{nv} + \dot{\varepsilon}^{vp} \tag{1-2}$$

All these models were developed in the framework of the general three-component model (Figure 1-5) (Di Benedetto, 1987; Di Benedetto and Tatsuoka, 1997; Di Benedetto and Hameury, 1991; Di Benedetto et al., 1999a, 1999b), in which the decomposition given in equation 1-2 is assumed.

Bodies EP1 and EP2 of Figure 1-5 have a non-viscous (sometimes called elastoplastic) behaviour. A great number of constitutive laws have been proposed to describe a non-viscous (or elastoplastic) behaviour, such as: elasticity, plasticity, elastoplasticity, hypoplasticity, interpolation type, among many others (cf. Figure 1-5). It can be shown (Darve, 1978) that the general form of the strain increment given by the body of EP type, is:

$$d\varepsilon = M\{h, dir(d\sigma)\} \, d\sigma \tag{1-3}$$

Or, if expressed with objective rates, as $\dot{\varepsilon} = d\varepsilon/dt$ and $\dot{\sigma} = d\sigma/dt$:

$$\dot{\varepsilon} = M(h, dir\dot{\sigma}) \, \dot{\sigma} \tag{1-4}$$

where these quantities are tensors and $dir(d\sigma) = d\sigma/\|d\sigma\|(= dir\dot{\sigma} = \dot{\sigma}/\|\dot{\sigma}\|)$ is the direction of the stress increment (or objective stress rate) whose norm is 1 ($\|dir(d\sigma)\| = \|dir\dot{\sigma}\| = 1$). The parameter h represents the whole history parameters, also called memory, hardening, state … parameters. M is the constitutive (or compliance) tensor, which depends on h and $dir(d\sigma)$.

The introduction of $dir\dot{\sigma}$ expresses the irreversibility and the parameters h, which may be scalars, vectors or tensors, describe the stress history dependence.

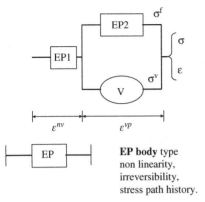

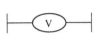

EP body type
non linearity,
irreversibility,
stress path history.

Examples : Elasticity , plasticity, elastoplasticity,
hypoelasticity, hypoplasticity, interpolation type , ...

V body type
viscous effect
non linearity, irreversibility,
stress path history.

Examples : Newtonian linear , newtonian non-linear:
parabolic creep , viscous evanescent , ...

Figure 1-5. Framework of the non-linear three-component model (Di Benedettto, 1987; Di Benedettto and Tatsuoka, 1997; Di Benedetto *et al.*, 2002; Tatsuoka *et al.*, 2002) (schematic).

Body V of Figure 1-5 creates the viscous property dependency of the material. It represents a specific time-dependent behaviour, which is expressed by the following equation 1-5:

$$\sigma = f(h, \dot{\varepsilon}) \qquad (1-5)$$

where f is the viscous tensor, which depends on h and $\dot{\varepsilon}$ (objective strain rate).

When summarisinging the properties of EP1, EP2 and V bodies and applying them to the three-component framework (Figure 1-5), the following general set of equations (1-6 to 1-9) are obtained:

$$\dot{\varepsilon}^{nv} = M^{nv}(h^{nv}, dir\dot{\sigma})\ \dot{\sigma} \qquad (1-6)$$

$$\dot{\varepsilon}^{vp} = M^{f}(h^{f}, dir\dot{\sigma}^{f})\ \dot{\sigma}^{f} \qquad (1-7)$$

$$\dot{\varepsilon}^{vp} = N^{v}(h^{v}, \sigma^{v}) \text{ or } \sigma^{v} = f(h^{v}, \dot{\varepsilon}^{vp}) \qquad (1-8)$$

$$\sigma = \sigma^{v} + \sigma^{f} \qquad (1-9)$$

where "nv", "f" and "v" stand for respectively the EP1, EP2 and V bodies. The history parameters of each of the three bodies may be different. σ^{v} is the "viscous" stress and σ^{f} is the "inviscid" or "creep" stress.

The proposed viscous component can be added to any non-viscous law formalism, which then imposes the form of the non-viscous tensor M^{nv} and of the inviscid (or creep) tensor M^{f}. If some identical parameters are introduced into the sets h^{nv}, h^{v} and h^{f}, a coupling between viscous and non-viscous part can be introduced.

It can be shown that this formalism is a three dimensional generalization of classical theories such as vicoelasticity and viscoplasticity.

If the time "t_{c}" (cf. section 1.2) is incorporated into the parameters h^{nv}, h^{v} or h^{f}, then ageing effect can be modelled. Observations of stress-strain-time behaviour of geomaterials, with positive ageing effects, reveals that strain rate (respectively stress rate) may or may not exist at constant stress (respectively strain) depending on the loading history. A more general form of equations 1-6 and 1-7 in the case with ageing effects is given in the paragraph 3.

It has to be noted that equations 1-6 and 1-7 can be written in the inverse way giving the stress rate in function of the strain rate. In this paper expressions having the strain rate as variable are sometimes used. For example, in that case equation 1-7 takes the form expressed in equation 1-10, where N^{f} is the inverse of M^{f} ($N^{f}.M^{f}$ = identity). The expression of equation 1-10 is more general than equation 1-7: it allows, in particular, treating softening aspect. Meanwhile the equations 1-6 and 1-7 are very commonly used. The formal adaptation of one type of equation to the other remains rather simple and is not systematically treated further.

$$\dot{\sigma}^{f} = N^{f}(h^{f}, dir\dot{\varepsilon}^{vp})\dot{\varepsilon}^{vp} \qquad (1-10)$$

The differences between the models developed by the authors, which are presented in section 3, come from different choices for bodies EP1, EP2, V. This choice is adapted to the specific observed behaviour of the considered geomaterial. In general EP1 body is considered as elastic. Then the strain decomposition (equation 1-2) becomes:

$$\dot{\varepsilon} = \dot{\varepsilon}^{e} + \dot{\varepsilon}^{vp} \qquad (1-11)$$

When considering from the simplest toward more sophisticated behaviour, the list of the proposed models (cf. section 3) consist of:

- the New Isotach model (Tatsuoka *et al.*, 2001a, 2001b; Di Benedetto *et al.*, 2002),
- the TESRA (Temporary Effects of Strain Rate and Acceleration) model (Tatsuoka *et al.*, 2002, Di Benedetto *et al.*, 2002),
- the general TESRA model (Tatsuoka *et al.* , 2002),
- the viscous evanescent (VE) model (Di Benedetto *et al.*, 1999b, 2001b; Tatsuoka *et al.*, 2002; Sauzéat, 2003),

– the combination of the New Isotach and general TESRA model (Hirakawa *et al.*, 2003c).

These models were successively developed to take into account the specific viscous behaviour observed on different types of geomaterial, including clay, silty sand, sand, gravel, crushed concrete, sedimentary soft rock, cement-treated soil and bituminous mixture. The New Isotach model is based on a classical formalism. That is, upon a sudden or gradual change in the viscoplastic strain rate, the stress-strain curve re-joins the curve that would be obtained for a monotinic loading performed continuously at the instantaneous viscoplastic strain rate. The next models have a specific feature that can describe a peculiar behaviour that viscous effects decay with an increase in the viscoplastic strain. This surprising type of behaviour is observed particularly with clean sand but also with other types of granular material in a less pronounced way. It is not possible to model this type of behaviour by any other laws proposed based on classical theories. This last remark shows an interesting feature and an originality of the proposed developments. The respectively adapted approaches of modelling for different types of geomaterial are shown in the next sections, where a great variety of experimental results for soils and soft rocks are presented.

The New Isotach model can be described as a simplified version of the TESRA one, while the TESRA model can also be described as a simplification of the general TESRA model (or as a specific expression of the VE model). This fact indicates the versatility of the TESRA and VE models, as they can be translated into simpler or more complicated ones following the considered characteristics for each of the elementary body EP1, EP2 and V (cf. Figure 1-5).

However, the ageing effect has not been incorporated in the constitutive models previously introduced. A general theory has been proposed for ageing linear viscoelastic materials (Salençon, 1983). Some non linear models are also proposed for concrete during setting (Feron, 2002). On the other hand, the ageing effect is rarely taken into account in the models developed for geomaterials.

An introduction of ageing effect into the three-component model is also given in this paper. Ageing influence is taken into account in one or several of the three bodies, EP1, EP2 and V. As a matter of fact, the non-viscous bodies EP1 and EP2 can be sensitive to time effects due to ageing (a simple example of such a case is given by an elastic body which exhibits modulus changing with time). This point also reveals the versatility of the proposed three-component framework, which appears as adapted to describe a wide and comprehensive variety of behaviours.

In the following paragraphs the behaviour observed on a wide range of geomaterials is first presented.

Homogeneous tests and physical models and full scale tests are considered.

Then, the simulation and associated issues are developed in the framework of the three-component model.

At least examples of FEM simulation using the specifically developed models, having the property of "viscous evanescent" effects or TESRA, are proposed.

2 OBSERVATIONS ON DIFFERENT GEOMATERIALS

2.1 *Homogeneous element tests*

In engineering design, it is often required to predict:

– load-displacement behaviour at different rates of construction or loading,
– creep and stress relaxation behaviour following construction or loading at different rates,
– load-displacement behaviour after construction or if loading is restarted at a certain rate following a long period of intermission,
– creep and stress relaxation behaviour at unloaded conditions, and
– evolution of the strength resistance, or more generally stress-strain response, after a long period of ageing.

For such a prediction as described above, the characterisation of time-dependent stress-strain behaviour of geomaterial corresponding to the following viscous aspects is essential:

– effects of constant strain rate on the stress-strain behaviour, including those on the peak strength,
– changes in the stress-strain behaviour when the strain rate suddenly or gradually increases or decreases from a certain value to another,
– creep deformation and stress relaxation,
– stress-strain behaviour when loading is restarted at a constant strain rate after a stage of creep or strain relaxation; and
– time-dependent stress-strain behaviour in the course of unloading and reloading, or more generally, the stress-strain-time behaviour for arbitrary general stress histories, including cyclic loading. For ageing-sensitive geomaterials, the start of loading could be at different periods of ageing.

All these aspects were evaluated by performing a comprehensive series of sophisticated homogeneous element tests (triaxial tests – two & three dimensional –, plane strain compression tests, one-dimensional loading tests (i.e., oedometer tests) and torsional tests) on a wide variety of geomaterial, for the last decade at the University of Tokyo, Japan, and ENTPE, France, particularly focusing on the stress-strain-time properties of sands and bituminous materials (Tatsuoka *et al.*,

1999a, 2000, 2001a, 2002; Cazacliu, 1996; Di Benedetto, 1997; Cazacliu and Di Benedetto, 1998; Ibraim, 1998; Di Benedetto et al., 1999a, 1999b; among others...). In order to give answers to the respective engineering needs, the characteristic features of this long-term research programme included the evaluation of; i) the elastic properties by precisely measuring strains of less than 0.001% at different levels of loading histories; ii) the viscous effects on the whole range of pre-peak strain-hardening behaviour; and iii) the peak strength and the post-peak strain softening behaviour.

The loading histories employed in the experimental investigations consisted of; i) monotonic loading and unloading tests performed at largely different strain rates (by a factor of up to 1000 in a single test); ii) monotonic loading tests in which the strain rate was changed stepwise; and iii) creep and stress relaxation tests for different periods.

The time effects were evaluated by the procedures described above not only under loading conditions but also under unloading, reloading and cyclic conditions.

The respective testing method, which depends on the type of geomaterial and the type of experimental device, is described in previous papers (Goto et al., 1991; Tatsuoka et al., 1994, 2000; Cazacliu and Di Benedetto, 1998; Santucci de Magistris et al., 1999; Pham Van Bang and Di Benedetto, 2003; ...) and will not be presented again in this paper. It is to be noted however that axial and radial strains were measured locally by using local deformation transducers in most of the tests (in particular with stiff specimens), and a wide variety of geomaterial, including undisturbed and reconstituted soft and stiff clays, sands, gravels, sedimentary soft rocks and cement-mixed clays, sands, gravels and bituminous mixes, were tested.

The main features of the observed stress-strain-time behaviour are presented in the following paragraphs. First, general stress-strain behaviour in terms of a single stress parameter and a single stain parameter. The considered stress can be the normal stress (σ), the shear stress (τ), the mean pressure (p), the deviator stress (q) or the principal stress ratio (R), while the considered strain can be the relative length variation (ε), the shear distortion (γ), the volume strain ($\varepsilon_{vol} = \Delta v/v$), etc. Then, the behaviour in the strain space (i.e., strain path) is presented.

2.1.1 General stress-strain behaviour (single stress and strain componentss)

2.1.1.1 Small strain domain

The time effects on the stress-strain response in the small strain domain (up to some 10^{-5} m/m) is generally evaluated by two types of test, monotonic loading tests at a fixed strain (or stress) rate and cyclic loading tests at a fixed frequency with a fixed strain (or stress) amplitude. Different shapes of cyclic stress or strain

(saw tooth, sinusoidal, trapezoidal ...) may be employed. By these two types of test, different moduli (E_{tan}, E_{sec}, E_{eq}), as defined in Figure 2-1, are obtained. Proper interpretation and comparison of the results obtained for different loading conditions are necessary to correctly evaluate these different types of modulus.

It has been confirmed by several authors (Kim and Stokoe, 1994; Tatsuoka et al., 1991a, 1999a; Hameury, 1995; Di Benedetto and Tatsuoka, 1997; Jardine, 1992; Lo Presti et al., 2001; among others) that, at small strains for a given type of geomaterial, the stress-strain curve can be rather accurately represented by a straight line having the same slope whatever the amplitude (ε_0 in Figure 2-1) and the loading rate (non viscous materials) are. In that case, the geomaterial is linear for the considered loading domain and the three types of modulus (tangent, secant and equivalent) are equal to each other (i.e., $E_{tan} = E_{sec} = E_{eq}$) and independent of the strain and strain rate values. For example, this practically linear and non-viscous behaviour is obtained

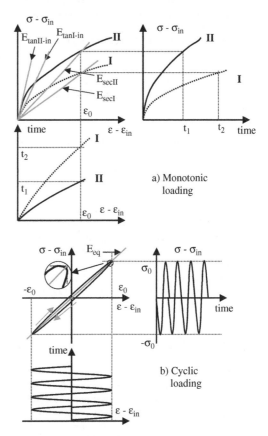

Figure 2-1. Different curves and moduli obtained for monotonic loading tests (a) and sinusoidal cyclic tests (b) (schematic). Loadings start at σ_{in} and ε_{in}. E_{tan-in} represents the tangent modulus at $\varepsilon = \varepsilon_{in}$.

with sands, gravels, mortar, hard rock (Tatsuoka *et al.*, 1999a) if the strain amplitude remains lower than about 0.001% and the range of strain rate (or frequency) remains within a factor of several hundreds.

For viscous materials the comparison is more complex. It can be easily shown that the linearity (the behavior is then linear viscoelastic) implies that:

– in the case of monotonic constant strain rate loading, the secant modulus at time "t" ($E_{sec}(t)$) is the same whatever the strain rate ($\dot{\varepsilon}$) and the strain (ε_0). In addition, the "real" tangent value at $\varepsilon = \varepsilon_{in}$ (E_{tan-in}) (which can be obtained only if the accuracy of the experimental device is high enough, while, if the accuracy is not good, it is a secant modulus that is measured) is a constant, which depends only on the previous history. The linear viscoelastic response for constant strain rate loading is plotted in Figure 2-2.
– in the case of sinusoidal loadings the equivalent modulus for a given frequency "fr" ($E_{eq}(fr)$) is independent of the strain amplitude (ε_0). This modulus corresponds to the norm of the so-called complex modulus (E*) that has the definition given in equations 2-1 and 2-2.

As the cyclic strain is sinusoidal the stress is also sinusoidal with the same frequency but with a phase lag "ϕ". It comes:

$$\varepsilon = \varepsilon_0 \sin(\omega t) \quad and \quad \sigma = \sigma_0 \sin(\omega t + \phi) \qquad (2\text{-}1)$$

$$E^* = \sigma_0/\varepsilon_0 \, e^{i\phi} = |E^*| \, e^{i\phi} = E_{eq} \, e^{i\phi} \qquad (2\text{-}2)$$

where ω is the pulsation (or angular frequency) ($\omega = 2\pi fr = 2\pi/T$). T is the period. Both the norm of the complex modulus ($|E^*| = E_{eq}$) and the phase lag (ϕ) depends only on the frequency (or ω, or T). "i" is the complex number: $i^2 = -1$.

For the linear viscoelastic materials, the phase lag "ϕ" is linked with the damping ratio "D" (Figure 2-3) by:

$$D = \frac{\pi \, \sigma_0 \, \varepsilon_0 \, \sin(\phi)}{2\pi \, \sigma_0 \, \varepsilon_0} = \frac{\sin(\phi)}{2} \qquad (2\text{-}3)$$

From the previous remarks it comes that, for monotonic constant strain rate loading, the "good" variable to express the modulus is the time "t", while for sinusoidal loadings the frequency "fr" (or $\omega = 2\pi fr$ or $T = 1/fr$) should be chosen. The first type of analysis is made in a temporal scale and the second one is in the frequency domain. The strain rate ($\dot{\varepsilon}$), often used in the literature, is not pertinent as it appears connected

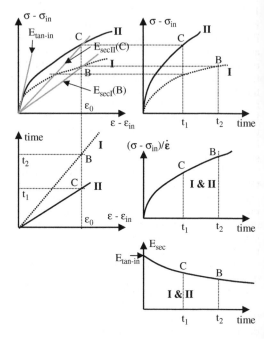

Figure 2-2. Stress-strain and modulus evolutions for linear viscoelastic material tested at constant strain rate. The state "0" (σ_{in}, ε_{in}) corresponds to a stable (or "quasi-stable") state (i.e.: stress and strain values don't vary (or vary extremely slowly) with time).

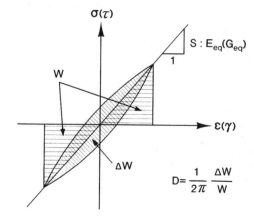

Figure 2-3. Definition of damping ratio "D" (figure 1 of Di Benedetto and Tatsuoka, 1997).

with the considered strain "ε_0" ($\dot{\varepsilon} = \varepsilon_0/t$ for constant strain rate tests and, $\max(\dot{\varepsilon}) = \omega\varepsilon_0$ for sinusoidal tests). The strain rate is relevant as the variable when comparing results from different tests only if the considered strain "ε_0" is the same for all the tests.

Then, the key questions when comparing and analysing results coming from different types of test are:

- What is the good (or relevant) criterion to confirm the linearity of the behaviour?
- How to plot and compare moduli obtained from the two kinds of tests (monotonic and cyclic)?
- What is the influence of the type of cycle on the secant modulus values?
- Is the value of E_{tan-in} unique (Figure 2-1 and Figure 2-2)?

These questions aim to give an interpretation of the data to answer to the two important points when modelling the behaviour, i) existence of a linearity domain of behaviour and ii) analysis and comparison of results from different origins (i.e., different types of test).

(1) Existence of a linearity domain

Of course, the first essential point is to check whether linear behaviour can be observed in a given loading domain around a given stress-strain state. As it is the case when viscous effects are negligible. This property is verified if the secant modulus (respectively equivalent modulus) is independent of the strain rate (respectively strain) for a given time (respectively frequency) during a monotonic constant strain rate test (respectively sinusoidal strain (or stress) test). Both types of test could be proposed (Di Benedetto and De la Roche, 1998), but the simpler one is the sinusoidal cyclic loading tests. For sinusoidal cyclic loading, the modulus remains constant when increasing (or decreasing) strain level inside the linearity domain for any given frequency. The linearity checking is then rather easy.

An example of this kind of "linearity test" is proposed in Figure 2-4 for a bituminous mix (Doubbaneh, 1995). It may be seen from Figure 2-4 that the behaviour of the bituminous mixture is essentially linear for strain amplitude below around some 10^{-5} m/m.

A great amount of experimental data is available that shows the existence of the linear domain for soils and rocks for a narrow range of frequency. However, only few results are available that confirms the linearity limit for a relatively wide range of frequency. In fact such tests are very delicate to perform. The limitation of accuracy and even feasibility of actual devices is rapidly reached if a wide range of frequency (or time) is realised (Abrantes and Yamamuro, 2003). Some data that is presented in Di Benedetto and Tatsuoka (1997) tends to confirm that the behaviour of all the types of geomaterial can be considered as linear (as the first approximation) for a strain amplitude lower than around 10^{-5} m/m. This practically linear behaviour is ether predominantly non-viscous or noticeably viscous. It is also a general trend that the stiffer the geomaterial, the larger the linear domain. This tendency is schematically illustrated in Figure 2-5 (from Tatsuoka and Shibuya, 1991a).

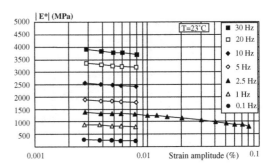

Figure 2-4. Norm of the complex modulus (equivalent modulus, cf. Figure 2-1), of a bituminous mix having a compacity of 96%, for sinusoidal loadings at different strain amplitudes and frequencies, linearity domain evaluation. (Doubbaneh, 1995).

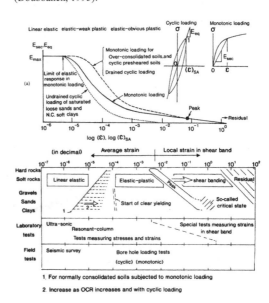

Figure 2-5. Limits of different types of behaviour for different geomaterials (from Tatsuoka and Shibuya, 1991a).

This feature can also be confirmed by the results on bituminous mixtures presented in Figure 2-6 from Doubbaneh (1995). The linear domain of the bituminous mixtures seems to become smaller in the strain space with an increase temperature, which corresponds to a decrease in the rigidity. Surprisingly, this last result is the opposite as the one observed on pure bitumen (Airey et al., 2003). This fact tends to indicate that the granular skeleton has a main role relatively to the linear domain.

Meanwhile the size of the linear domain is strongly influenced by the previous loading history. This statement is confirmed in Figure 2-7 and Figure 2-8. Figure 2-7, from Park and Tatsuoka (1994), presents

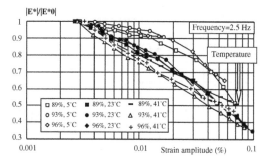

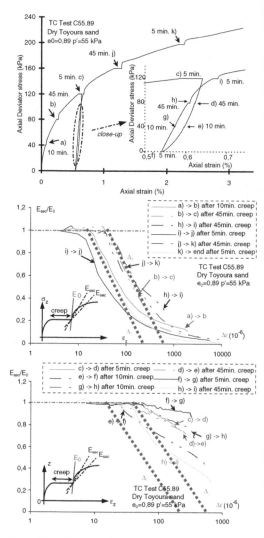

Figure 2-6. Complex modulus (equivalent modulus, cf. Figure 2-1) divided by the modulus at a strain amplitude of 0.003% ($E*_0$|), of bituminous mixes having different compacities, for sinusoidal loadings at different strain amplitudes and frequencies. The linearity domain becomes smaller when increasing temperature, which correspond to a decrease of rigidity (from Doubaneh, 1995).

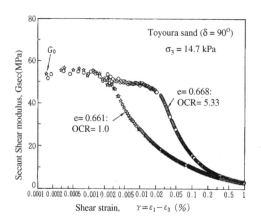

Figure 2-7. Influence of a previous isotropic cyclic loading on the size of the linear domain for PSC test on Toyoura sand (from Park and Tatsuoka, 1994).

the secant modulus evolution for two PSC tests on Toyoura sand. It shows that the linear domain is notably enlarged when a previous isotropic cyclic loading (up to OCR = 5.33) is applied. In Figure 2-8, a TC test on dry Toyoura sand with creep periods of different times is considered. The general axial strain versus axial stress curve is presented in part (a). The secant modulus during the reloading steps just after the creep periods are plotted, in function of the logarithm of the axial strain, in parts (b) and (c). As can be seen the linear domain is larger after the creep periods of 45 minutes than the one after the creep periods of 5 or 10 minutes. This result seems also verified for the creep periods applied during the cyclic loading of large amplitude, but this large cycle and the distance from the stress reversal also seem to affect the size of the linear domain.

Figure 2-8. CT test on dry Toyoura sand with creep periods of different times. The influence of the creep period duration on the size of the «quasi» elastic domain is visible in part b and c (Duttine, ENTPE unpublished data 2003).

The following conclusion could be obtained with respect to the existence of the domain of linear behaviour, which corresponds to viscoelastic behaviour for rate-dependent materials, in the neighbourhood of a "stable" stress strain state (i.e., stress and strain values don't vary, or vary very slowly, with time). The recent experimental investigations tend to confirm that the traditional approximation to non-viscous materials can be adopted in the general case. The linearity limit for cyclic loading is found at a strain of about 10^{-5} m/m. Of course as already mentioned earlier, this linearity must only be considered as the first, but justified, approximation. Results from a series of experimental and modelling researches on sands, shown later in this paper, indicate that the viscous part remains non-linear in the strain range that could be reached experimentally (around 10^{-6} m/m). Yet the viscous effect remains rather small in the time domain (for monotonous loading tests) and frequency domain (for cyclic loading tests) that can be experimentally examined. Then, the (non-viscous) elastic behaviour becomes predominant. This last assumption is not verified for all the loading paths. For example during and after creep and relaxation periods viscous effect can be of primary importance on dry sand.

(2) Comparison of data in the linear domain from different types of test

It is important to compare the moduli obtained from different types of test (cf. Figure 2-1). The relationship among these moduli is can be clearly determined because of the linearity of behaviour when this hypothesis is verified (i.e. in the very small strain amplitude range).

For viscous materials the equivalent modulus (Eeq) is dependent of the shape of the applied cycle. The Figure 2-9, from Di Benedetto and Tatsuoka (1997), shows the calculated stress-strain loops for a SAB model (spring in series with a spring in parallel with a dashpot, cf. Figure 3-1) for sinusoidal and constant absolute strain rate cyclic loading tests. Despite that the considered cycles have the same strain amplitude and frequency, the curves are slightly different. The magnitude of this difference is a function of the constitutive behaviour. As can be seen from the example presented in Figure 2-9, the values of secant modulus and damping ratio in these two cases are rather close to each other. This simulation result is confirmed by the experimental data from Nakajima et al. (1994), which are presented in Figure 2-10. Even if no global proof is given, it seems that the moduli obtained from different cyclic shapes, classically used for experimental investigation on geomaterials, are very close and can be directly compared.

The comparison between secant and equivalent moduli is more controversial. The theoretical relations giving these two moduli for linear viscoelastic materials (cf. Figure 2-2) is given in equations 2-4 & 2-5 for sinusoidal loading and in equation 2-6 for constant strain rate loading (Mandel, 1966; Salençon, 1983):

$$E^*(\omega) = i\omega \int_0^\infty e^{-i\omega t} R(t)\, dt \qquad (2\text{-}4)$$

$$E_{eq}(\omega) = |E^*| \qquad (2\text{-}5)$$

$$E_{sec}(t) = \frac{\int_0^t R(\tau)\, d\tau}{t} \qquad (2\text{-}6)$$

where R(t) is the relaxation function ($\sigma(t) = R(t)\varepsilon_0$; if ε_0 is constant).

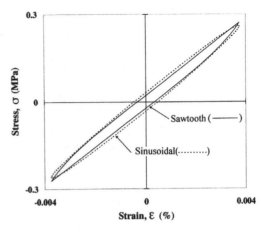

Figure 2-9. Calculated loop with the SAB model (in Figure 1-1-5, EP1 and EP2 are springs and V is a dashpot, cf. Figure 3-1) for sinusoidal and constant absolute strain rate cyclic tests (frequency = 0.07 Hz, K_0 = 100 MPa, K = 200 MPa, η = 200 MPa.s) (from Di Benedetto and Tatsuoka, 1997).

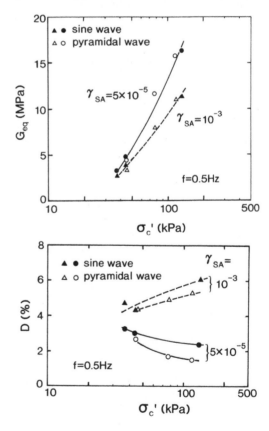

Figure 2-10. Influence of cyclic shape on (a) shear modulus and (b) damping from tests on intact clay samples (Nakajima et al., 1994).

The plot on a same figure of E_{eq} and E_{sec} requires that correct couples of "ω" and "t" be chosen. In fact, the relation (w) between ω an t; $\omega = w(t)$, assuring $E_{sec}(t) = E_{eq}(w(t))$ depends on the whole time dependent behaviour. In addition, this relation is not mathematically easy to obtained (Marasteanu, 1999; Olard et al., 2003). Simple commonly used relations are, $1/2\pi \leqslant fr^*t \leqslant 1$. It can be shown (Di Benedetto and De la Roche, 1998) that these relations can give wrong values up to some 10%. Such levels are obtained if the material is very sensitive to rate effects, which is not the case for most of the geomaterials except for bituminous mixes as it is evident on Figure 2-12, as explained further.

Another important information is the value(s) of the modulus "$E_{tan\text{-}in}$" (cf Figure 2-1). It can be shown for linear materials that $E_{tan\text{-}in}$ is a unique material parameter, which can be obtained by:

$$E_{tan\text{-}in} = \lim_{t \to 0}(E_{sec}(t)) = \lim_{w \to \infty}(E_{eq}(w)) \qquad (2\text{-}7)$$

From equations 2-4, 2-5 and 2-6 it comes:

$$E_{tan\text{-}in} = R(0) \qquad (2\text{-}8)$$

Equation 2-7 means that a "small" limit time t_{lim} (respectively a "high" limit frequency fr_{lim}) can be introduced such as for "small" time; if $t < t_{lim}$ (respectively for "very high" frequency; if $fr > fr_{lim}$), in the case of monotonic loadings (respectively cyclic loadings), the behaviour can be considered as (quasi)elastic. In this case, the material becomes rate-independent and the shape of the cycle as well as the strain rate has no influence on the moduli, which then respect the equality:

$$E_{tan\text{-}in} \approx E_{sec} \approx E_{eq} \qquad \text{if } t < t_{lim} \text{ or } fr > fr_{lim} \qquad (2\text{-}9)$$

From the previous explanations one should distinguish the linear behaviour domain, which is probably dependent of the strain amplitude (as presented in section 2.1.1.1) and, inside that linear domain, the elastic (or quasi-elastic) domain, which is limited in time for monotonic loading and frequency, for cyclic loading.

Figure 2-11 presenting triaxial undrained test results on NSF clay, from Shibuya (2001), tends to confirm these two types of behaviour. On this figure it can be estimated from the cyclic tests (at 0.1 Hz and 0.01 Hz) that the linear domain is limited at about 8 10^{-5} m/m. Inside that domain, the E_{eq} modulus is frequency dependent as $E_{eq}(0.01\,\text{Hz})$ is smaller than $E_{eq}(0.1\,\text{Hz})$. The frequency of 0.01 Hz is not high enough to respect equation 2-9. Then, the elastic behaviour is not reached at 0.01 Hz for the tested NSF clay. As a general remark, this inequality ($E_{eq}(0.01\,\text{Hz})$

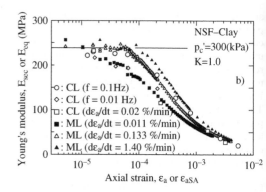

Figure 2-11. Esec and Eeq moduli from triaxial undrained monotonic and cyclic tests on NSF clay (Shibuya, 2001). See Figure 2-1 and Figure 2-2 for modulus definition. ML and CL mean respectively monotonic and cyclic loading.

smaller than $E_{eq}(0.1\,\text{Hz})$) is also verified in the non linear domain (ε greater than $8\ 10^{-5}$ m/m) up to strain amplitude of the order of one percent, where the viscous effects are certainly hidden by the non linearity. The 2 monotonic loading tests, respectively at 0.011 and 0.133%/min, give the value of the strain below which the modulus E_{sec} remains constant and equal to $E_{tan\text{-}in}$ (close to 250 Mpa). The 2 limit strain values are respectively: $\varepsilon_{lim}(0.011) = 9\ 10^{-6}$ m/m and $\varepsilon_{lim}(0.133) = 7\ 10^{-5}$ m/m. The time (t_{lim}) at which these limit strains are reached can be calculated from the strain rate. It comes: $t_{lim}(0.011) = 4.9$ s and $t_{lim}(0.133) = 3.2$ s. As a confirmation of the theory for linear viscoelastic materials, these two limit times are rather close and can be considered as equal, when taking into account the accuracy of the measurements. While it is not the case for the 2 limit strains ($\varepsilon_{lim}(0.011)$ and $\varepsilon_{lim}(0.133)$), which are very different. The monotonic test at a strain rate of 1.4%/min does not respect the previous order of magnitude as $t_{lim}(1.4) = 0.4$ s. This result is also in conformity with the theory because in that case the E_{sec} modulus decrease is not due to time limit but to strain amplitudes becoming higher than the linear domain.

The cyclic loading test at constant absolute strain rate of 0.02%/min and strain amplitude of $16\ 10^{-6}$ m/m (inside the linear domain) is performed at a frequency of 0.05 Hz. The obtained secant modulus is higher than the one from CL test at 0.01 Hz and close to the modulus of the CL test at 0.1 Hz. Which again confirm what is expected if linear behaviour is respected. The analysis of the cyclic data at a strain amplitude of 220 m/m (corresponding to a frequency of 0,004 Hz for the CL test at 0.02%/min) gives the following ranking for the equivalent moduli: $E_{eq}(0.1) > E_{eq}(0.01) > E_{eq}(0.02\%/\text{min})$, which respect the expected viscous influence in the non linear domain.

The data of Figure 2-11 give an evaluation of the two limits (t_{lim} and fr_{lim}) limiting the validity of equation 2-9, corresponding to an elastic behaviour. The already calculated limit time values (4.9s and 3.2s) for ML tests is of the order of $t_{lim} = 4$ seconds. The cyclic test at 0.1 Hz gives, in the linear domain, a constant E_{eq} modulus lower but close to the maximum modulus (about 250 Mpa), which is equal to E_{tan-in} for linear materials. Then it can be considered that fr_{lim} is of the order of 0.1 Hz or slightly higher. As already said, the relation between t_{lim} and fr_{lim} is not mathematically easy to obtain. Meanwhile the previously given equations ($1/2\pi \leqslant fr*t \leqslant 1$) give a value of fr_{lim} deduced from t_{lim} between 0.04 Hz and 0.25 Hz. The order of magnitude is quite correct.

The presented analysis confirms that both a linear domain, which is strain amplitude dependent, and an elastic domain, which is time or frequency dependent, can be postulated for NSF clay, as a first approximation. This tends to validate the existence of a domain ("around" a given stress-strain state) where the behaviour is linear viscoelastic. It has to be underlined that the behaviour is not globally linear as E_{tan-in} depends on the previous loading history. For example, in the case of sand, E_{tan-in}, is a function of the void ratio and the actual stress (at least).

Another interesting point to investigate is the value of the lower modulus E_{min}, which corresponds to the limit of the secant and the equivalent moduli:

$$E_{min} = \lim_{t \to \infty}(E_{sec}(t)) = \lim_{w \to 0}(E_{eq}(w)) \qquad (2\text{-}10)$$

It can be easily shown for linear viscoelastic materials that the E_{min} value is not null if the two following tendencies are observed, i) the amplitude of strain during a creep test (i.e. at constant stress) is limited or, ii) the stress does not tend toward 0 during a relaxation test (i.e. at constant stress). It is physically admitted that most of the "classical" geomaterials (soils and rocks) respect these tendencies. Only some very soft clays and muds don't respect these tendencies and have probably a minimum modulus (E_{min}) equal to zero (as it is the case for liquid type behaviour).

The experimental investigation of the E_{min} value is not easy as the time (respectively the frequency) for monotonous tests (respectively for cyclic tests) must be very high (respectively very low). The limit of the actual devices is not always high enough to draw clear conclusion. In addition as the strain should remain inside the linear domain, the needed high time value limits the strain rate at a very low value for monotonic constant strain rate test. If this last condition is not respected, the modulus does not decrease because of rate effect but because of non linearity.

As equations 2-9 and 2-7 are formally identical, similar conclusions can be drawn for limiting value

E_{min} than for limiting value E_{tan-in}. Equation 2-10 means that a "high" limit time t_{limmin} (respectively a "low" limit frequency fr_{limmin}) can be introduced such as: for "high" time; if $t > t_{limmin}$ (respectively for "low" frequency; if $fr < fr_{limmin}$), in the case of monotonic loadings (respectively cyclic loadings), the behaviour can be considered as (quasi)elastic. In that case the material becomes rate independent and the shape of the cycle as well as the strain rate has no influence on the moduli, which then respect the equality:

$$E_{min} \approx E_{sec} \approx E_{eq} \qquad \text{if } t>t_{limmin} \text{ or } fr<fr_{limmin} \qquad (2\text{-}11)$$

Then, another elastic (elastic "2") domain can be identified inside the linear behaviour domain. The elastic "2" (or quasi-elastic "2") domain, is also limited in time for monotonic loading and frequency, for cyclic loading (equation 2-11).

From equation 2-6, it comes:

$$E_{min} = R_{min} = \lim_{t \to \infty}(R(t)) \qquad (2\text{-}12)$$

Then, all the measured moduli respect the following inequality:

$$R(\infty) = E_{min} \leq E_{sec} \text{ or } E_{eq} \leq E_{tan-in} = R(0) \qquad (2\text{-}13)$$

To summary, a linear domain of behaviour, which is limited in strain amplitude, can be considered for geomaterials. Inside that domain, two domains, limited in time or frequency, where the behaviour is elastic, can be isolated. The Elastic properties inside each of these domains can be very different. This conclusion is schematised in Figure 2-12, and translates a behaviour as a first approximation. It is quite probable that non linearity appears at very small strain but can be considered as negligible in this domain.

An interesting and complementary information on the rate dependent behaviour of geomaterials is given by the 2 time values: t_{limmin} and t_{lim} (respectively 2 frequency values: fr_{limmin} and fr_{lim}). These 2 values indicate the limitation of two elastic behaviours. From results presented in Figure 2-11, we obtained a value of t_{lim} of the order of 4 seconds for NSF clay (see previous paragraph). The value of t_{limmin} was not possible to evaluate for this material due to the lack of data.

Investigation on the bituminous mix presented in Figure 2-13 gives the following limits at a temperature of 10°C. The limit frequencies are calculated for modulus at around 25% of respectively E_{min} and E_{tan-in}: $E_{min} = 65$ Mpa, $E_{tan-in} = 34\,000$ Mpa, $fr_{limmin} = 3 \ 10^{-5}$ Hz, and $fr_{lim} = 10^{+4}$ Hz. From these results the expected limited times are around: $t_{limmin} = 3 \ 10^{+4}$s, and $t_{lim} = 10^{-4}$s. These values, which could not be obtained from experiments at 10°C, are deduced

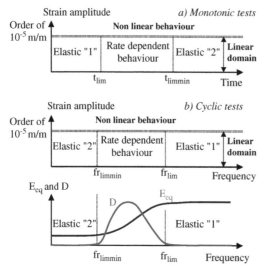

Figure 2-12. Schematic representation of the linear and non linear domains of behaviour and of the 2 elastic domains inside the linear domain. t_{lim}, t_{limmin}, fr_{lim} and fr_{limmin} values are very different following the type of geomaterials.

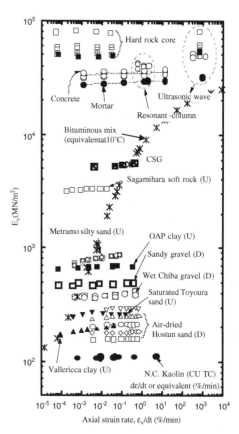

Figure 2-13. Vertical modulus, at a strain amplitude close to 10^{-5} m/m, in function of the strain rate (or equivalent strain rate for cyclic tests) for different geomaterials (most data from Kohata et al., 1997 and Tatsuoka et al., 1999a)

Test conditions for the data presented are:

1) cyclic triaxial tests (U; undrained and D; drained)

a) Saturated Sgamihara soft mudstone, undisturbed saturated OAP clay and moist Chiba gravel (Di Benedetto and Tatsuoka, 1997);

b) Air-pluviated Toyoura sand (e = 0.658 and $\sigma'_v = \sigma'_h = 1.098$ kPa; Santucci di Magistris et al., 1999), air-pluviated Hostun sand (e = 0.72 and $\sigma'_v = 78 \sim 245$ kPa and $\sigma'_h = 78$ kPa, Hoque, 1996);

c) Saturated compacted Metramo silty granite sand ($\sigma'_c = 392$ kPa; Santucci de Magistris et al., 1999),

d) Saturated undisturbed Vallericca clay (Material properties reported in Tatsuoka et al., 2000; $\sigma'_c = 98$ kPa, e = 0.819, Consolidation time = 2780 min)

e) Moist CSG (cement-mixed sandy gravel; $D_{50} = 1.9$ mm; Uc = 4.6; w = 5%; cement content = 60 kg/m3; $\sigma'_v = \sigma'_h = 200$ kPa; & q_{max} by drained CT = 500 kPa) (Omae et al., 2003)

2) CU TC tests; saturated NC kaolin ($pc' = 296$ kPa; $K_c = 0.6 \sim 1.0$) (Tatsuoka et al., 1994),

3) Unconfined cyclic tests and ultrasonic tests on rocks, concrete and mortar (Sato et al., 1997a and 1997b), and

4) Compression tension tests on bituminous mixtures (Olard et al., 2003).

from results at different temperatures with the hypothesis of thermorheologically simple behaviour (Ferry, 1980; Di Benedetto and De la Roche, 1998).

Figure 2-13 summarises the data obtained from a comprehensive series of test on different types of geomaterial (from hard rocks to soft clays) performed along different loading paths. Most of the tests were performed at the University of Tokyo. The values of vertical modulus (E_v) presented in this figure are either the statically measured secant (E_{sec}) or equivalent (E_{eq}) modulus from, respectively, monotonic triaxial compression tests and cyclic triaxial tests, or the dynamically measured modulus (E_{dyn}) obtained by wave velocity measurements or resonant tests. It had been shown previously by many researchers that the quasi-static and dynamic measurements give similar results for non-viscous materials and these results should be considered as above. It can also be extrapolated to linear viscoelastic materials if plotted in correct axes. The E_{dyn} modulus (which is measured for small enough strain amplitudes to consider linear behaviour) is equal to the E_{eq} modulus that would be obtained at the same frequency. This last assumption is verified for all the geomaterials, presented in Figure 2-13, except for concrete, for which an experimental problem could be expected: i.e., the wave velocity measurement could provide a higher modulus than static measurements due to a significant heterogeneity of the material in the scale of wave length.

In Figure 2-13, the chosen horizontal axis is the axial strain rate, which corresponds to the respective actual value for the monotonic tests at a constant strain rate and to an equivalent value for the cyclic tests, in which the stress rate was constant. Considering the previous developments, the chosen equivalent strain rate for the cyclic tests is:

$$\dot{\varepsilon}_{eq} = 4\varepsilon_0 \ fr \qquad (13)$$

As the strain amplitude (ε_0) was similar for all the tests, close to 10^{-5} m/m, it is possible to make a direct comparison between the monotonic and cyclic loading test data as explained before. The results obtained at different temperatures (from -20 to $40°C$) and the application of the "time-"temperature" principle (Ferry, 1980; Di Benedetto and De la Roche, 1998) allow to plot the data at a reference temperature of $10°C$ for the considered bituminous mix. If the data for the bituminous mix and the data from dynamic tests are excluded, the range of frequencies (or equivalent frequencies) applied in Figure 2-13 is between 0.0004 Hz and 4 Hz. This range is rather wide but probably not large enough to cover all the variety of behaviour obtained in practice and to draw a conclusion on the fr_{limmin} and fr_{lim} values.

From the data presented in Figure 2-13, one can distinguish three groups of material regarding the sensitivity to rate effects in the considered linear domain as follows:

– the bituminous mix, which is extremely sensitive to rate effects, with a modulus varying by a factor of more than 100 in the considered range,
– Vallericca clay, wet Chiba gravel, Metramo silty sand, the mortar, cement-mixed sand gravel (CSG), and concrete, showing a range of variation of some 10%, which can be considered as slightly viscous,
– the other types of geomaterial, exhibiting a very small rate-sensitivity. Their behaviour at small strains can be treated as non-viscous materials in the considered domain of loading: i.e., the modulus of these materials is not influenced by the strain rate for these tests. Note however that the above does not imply that the viscous effects never exist with these types of geomaterials. Some examples are given in the following sections. In particular, it will be shown how important viscous effects could be with dry sands (even for practical purposes) under other loading conditions including different time and frequency ranges.

From the previous developments, on modulus evolution in the linear domain, the observed small sensitivity to rate effects for some materials can have two interpretations:

– the material is rate-independent or non viscous,
– for the applied frequency range the material is either in elastic domain "1" (if $fr_{lim} < 4$ Hz or 4000 Hz if

dynamic test are included) or in elastic domain "2" (if $fr_{limmin} > 0.0004$ Hz) (cf. Figure 2-12).

The correct answer can only be obtained if investigation in a wider time (or frequency) range is performed. Unfortunately, the applied range is limited in the laboratory by experimental constraints. An extreme strain rate value of more that 1000% per minutes is, for example, obtained by Abrantes and Yamamuro (2003) using impact tests. For practical purpose, it is needed that the considered range corresponds to time scale observed for civil engineering structure design.

2.1.1.2 Pre-peak, peak and post-peak domains

(1) Monotonic loadings at different loading rates
The loading path, which is considered in this section, is monotonic loading at constant strain (or stress) rate. Depending on the respective geomaterial type, the stress-strain response curves can exhibit different loading rate effects:

(a) Negligible influence of strain rate and "pure" evanescent viscous behaviour.
This behaviour is illustrated in Figure 2-14 and Figure 2-15 from Tatsuoka *et al.* (2000). A set of drained Plane Strain Compression (PSC) tests were performed at a constant confining pressure of $\sigma'_h = 392$ kPa on saturated Hostun sand at different constant strain rates. In this figure, $\dot{\varepsilon}_0$ means the basic axial strain rate, equal to 0.0125%/min. The initial void ratios e_0, measured at an isotropic stress state of $\sigma'_c = 29$ kPa, were nearly the same among the specimens. The following trends of behaviour can be seen from these figures:

i) The stress ratio $R = \sigma'_v/\sigma'_h$ and shear strain $\gamma = \varepsilon_v - \varepsilon_h$ relationships for the constant axial strain rates $\dot{\varepsilon}_v$ that are different by a factor of up to 500, obtained from tests H302C through H307C, are essentially independent of the axial strain rate $\dot{\varepsilon}_v$. Note that the $\dot{\varepsilon}_v$ value was kept constant in each test. Some "small" effects of $\dot{\varepsilon}_v$ may be seen only in the stress-strain behaviour immediately after the start of loading from a stress ratio R equal to 3.0 (Matsushita *et al.*, 1999) (cf. Figure 2-14 for focus).

ii) Despite the above, significant creep deformation and stress relaxation can be seen when the deviatoric stress and axial strain are kept constant (Figure 2-15).

iii) In addition, in test HOS02 (Figure 2-16), the $R = \sigma'_v/\sigma'_h$ value (i.e., the shear stress) increases/decreases at a very high rate immediately after the strain rate $\dot{\varepsilon}_v$ increases/decreases stepwise by a factor of 100. Note that Matsushita *et al.* (1999) showed that the effects of the inertia of the testing system are negligible and cannot be the origin of this behaviour. Then, with an increase in the strain, the stress-strain curve exhibits a marked change in the tangent modulus and gradually converges into an essentially unique stress-strain curve that would have been obtained if

73

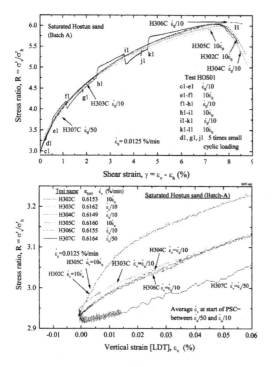

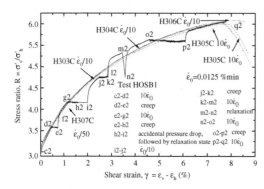

Figure 2-14. Relationships between $R = \sigma'_v/\sigma'_h$ and $\gamma = \varepsilon_v - \varepsilon_h$ from a series of drained PSC tests ($\sigma_h = 392$ kPa) at different constant axial stain rates and a test with step changes in the constant strain rate on saturated Hostun sand (batch A); (Matsushita et al., 1999; Tatsuoka et al., 2000).

Figure 2-15. Relationships between $R = \sigma'_v/\sigma'_h$ and $\gamma = \varepsilon_v - \varepsilon_h$ from a series of drained PSC tests at different constant axial stain rates and a test including creep and stress relaxation stages on saturated Hostun sand (batch A) (Matsushita et al., 1999; Tatsuoka et al., 2000).

the strain rate had been kept constant without a step change. This observation obtained under drained condition is also true in undrained tests, as can be seen from Figure 2-16 (Tatsuoka et al., 2002).

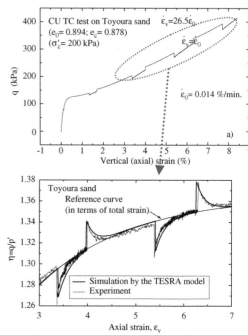

Figure 2-16. Stress-strain response for undrained TC test on Toyoura sand (Tatsuoka et al., 2002). Simulation presented in the zoom up part is introduced in section 3.

The observations, described in section i) above, may lead to the conclusion that the behaviour is time-independent or non-viscous. In fact, the results underlined in sections ii) and iii) above mean that viscous effects do exist and are not negligible for the considered loading conditions.

Similar trends of behaviour have been observed in PSC tests on air-dried Hostun sand (Di Benedetto et al., 2002), in the torsional shear tests on air-dried Hostun sand (Di Benedetto et al., 2001a and 2001b) and in the triaxial compression tests (TC tests) on water-saturated and air-dried Toyoura sand (Matsushita et al., 1999). The Figure 2-17 from Di Benedetto et al. (2002) shows results from a drained PSC test on dry Hostun sand realised in the same condition as the tests presented in Figure 2-14 in and Figure 2-15. A very small different response between the dry and saturated specimens shown respectively in Figure 2-16 and Figure 2-17 seems to be due to the different batches (batch A and B) and is not due to the different wet conditions. The trends of behaviour described above are therefore not unique for these particular drained PSC tests on saturated Hostun sand and are not due to the presence of water. They have also been observed for cement-mixed soils (Kongsukprasert et al., 2001) and cement-mixed gravels (Kongsukprasert and Tatsuoka, 2003a and

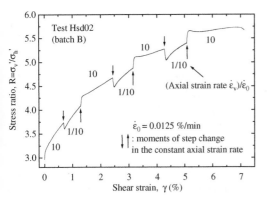

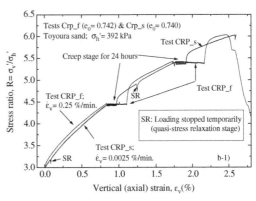

Figure 2-17. Relationships between $R = \sigma'_v/\sigma'_h$ and $\gamma = \varepsilon_v - \varepsilon_h$ from a drained PSC test ($\sigma_h = 392$ kPa, $e_0 = 0,7$) with step changes in the constant strain rate on dried Hostun sand (batch B); (figure 16 of Di Benedetto et al., 2002) same trend as Figure 2-14.

Figure 2-18. Results from two drained PSC tests on saturated Toyoura sand in the R versus axial strain axes (Tatsuoka et al., 2002). Simulations by the TESRA model are presented in 3-9.

2003b), these materials also include ageing effects (see section 2.1.1.2(7)).

It is important to note that these apparently contradictory stress-strain behaviours described above are very difficult to be simulated by any existing theory as far as the authors know. The new models ("Temporary Effect of Strain Rate Acceleration" (TESRA) and "viscous evanescent" (VE)), developed by the authors to translate this behaviour, are presented in section 3.

Another feature of the TESRA or VE behaviour is seen from the data presented in Figure 2-18 (Tatsuoka et al., 2002). In this figure, the relationships between the stress ratio and the vertical strain versus from two drained PSC tests on saturated Toyoura sand are represented. The axial strain rate for the fast rate test (CPR-f) is 0.25%/min and 100 times lower for the low rate test (CPR-s). Creep loading was applied at the same stress levels in the two tests. Due to the viscous evanescent properties, the stress and the stress values had become very close at the start of the respective creep loading stage in the two tests. Surprisingly, even though nearly the same stress and stress state was reached before the start of creep loading, the creep evolution was significantly different from one test to the other. Creep deformation increment was more than twice in the test in which monotonic loading was faster. This peculiar trend of behaviour could also be well simulated with the laws developed by the authors. A simulation of these tests with the TESRA model is presented in Figure 3-9.

(b) Clear influence of strain rate and isotach type behaviour.

Some geomaterials, such as some kinds of clay, sedimentary soft rock, bituminous mixtures, show a clear and persistent influence of strain (or stress) rate on

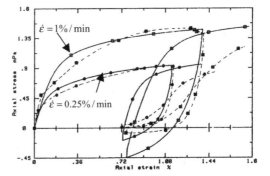

Figure 2-19. Stress-strain curves obtained for simple compression tests on bituminous mixtures at two different strain rates (from Di Benedetto, 1987), dashed curves are data and full curves model simulations (cf. section 3).

the stress-strain relation and each stress-strain curve is associated with a strain rate. Even if the curves for different strain rates have the same shape, a clear distinction can be made among them. The faster the loading, the stiffer the behaviour becomes. This behaviour is observed in Figure 2-19, where results from two unconfined compression tests on bituminous mixtures performed at different axial strain rates (from Di Benedetto, 1987) are plotted. The simulation curves presented in this figure are explained in section 3.

In addition, this "isotach" type behaviour is characterised by a sudden jump in the stress (or strain) upon a stepwise change in the strain (or stress) rate and the stress-strain relation after such a jump is persistent as far as the monotonic loading continues at a constant strain rate while rejoining the monotonic curve corresponding to the new rate. An example of such behaviour obtained from a undrained triaxial

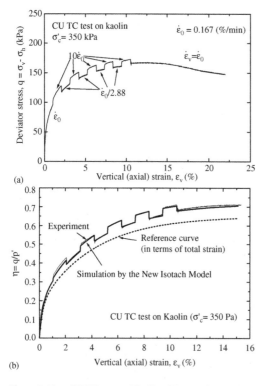

(a)

(b)

Figure 2-20. CU TC test on Kaolin with step changes in $\dot{\varepsilon}_v$ (Tatsuoka *et al.*, 2002); a) q vs ε_v relationship; b) $\eta = q/p'$ vs ε_v relationship (with simulated one, cf. section 3).

compression test on kaolin is presented in Figure 2-20 (Tatsuoka *et al.*, 2002).

(c) Clear influence of strain rate with transition from isotach to more viscous evanescent type behaviour. Behaviour corresponding to an intermediate trend between the two extreme cases described in (a) and (b) is also observed. Figure 2-21, from Tatsuoka *et al.* (2002), shows the stress-strain relationship from three consolidated undrained triaxial compression tests (CU TC tests) on saturated reconstituted Fujinomori clay, isotropically consolidated at $\sigma'_c = 200$ kPa. Two tests were performed at a constant axial strain rate (respectively 0.005%/min and 0.5%/min) and, in the other test, the axial strain was changed stepwise from one rate to another at different strain levels during otherwise monotonic loading at a constant strain rate. It may be seen that, at ε_v less than about 1.0%, the stress-strain relation is essentially a unique function of instantaneous $\dot{\varepsilon}_v$ for these largely different strain rate histories. It is also the case with the effective stress paths (Figure 2-21c, Tatsuoka *et al.*, 2002). This behaviour is of the isotach type discussed in (b). At ε_v larger than about 1%, however, as monotonic loading continues at a constant $\dot{\varepsilon}_v$, the effects of $\dot{\varepsilon}_v$ on the stress-strain rela-tion become gradually smaller. That is, at ε_v larger than

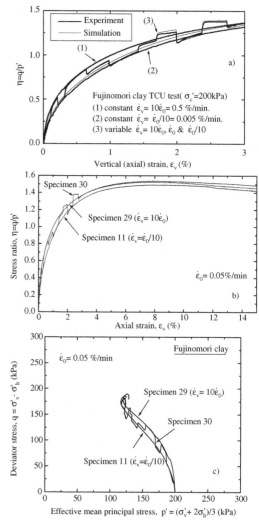

a)

b)

c)

Figure 2-21. Stress-strain relationship for three CU TC tests on Fujinomory clay (Tatsuoka *et al.*, 2002); a) experiment and simulations with the general TESRA model (cf. section 3); b) data up to an axial strain of 15%, c) representation in a p–q plot.

about 1%, although the change in the $\eta = q/p'$ value by a step change in $\dot{\varepsilon}_v$ is still significant as it is at smaller strains, the stress-strain curve over-shoots the one during continuous monotonous loading and decays as loading continues at a constant $\dot{\varepsilon}_v$. A similar change in the behaviour was observed in the effective stress paths (Tatsuoka *et al.*, 2002) (cf Figure 2-21c).

The partial decay of viscous effect, as described above, was observed on well graded gravels (Ahn Dan *et al.*, 2004), Saturated Metramo silty sand in undrained TC (Santucci de Magistris *et al.*, 2004) and

76

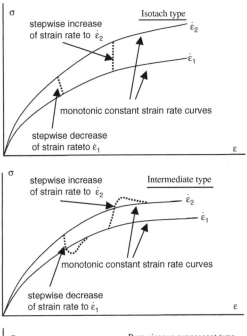

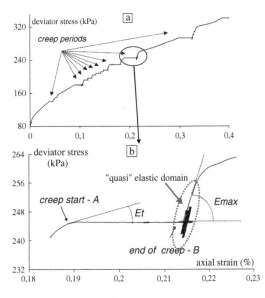

Figure 2-23. Drained hollow cylinder (T4C Stady) triaxial test on dry medium density RF Hostun sand with creep periods ($\sigma_3 = 80\,\text{kPa}$) from Di Benedetto (1997).

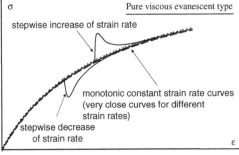

Figure 2-22. Types of stress-strain behaviour (schematised).

low plasticity clay. When considering only stress-strain response, as it is the case in this chapter, this type of behaviour is obtained for very different loading paths, such as: 1D compression, triaxial, plane compression, shear ...

The two types of extreme behaviour, called respectively "isotach" and "pure viscous evanescent", and the intermediate one are schematically described in Figure 2-22.

(2) Analysis of creep and relaxation stages
Other types of classical loading paths that have been analysed to evaluate the viscous properties of materials are creep and relaxation periods.

Creep periods during drained triaxial compression (TC) test performed on dry Hostun sand with a hollow cylinder apparatus (T4CstaDy) are presented in Figure 2-23, from Di Benedetto (1997).

During loading after a creep at the same strain rate as before the start of creep, the stiffness is very high (E_{max} in Figure 2-23b). If the creep period is long enough (more than one hour for dry sand), it was experimentally shown (Di Benedetto, 1997; Cazacliu and Di Benedetto, 1998; Di Benedetto et al., 1999a and 1999b) that the behaviour for small cycles becomes "quasi" elastic. Then the behaviour can be modelled by a hypoelastic law inside this "quasi" elastic domain. For example, a specific hypoelastic formulation has been proposed at ENTPE: the DBGS law (Di Benedetto et al., 2001a and 2001b; Geoffroy et al., 2003; Sauzéat, 2003). It appears that creep (as well as the "small" cyclic loading, which is not clearly visible in Figure 2-23) modifies only locally the response, while general monotonic loading curve is closely rejoined after reloading. The previous tendencies are confirmed by the results on air-dried Toyoura sand in drained TC presented in Figure 2-24. The test performed at a confining pressure of 400 kPa includes a series of creep and relaxation periods. The similarity with the trends obtained on PSC tests (Figure 2-15) can be underlined.

From these experimental observations it can be concluded that:

i) The dry sand shows viscous behaviour,
ii) The loading stress-strain curve is not a stable state for sand, in the sense that the stress and strain values on that curve cannot be kept constant: i.e., the stress-strain values do not stay on that curve if

77

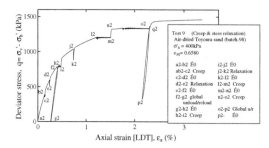

Figure 2-24. Series of creep and relaxation periods on air dried Toyoura sand in drained TC from Matsushita *et al.* (1999).

loading is stopped or if loading rate is stepwise changed. One could imagine that a monotonic loading at an infinitely slow strain rate could give a stable state. This question remains open, since even at the slowest strain rates allowed by the actual devices, creep has always been observed.

iii) After a rather "long" period of creep, which depends on the type of materials, an elastic domain seems to exist around the actual stress-strain state. The size of this domain, which remains very small, depends on the previous history and is rapidly covered and forgotten after the restart of monotonic loading. This conclusion is confirmed by the results presented in Figure 2-8 already discussed in section 2.1.1.1.

This qualitative trend of behaviour obtained on dry sand has been observed also on some other and very different types of geomaterials. Other examples are given latter in this paper.

When plotting the creep strain evolution of sand with time a rapid variation, which decreases rapidly is usually observed. Nevertheless, even after some thousands seconds, the strain is not totally stabilised and follows a rather linear evolution when the logarithm of time is considered (Figure 2-25 from Sauzéat *et al.*, 2003). It has to be noted that due to the rotation of principal axes during the test presented in Figure 2-25, shear strain, axial strain, orthoradial strain and radial strain had to be considered (Di Benedetto *et al.*, 1999b) (cf. § 2.1.2).

It is important to note that the direction of creep strain increment (in the strain space) is not uniquely linked to the instantaneous stress level but it depends on the latest reversing mode of strain (or stress). This feature can be, for example, confirmed by the analysis of creep strain direction at point 4 in Figure 2-25. This feature can also be clearly confirmed by the behaviour at creep loading performed on the unloading branch in test DR2 presented in Figure 2-26. This result was obtained from a drained TC test on moist Chiba gravel (Ahn Dan *et al.*, 2004). The creep strain

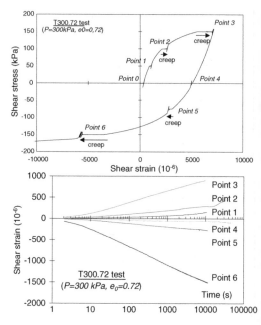

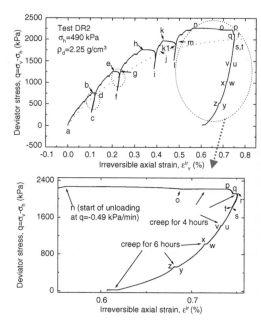

Figure 2-25. Pure torsion test $(\sigma_{rr} = \sigma_{\theta\theta} = \sigma_{zz} = 300\,kPa)$ T300.72 (T4CStaDy) on dry Hostun sand: a) Evolution of shear stress $(\sigma_{\theta z})$ and shear strain $(\varepsilon_{\theta z})$ during test, b) Evolution of shear strain with time, during creep periods applied at different stress levels (from Sauzéat *et al.*, 2003) (cf. Figure 2-52 for strain evolution).

Figure 2-26. Drained TC test on moist Chiba gravel from Ahn Dan *et al.* (2004).

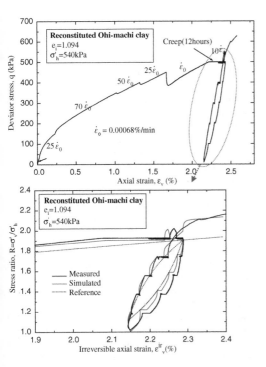

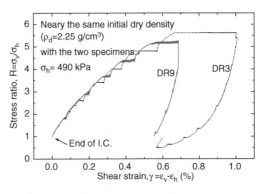

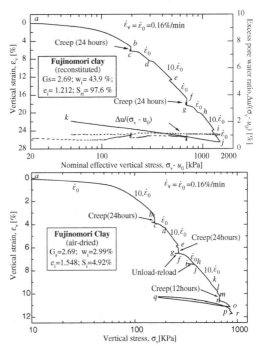

Figure 2-27. Drained TC test on reconstituted saturated Ohimachi clay from Li *et al.* (2003). Simulations presented in the zoom up part use the new isotach model and are introduced in section 3.

axial increment that developed at creep loading stages denoted as "UV, WX, YZ", applied at positive values of the deviator "q", was negative. Figure 2-27 shows results from a drained TC test on reconstituted saturated Ohimachi clay (Li *et al.*, 2003). The test results presented in this figure confirm the test results described above with respect to the direction of creep strain (i.e., negative creep axial strain increments on the unloading branch). Another conclusion that can be derived from Figure 2-26 and Figure 2-27 is that the general trend of viscous behaviour that was previously described for sand is also relevant to other types of geomaterial.

An example of the influence of the water content for Chiba gravel on the stress-strain response curve is given in Figure 2-28. A pair of saturated and moist specimens (respectively DR9 and DR3) was subjected to drained TC with several intermediate creep loading stages. It was found that the viscous effects, that induced creep strain increments, are very similar for the two types of samples. It appears that the viscous properties are independent of water content within the limit of this test condition.

Effects of a more drastic water content difference on the viscous effects were examined by means of drained one-dimension compression œdometer tests

Figure 2-28. Drained TC tests on saturated (test DR9; $\rho_d = 2.25$ g/cm³, $e_0 = 0.19$, $S_r = 100\%$) and moist (test DR3, $\rho_d = 2.25$ g/cm³, $e_0 = 0.19$, and $S_r = 77.29\%$) samples of Chiba gravel (Ahn Dan *et al.*, 2001) creep period at each step is about 6 hours (cf. Figure 2-53 for strain evolution).

Figure 2-29. One dimension compression drained œdometer tests (CDO tests) on Fujinomori clay. Creep periods and stepwise strain rate changes are applied during loading on a saturated (w = 43.9%) and an air dried (w = 2.99%) sample (Li *et al.*, 2004).

(CDO tests) on Fujinomori clay, as presented in Figure 2-29. Several creep loading sages and stepwise strain rate changes were applied during otherwise monotonic loading on a saturated (w = 43.9%) specimen that was trimmed from a large cake made by consolidated slurry and an air-dried (w = 2.99%) specimen that was prepared by the tamping method. Despite that the stress-strain curves are quantitatively different, the trend of viscous phenomenon seen in the test tests is quite similar and in a good agreement with the previous observations of viscous properties. In particular, the stiffened stress-strain behaviour observed when loading is restarted following a creep loading period is due to the viscosity-created "quasi" elastic domain (cf Figure 2-23) but it cannot be explained by a phenomenon other than the viscous behaviour. Ageing, in particular, does not seem to be the relevant explanation for this behaviour.

The viscous effects are also the origin of effective mean pressure (p′) decrease during an undrained TC creep loading with a constant deviator value (q). This testing phase is usually called "undrained creep". Undrained creep could result in instable behaviour for sand (Leong and Chu, 2002; Lade and Yamamuro, 1996). If the material had no viscous properties, no changes in the effective mean pressure would take place as no creep strain takes place during drained creep loading. Figure 2-30 presents results from a TC test in which a series of drained and undrained creep

loadings were applied alternatively to a saturated loose specimen of silica sand (D_{50} = 0.077 mm; U_c = 2.43; G_s = 2.655; e_{max} = 1.335; and e_{min} = 0.73) performed by Kiyota (2004). At relatively high deviator stresses, the shear strain increment is noticeably smaller at a drained creep loading stage than at a undrained creep loading stage for the same stress-strain condition at the start of creep loading. This result gives a hint for the shape of yield surface (or locus) of the EP2 body (cf. Figure 1-5) when a classical elastoplastic formalism is considered for this body. The two traditional choices are a capped type or an open type in the framework of a double hardening model. This issue is, for example, discussed in Nawir *et al.* (2003b and 2003c) for shear yielding of sand; they showed that an open type surface seems more relevant for shear yielding.

(3) A specific effect of drained creep, which might be mixed up with anisotropy

It seems that the viscous properties of sand can explain the different behaviours obtained from undrained triaxial tests performed on isotropically and anisotropically consolidated samples. Figure 2-31 shows undrained TC test responses after point A, obtained on five very loose Hostun sand specimens. These five specimens have different stress-time histories before point A. The following two types of undrained behaviour are observed: i) the first type for tests 1 and 4 (slope V_1 when undrained TC is applied); and ii) the second type, corresponding to a higher slope in $p' - q$ plane (slope V_2 when undrained TC is applied), for tests 2, 3 and 5. For tests 1 and 4, no creep is allowed, while drained creep loading for a period of about two hours is applied for tests 2 and 3 at point A. Considering the very small loading rate, during anisotropic consolidation in test 5, it is also considered that some larger creep deformation could occur

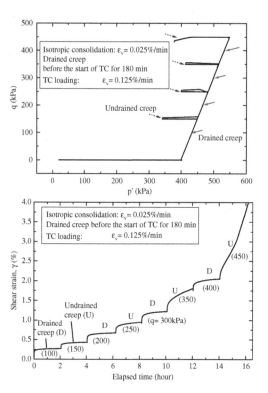

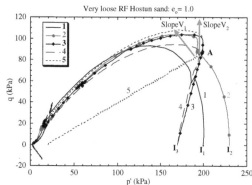

Figure 2-30. Series of drained and undrained creep periods for saturated silica sand in TC test performed by Kiyota (2004) (cf. Figure 2-56 for strain evolution).

Figure 2-31. Undrained TC behabviour after point A of five very loose Hostun sand specimens having different stress-strain histories before point A: drained creep is allowed at point A only for tests number 2, 3 and 5 (Di Benedetto, 1997; Ibraim, 1998).

before reaching point A in that test. One can conclude that the a large difference in the stress path direction after point A between the two types of tests (slopes V_1 and V_2) is not due to anisotropy (particularly in case of test number 5) but can be considered as a rate-dependent behaviour effect. The slope difference can be explained by the existence of the "quasi" elastic domain previously identified (cf. Figure 2-23). In addition it should be underlined that this elastic behaviour is probably anisotropic.

A result similar to the above was obtained from a series of undrained triaxial compression test on reconstituted specimens of Fujinomori clay (Figure 2-32, Tatsuoka *et al.*, 2000). Specimen 8 was isotropically consolidated to $p' = 200$ kPa, where it was subjected to drained creep for one day, followed by undrained TC loading at a constant axial strain rate $\dot{\varepsilon}_v = 0.05\%$/min ($= \dot{\varepsilon}_0$) to ultimate state without a pause at stress state A. Specimen 9 was anisotropically consolidated at a stress ratio $\sigma'_h/\sigma'_v = 0.5$ and at an axial strain rate $\dot{\varepsilon} = 0.15\,\dot{\varepsilon}_0$ up to stress state A, which is located on the undrained TC effective stress path of specimen 8. Then, without a pause at stress state A, undrained TC loading was started at an axial strain rate $\dot{\varepsilon}_v = \dot{\varepsilon}_0$, which was $1/0.15 \approx 7$ times faster than that during the anisotropic consolidation. Specimen 16 has the same loading history as specimen 8 until point A. Specimen 16 was subjected to drained creep for two days at stress state A, followed by restart of undrained TC loading at a constant axial strain rate $\dot{\varepsilon}_v = \dot{\varepsilon}_0$ as before. Specimen 14 has the same loading history as specimen 9 until point A. Specimen 14 was subjected to drained creep for two days at stress state A before starting undrained TC loading at $\dot{\varepsilon}_v = \dot{\varepsilon}_0$ to ultimate state. Finally, specimen 28 was isotropically consolidated to $p' = 121$ kPa, where it was subjected to drained creep for one day, followed by *drained* TC at $\dot{\varepsilon}_v = \dot{\varepsilon}_0/10$ to stress state A. Then, after having been subjected to drained creep for two days at stress state A, the specimen was subjected to undrained TC loading at $\dot{\varepsilon}_v = \dot{\varepsilon}_0$. It may be seen from Figure 2-32(a) & (b) that only the specimens that were subjected to drained creep for two days at stress point A have a large stress range where the stress path is nearly vertical with slope V_2 irrespective of loading histories until stress point A (i.e., whether anisotropic compression, drained triaxial compression and undrained triaxial compression). On the other hand, the stress path following stress point A of specimen 9, which has been anisotropically consolidated until point A, is nearly vertical with slope V_2 for a much smaller stress range.

The first conclusion that can be derived from the test results shown above is that the conventional anisotropic elasto-plastic approaches are then not relevant to correctly model the observed behaviour. More generally, any modelling using a non-viscous framework will not be able to explain such test results shown above by any adaptation.

The second conclusion is linked to the creep effects. As underlined previously (cf. Figure 2-23), an elastic domain develops by drained creep around the stress state where the drain creep has taken place. Then, the slope V_2 (Figure 2-31) corresponds to an elastic behaviour response, which also explains the stiffened behaviour observed after creep seen in Figure 2-23.

(4) Analysis of stress jumps when stepwise changes in loading rate

A specific stress path including stepwise changes in the loading rate was more particularly chosen to evaluate and model the viscous behaviour (cf. section 3). These changes can be applied to any general stress path with or without rotation of axes, for any loading devices such as: triaxial, plane strain compression, œdometer, hollow cylinder, among others, in drained

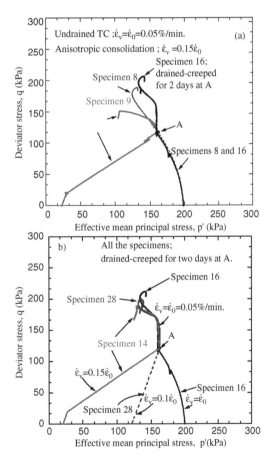

Figure 2-32. Undrained TC behaviour after point A of sample of clay kaolin having different stress-strain histories before point A: creep is applied at point A except for tests 8 and 9 (Tatsuoka *et al.*, 2000).

or undrained conditions. As can be seen in a number of figures presented in this paper, stress jumps appear. It is positive when the rate is suddenly increased and negative in the opposite case (cf. for example Figure 2-14 and Figure 2-17, Figure 2-33, Figure 2-34, Figure 2-35). The stress jump measurement is shown in Figure 2-36. The stress ratio difference $\Delta R = \Delta(\sigma'_v/\sigma'_h)$, caused by each step change in the strain rate, is considered. As ΔR is measured after correction of the monotonic evolution, it gives a good approximation of the stress jump caused solely by the strain rate increase, which is a viscous property only (as shown in section 3, it is connected with the body V, cf. Figure 1-5). Figure 2-33 to Figure 2-35 show some results obtained on Toyoura sand in drained TC tests including stepwise changes (Nawir et al., 2003a). The influences of the confining pressure (σ_h), the void ratio (e) and the saturated or dry state are respectively studied. The stress jump values ΔR for the tests presented in these 3 figures (Figure 2-33 to Figure 2-35), are summarised in Figure 2-36. This figure represents the data in the axes:

$$\log\left(\frac{(\dot{\varepsilon}^{vp})_{after}}{(\dot{\varepsilon}^{vp})_{before}}\right) \; versus \; \frac{\Delta R}{R}$$

where $(\dot{\varepsilon}^{vp})_{before}$ [respectively $(\dot{\varepsilon}^{vp})_{after}$] is the viscous (or deferred) strain rate (cf. equation 1-2) before [respectively after] the rate change. The (hypo-)elastic part (ε^e of EP1) has in general very few effect. Then, the viscous strain rate ratio is generally very close to the strain rate ratio so that it can be written as:

$$\log\left(\frac{(\dot{\varepsilon}^{vp})_{after}}{(\dot{\varepsilon}^{vp})_{before}}\right) \approx \log\left(\frac{(\dot{\varepsilon})_{after}}{(\dot{\varepsilon})_{before}}\right) \qquad (2\text{-}14)$$

For the data reported in Figure 2-36, a linear relation can be considered:

$$\frac{\Delta R}{R} = \beta \log\left(\frac{(\dot{\varepsilon}^{vp})_{after}}{(\dot{\varepsilon}^{vp})_{before}}\right) \qquad (2\text{-}15)$$

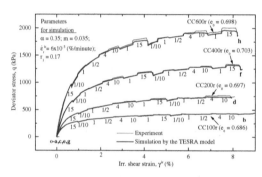

Figure 2-33. Drained CT tests on saturated Toyoura sand at different confining pressures (from 100 to 600 kPa) including stepwise changes in the strain rate (Nawin et al., 2003a) (cf. section 3 for simulations).

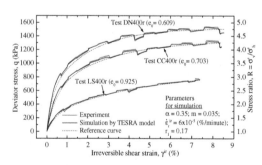

Figure 2-34. Drained CT tests on saturated Toyoura sand at different densites, including stepwise changes in the strain rate (Nawir et al., 2003a) (cf. section 3 for simulations).

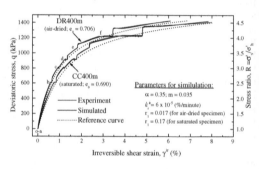

Figure 2-35. CT tests on Toyoura sand at different water contents, including stepwise changes in the strain rate (Nawir et al., 2003a) (cf. section 3 for simulations).

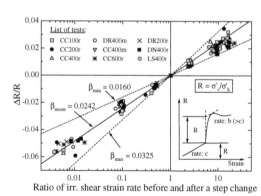

Figure 2-36. Stress jumps and rate sensitivity coefficients β obtained on Toyoura sand for tests presented in Figure 2-33 to Figure 2-35 (Nawir et al., 2003a).

with β, called the "rate sensitivity coefficient", close to 0,024.

A noticeable result is that no distinction can be made in these axes among the different test conditions. In particular dry specimens (Figure 2-35) reveal a slightly stiffer stress-strain response curve than saturated ones, while the plots made to evaluate the viscous properties (Figure 2-36) are essentially the same.

Figure 2-37 represents the data obtained in a series of such TC tests on dry Hostun sand by Pham Van Bang and Di Benedetto, 2003. The experimental stress overshoot (resp. undershoot), is observed when the strain rate is multiplied by a factor 100 (resp.1/100) from an initial value (or $\dot{\varepsilon}_{before}^{vp}$) equal to 0,006%/min (resp. 0,6%/min). Tests were performed on dense and loose ($e_0 = 0.72$ and 0.93) samples at confining pressures of 80, 200 and 400 kPa. When comparing Figure 2-36 and Figure 2-37 one can conclude that the β coefficient (equation 2-15) is the same for both sands and the various considered experimental conditions.

More extensively, equation 15 was evaluated for the following materials by drained loading tests, listed in Appendix 2:

1) original Chiba gravel (moist in drained TC);
2) model Chiba gravel (air-dried in drained TC and 1D compression);
3) Hostun sand (saturated and air-dried in drained PSC and TC);
4) Toyoura sand (saturated in drained PSC; and saturated and air-dried in drained TC);
5) Silica sand No.8 (saturated in drained TC):
6) Jamuna River sand (air-dried in drained PSC);
7) crushed concrete (moist in drained TC);
8) Fujinomori clay (saturated, wet and air-dried in drained TC; saturated, air-dried and oven-dried in 1D compression; air-dried and oven-dried specimens were prepared by tamping the clay powder);
9) kaolin (oven-dried in 1D compression: the specimens were prepared by tamping the clay powder);
10) undisturbed Kitan clay (saturated in drained TC and 1D compression);

11) reconstituted Kitan clay (saturated in drained TC);
12) undisturbed Ohmachi clay (saturated in drained TC and 1D compression):
13) reconstituted Ohmachi clay (saturated in drained TC and 1D compression);
14) undisturbed Pisa clay (saturated in 1D compression); and
15) reconstituted Pisa clay (wet and air-dried in 1D compression).

The rate sensitivity coefficient β values for drained TC and PSC tests at constant confining pressure σ_h' were obtained from the slopes of the relationship between $\Delta R/R$ and $\log\{(\dot{\varepsilon}_v^{ir})_{after}/(\dot{\varepsilon}_v^{ir})_{before}\}$, where $R = \sigma_v'/\sigma_h'$. In so doing, it was assumed that the shear deformation is controlled by the stress ratio, $R = \sigma_v'/\sigma_h'$. On the other hand, the β values for 1D compression tests were obtained from the slopes of the relationship between $\Delta\sigma_v'/\sigma_v'$ and $\log\{(\dot{\varepsilon}_v^{ir})_{after}/(\dot{\varepsilon}_v^{ir})_{before}\}$. Some of the results from these tests are presented in Figure 2-38 through Figure 2-40. The general trends of behaviour seen from these and other similar figures are summarised below:

1) The range of β for a large variety of geomaterial types under largely different test conditions (loose to dense, dry to saturated, drained to undrained, low to high confining pressures and so on) from different testing methods is not large, from about 0.02 to about 0.08, as seen in Figure 2-38. Figure 2-38 compares the β values of clays (including air-dried Fujinomori clay) in TC with the β values of undisturbed and reconstituted Kitan clay in drained TC as well as silica No. 8, Jamuna River, Toyoura and Hostun sands, crushed concrete, model Chiba gravel and Chiba gravel in drained TC or PSC.

2) Yet the following trends of behaviour were found by carefully examining the data of the tests listed in Appendix 2.

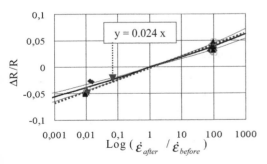

Figure 2-37. Rate sensitivity coefficient β for dry Hostun sand (from Pham Van Bang et al., 2003).

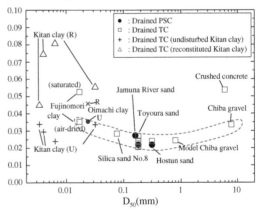

Figure 2-38. Effects of D_{50} on the rate sensitivity coefficient β from drained TC and drained PSC of clays, sands and gravels.

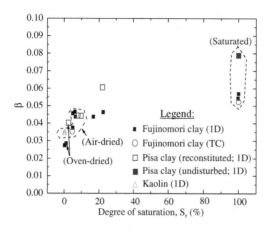

Figure 2-39. Effects of saturation degree on the rate sensitivity coefficient from oedometer and drained TC tests on Fujinomori clay (Li *et al.*, 2004).

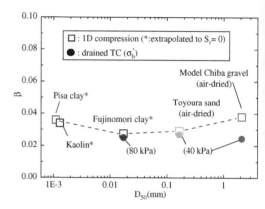

Figure 2-40. Effects of D_{50} on the rate sensitivity coefficient from oedometer tests on a wide variety of geomaterials.

a) The β value is slightly larger in 1D compression than in drained TC at a fixed confining pressure under otherwise the same test conditions. This difference should be due to different stress paths in these two types of tests and should be explained by a model that is extended to be applicable to the two or three-dimensional stress conditions.

b) The β value of the reconstituted clay specimens is noticeably larger than the value of the corresponding undisturbed clay samples under otherwise the same test conditions. More detailed discussions on this issue with Kitan clay is given by Komoto *et al.* (2003).

c) Figure 2-39 shows the effects of the degree of saturation on the β value of Fujinomori clay in dimensional-compression tests and drained TC tests (Li et al., 2004). The oven-dried and air-dried specimens were prepared by tamping clay powder in the compression ring. It is seen that the β value of the nearly perfectly dried specimens is still more than a half of the value of the saturated specimen.

d) In Figure 2-40, the β value of the oven-dried specimen of Fujinomori clay in 1D compression, together with those of kaolin and Pisa clay, are compared with the β values of the other two geomaterials (model Chiba gravel and Toyoura sand) in 1D compression. It may be seen that the β value of the oven-dried specimens of the three types of clay are well comparable with those of the sands and gravels, showing that the β value when free from the effects of pore water is rather independent of particle size. It may also be seen from Figure 2-38 that the β value of air-dried Fujinomori clay in TC is similar to the β values of reconstituted Kitan clay in drained TC as well as silica, Jamuna River, Toyoura and Hostun sands, model Chiba gravel and Chiba

gravel in drained TC or PSC. These facts indicate that the viscous properties of saturated clay are not due totally to the presence of pore water, but due also to the same mechanism as the one that is relevant to the viscous properties of sands and gravels.

Interpretation and interest of the rate-sensitivity coefficient β in the framework of the 3 component model is given in section 3, where simulation is considered.

(5) Cyclic loading
The influence of the viscous effects on the stress-strain behaviour during loading and unloading cycles with a large stress or strain amplitude and those with a relatively small stress or strain amplitude at a given stress-strain state is discussed in this section. It is to be noted that only the stress-strain response curve is considered in this paragraph. The behaviour in the strain space, which is also of primary importance, is discussed in the section 2.1.2.

(a) Small stress or strain amplitude cyclic loading
When applying cyclic stresses or strains with a relatively small or moderate stress or strain amplitude applied at a given stress-strain state, rate-independent cyclic loading effects and "creep" (i.e.viscous) deformation are both active. For moderate stress level and for most of the geomaterials, the creep deformation asymptotically approaches its limit as the time elapses (and with an increase in the number of cycles). Once the development of creep deformation is stabilised, the stress-strain behaviour becomes elastic (or "quasi" elastic). This feature is visible in Figure 2-41, where the behaviour of loose Hostun sand is progressively more recoverable with the number of cycles but also with the creep deformation increase. In that sense creep and small cyclic loadings lead to the same effect, but the viscous aspect is predominant and viscous deformation should be "consumed" before the elastic behaviour

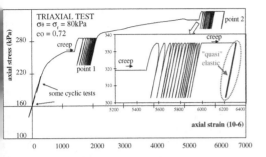

Figure 2-41. Drained TC test on loose Hostun sand with small cyclic loading and creep deformation at intermediate loading stages (Cazacliu, 1996).

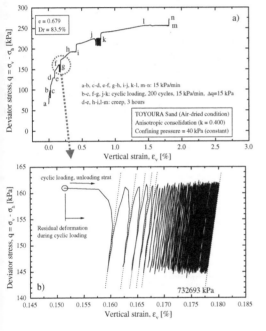

Figure 2-42. Cyclic loadings and creep stages during drained CT test on Toyoura sand (cf. Figure 2-58 for strain evolution) (Hirakawa, 2003).

appears nearly exclusively. Once again it is to be noted that the monotonic loading stress-strain curve is not a "stable" state as the stress or strain level cannot be maintained with time at a constant strain or stress. The statement given above can be reconfirmed by the results from a TC test on Toyoura sand plotted in Figure 2-42.

Figure 2-43 shows the time histories of shear strain during a pair of creep and cyclic loading tests on Toyoura sand starting from nearly the same stress strain state with nearly the same previous loading histories

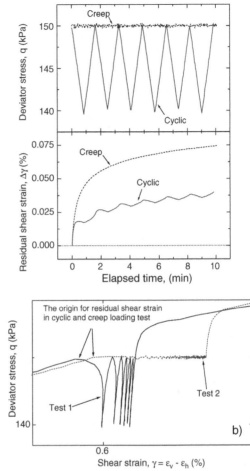

Figure 2-43. Time history of "q" and shear strain, and stress-strain evolution (b). Part of TC tests 1 and 2 on air-dried Toyoura sand presented in Figure 2-59, from Ko et al. (2003) (cf. Figure 2-59 for global stress-strain evolution).

(Ko *et al.*, 2003) (cf. Figure 2-59 for more details). The stress value during the small unload/reload cycles is always equal to or smaller than the constant creep stress. The development of residual strain during the cyclic loading test is obviously lower than the creep strain, which confirms that the viscous strain is the predominant factor in the development of residual strain during such a cyclic loading. An opposite situation can be observed in some cases. For example, Figure 2-44 reveals that a cyclic loading, during simple compression on bituminous mixes, can create a larger axial deformation than a creep test at a stress level equal to the maximum of the cyclic stress (Neifar *et al.*, 2003). This last result is explained by the irreversible strain after each cycle, which by accumulation becomes more important than the viscous strain. As the material

85

has a marked viscous behaviour, the effect of frequency during cycles is noticeable.

(b) Large stress or strain amplitude cyclic loading
As previously stated, the direction of creep deformation after a large stress or strain amplitude cyclic loading history is not only linked to the stress value but is also mainly affected by the previous history of stress reversal(s) (cf. Figure 2-25 to Figure 2-27). It seems that the last stress reversal value plays an essential role in the amplitude and the direction of the creep strain.

This statement is also valid for one-dimension œdometer tests. Figure 2-45 shows the relationships between the effective vertical stress and the vertical strain from a constant rate of strain oedometer test,

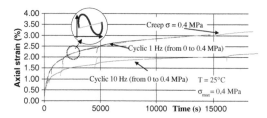

Figure 2-44. Simple compression cyclic and creep tests on bituminous mixes. Evolution of permanent axial strain with time (Neifar *et al.*, 2003).

including a large amplitude of loading and unloading with creep periods and stepwise changes in the strain rate, on undisturbed Kitan clay. The specimen was 6 cm in diameter and 2 cm in height with the top drained and the bottom undrained and the pore water pressure was measured at the bottom end. The developed excess pore pressure was very low throughout the test. The direction of creep vertical strain is either positive (compressive) or negative (expansive) depending on the loading branch (loading or unloading or reloading) and the distance from the stress value where the latest load reversal is made. In addition the rate-sensitivity coefficient β measured during reloading (denoted as 'reloading') (cf. section 2.1.1.2(4)) seems slightly higher than the values measured before and after the yielding point during loading (denoted as "before" and "after". For such "aged" clay, it seems that the β value increases by a history of loading exceeding the yield point. It seems that, except for such "aged" clay, for the first approximation, it can be considered that the β value is not largely affected by a large stress or strain amplitude cyclic loading history.

The previous conclusion for creep periods is also extended to relaxation periods (period where a constant strain is maintained). Figure 2-46 reports the results

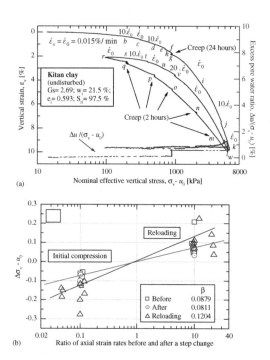

Figure 2-45. Creep periods during a cyclic œdometer test on undisturbed Kitan clay. Stress-strain curve and, (b) β parameter plot (cf § 2.1.1.2(4)) (Acosta_Martinez *et al.*, 2003).

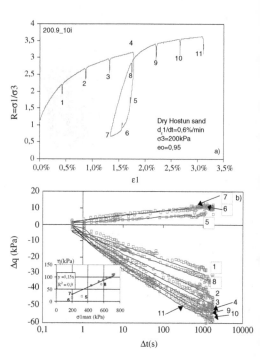

Figure 2-46. Drained TC tests with a large cycle on dry Hostun sand including 11 relaxation periods. The model and simulations of figure b) are performed with the ENTPE viscous evanescent model. η is the viscosity parameter of the model, which is presented in section 3.2.3 (Pham Van Bang, 2004).

of a drained CT test on dry Hostun sand with a large stress amplitude cycle up to negative values of the deviatoric stress. At different steps of the stress-strain curve, relaxation periods were applied (Figure 2-46a). For each of these 11 periods the variation of the deviatoric stress is plotted against the logarithm of time in Figure 2-46b. The relaxation period number 5 is particularly interesting as the stress increases even though the stress ratio is larger than one (positive deviatoric stress). A relaxation at the same level during monotonic loading would result in a decrease of the stress (cf. points 1, 2, 3, 9, 10, 11). Simulations with the ENTPE viscous evanescent model (cf. section 3) are also reported. The value of the viscosity parameter η (cf. equation 3-25) obtained by optimisation of the simulations for each relaxation periods is also indicated in Figure 2-46b.

(6) Shear band analysis
Figure 2-47 shows the results from two pairs of drained PSC tests on air-dried Jamura River sand, which includes a small amount of mica (about 5% by volume) (Yasin and Tatsuoka, 2003). Each pair of tests was performed at confining pressure equal to either 100 kPa or 400 kPa, and, at each confining pressure, two tests were performed at different strain rates. Figure 2-48 shows the results from a pair of drained PSC tests on air-dried gravel of Chert performed at a confining pressure of 196 kPa but at two different strain rates (Oie et al., 2003). It is interesting to note that, as the strain rate becomes larger, the axial strain at the peak becomes smaller, while, in the post-peak regime, the decreasing rate of stress with an increase in the axial strain averaged for the whole height of specimen decreases in both cases of PSC tests. These two phenomena, pre-peak and post-peak, should be due to different mechanisms. It seems that the latter post-peak phenomenon corresponds to the effects of strain rate on the manner of strain localization. At

least, it was confirmed with the specimens of Chert that the shear band thickness was larger in the test at a faster strain rate (Oie et al., 2003). If the shear band thickness increases with an increase in the strain rate because of material viscous properties and the relationship between the shear stress and the local shear strain within a shear band is independent of strain rate in monotonic loading at constant strain rates, the trend of post-peak behaviour seen in these figures is understandable.

(7) Ageing effects
The following two different cases with respect to the ageing effects are discussed herein:1) The material has been noticeably aged, but no additional effect of ageing can be noticed during a period of loading Δt (cf Figure 1-2): e.g., aged & reconstituted specimens of clay.2) The ageing effects is active during loading while the material exhibits noticeable viscous effects during loading (i.e., true coupling between the ageing and viscous effects): e.g., cement-mixed soils at a relatively young age.

With respect to the first case, Figure 2-49 compares the stress-strain relations of undisturbed and reconstituted specimens of Kitan clay from drained TC tests at constant confining pressure (Komoto et al., 2003). The following trends of behaviour may be seen:

1) Despite that the void ratio of the specimen that was reconstituted from slurry and normally consolidated to the same confining pressure as the undisturbed specimen was smaller than the undisturbed specimen, the pre-peak stiffness and peak strength of the reconstituted specimen was smaller than the undisturbed specimen. Moreover, the reconstituted specimen exhibited more contractant behaviour.

2) When monotonic loading was restarted at a constant strain rate following a creep loading stage, the stress-strain tended to rejoin the stress-strain curve that would have been obtained if the continuous monotonic loading had continued at the same strain rate

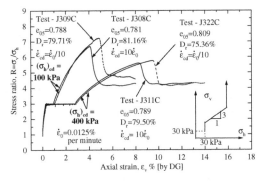

Figure 2-47. Drained PSC tests on air-dried Jamura River sand (Yasin and Tatsuoka, 2003): effect of strain rate on the shear band.

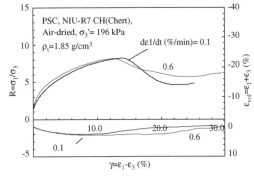

Figure 2-48. Drained PSC tests on air-dried gravel of Chert (Oie et al., 2003): effect of strain rate on the shear band.

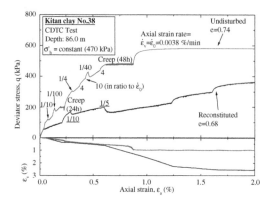

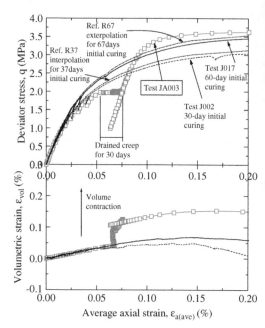

Figure 2-49. Comparison of stress-strain relations of undisturbed and reconstituted specimens of Kitan clay from drained TC at constant confining pressure (Komoto et al., 2003).

without an intermission of creep loading, showing that the stress-strain-properties have not changed during the creep loading stage.

Materials for which the ageing effects are noticeably active in the course of loading (i.e., case 2) show considerably different behaviour from the above when subjected to creep loading during otherwise monotonic loading. Figure 2-50 shows the relationships among the deviator stress and the axial and volumetric strains of moist cement-mixed Chiba gravel ($G_s = 2.71$, $C_u = 12.9$ and $D_{max} = 10$ mm) from drained TC tests in which different loading histories were applied to different specimens (Kongsukprasert and Tatsuoka, 2003a and 2003b). The gravel was mixed with ordinary Portland cement in a gravel to cement ratio of 2.5% by weight and compacted at water content of about 8.8% (nearly the optimum water content) to a dry density of 2.0 g/cm^3. The TC tests were performed at confining pressure equal to 19.7 kPa and an axial strain rate of 0.03%/min. In tests J002 and J017, the specimens were cured under the atmospheric pressure at constant water content for, respectively, 30 days and 60 days before the start of TC (i.e., at q = 0). In test JA003, the specimen was aged first under the atmospheric pressure at constant water content for 37 days and then subjected to drained creep at an anisotropic stress state (q = 2 Mpa) for 30 days under otherwise drained ML. The following trends of behaviour may be seen.

1) The results from tests J002 and J017 show that the respective deviator stress at different strains increased rather proportionally by curing under the atmospheric pressure.

2) In test JA003, the specimen exhibited noticeable creep strain, which actually stopped very fast. The volume contraction seen in this figure took place by effects of cement hydration at a later stage of creep

Figure 2-50. Stress-strain relation of cement mixed gravel from drained TC tests for different loading histories (Kongsukprasert and Tatsuoka, 2003a and 2003b).

loading. In this test, the stress-strain behaviour during cyclic loading followed by ML at a constant strain rate applied after curing at an anisotropic stress state was very stiff and nearly elastic for a large stress range. Subsequently, after having exhibited noticeable stress-overshooting and clear yielding, the stress-strain relation tended to join the one obtained from continuous ML of a specimen cured for the same total period under the atmospheric pressure.

The phenomenon described in the second term (noted also by Barbosa-Cruz and Tatsuoka, 1999 and confirmed by Sugai et al., 2003) can be explained only by taking into account the effects of ageing and viscous properties in a relevant coupled way (Tatsuoka et al., 2003). Kongsukprasert and Tatsuoka (2003a and 2003b) also reported interesting data from a drained TC test on the same type of cement-mixed gravel in which monotonic loading was performed at an extremely low strain rate. In this test, the ageing effects developing during loading exceeded negative effects due to viscous effects because of a very low strain rate.

2.1.2 Behaviour in the strain space

In the previous sections only the one-dimensional stress-strain response is introduced. A general constitutive formalism also needs to consider the other components. Some experimental results on the evolution of the other response components (in general, those represented in the strain space), in relation with viscous effect, are presented in the following sections.

2.1.2.1 Monotonic loading including different loading rates

The axial strain-volumetric strain curves obtained from six drained PSC tests on air-dried Jamuna river sand at confining pressures of 100 and 400 kPa and at different strain rates that are either constant or stepwise changed in each single test, are plotted in Figure 2-51b. The relationships between the principal stress ratio and the axial strain of the four monotonic loading tests (J308C, J309C, J322C and J311C) are presented in Figure 2-47 and commented in section 2.1.1.2(6), while those of the other two tests (J306C and J316C) is presented in Figure 2-51a. The following trends of behaviour may be seen from these figures:

1) This material is rather contractive due likely to the inclusion of mica.
2) The axial strain-volumetric stain curves of the three tests at a confining pressure of 100 kPa are rather close to each other despite that the strain rate is either constant, while being different by a factor of 100 (i.e., axial strain rates of 0.125%/min and 0.00125%/min), or changes by a factor of 100 during otherwise monotonic loading. Similar behaviour is observed at a confining pressure of 400 kPa.
3) More rigorously, the material is slightly more contractive (or less dilative) at lower loading rate.
4) In the tests in which the strain rate was changed stepwise (tests J306C and J316C), the axial strain-volumetric strain curves slightly deviate toward the dilative direction during the fast loading rate periods. This observation is consistent with the above mentioned trend of behaviour that the material is more contractive during monotonic loading at a low constant strain rate.

As described above, although the effects of strain rate and its change on the strain path are noticeable, they are very small and, for the first approximation, it could be assumed that the strain rate has no effect on the strain space evolution.

The accuracy and repeatability of the constant loading rate monotonic tests are usually not high enough to draw a clear conclusion on the effects of strain rate on the direction of strain path. This point could be generally better investigated by tests including creep and/or relaxation loading stages and/or stepwise changes in the strain rate during otherwise monotonic loading at constant strain rate.

2.1.2.2 Analysis of creep and relaxation stages and stress jumps when stepwise changes in loading rate

The direction of strain evolution during creep periods was investigated by means of hollow cylinder tests including or not rotation of principal axes by Sauzéat

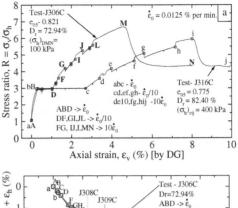

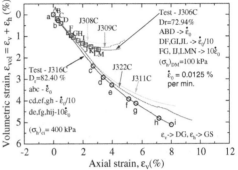

Figure 2-51. Drained PSC tests on air-dried Jamuna river sand at confining pressures of 100 and 400 kPa, and different strain rates. The stress-strain response of the 4 monotonic loading tests (J308C, J309C, J322C and J311C) is given in Figure 2-47 (from Yasin and Tatsuoka, 2003).

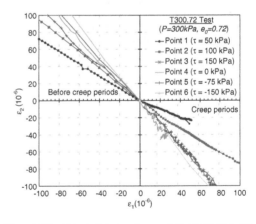

Figure 2-52. Principal strain values evolution just before and during creep periods for the pure torsion test on dry Hostun sand T300.72 (cf. Figure 2-25) from Sauzéat et al., 2003.

(2003). Figure 2-52 presents the paths in the principal strain axes "ε_1-ε_2" of the plane "z-θ", just before and during the creep periods applied during the test T300.72 on dry Hostun sand (cf. Figure 2-25).

The strain path for each point seems to have a continuous derivative at the origin of the creep period and the curves seem to keep a similar direction before and during creep (very close strain increments before and just after creep started). This observation means that the strain path followed during creep periods is very close to the one that would exist if the loading were not stopped. The results exposed further show that these two paths are, in fact, slightly different: the strain path direction change during creep.

The relationship between the shear strain and volumetric strain obtained from the two drained TC tests on moist and water-saturated specimens of Chiba gravel are presented in Figure 2-53. The relationship between the deviator stress and the irreversible strain are presented in Figure 2-26. The following trends of behaviour may be seen:

1) For the first approximation, the strain path is similar during monotonic loading and creep loading. More rigorously, however, the material becomes more contractive (or less dilative) during creep loading periods. This observation is consistent with the one with Jamuna river sand described above.
2) The water content, which is very different between the two specimens, has no clear visible influence on the trend of strain path during creep loading periods. Again, this fact confirms that creep, and more generally viscous effects, of granular material **do exist irrespective of water content**. The micro-physical explanation of the phenomenon not including effects of pore water should be found.

Results from other series of drained TC tests including creep periods are presented in Figure 2-54 and Figure 2-55 for other types of materials (Hostun and Toyoura sands). The previous conclusion can be confirmed. In addition, it may be seen that even at rather high shear stress levels where the material

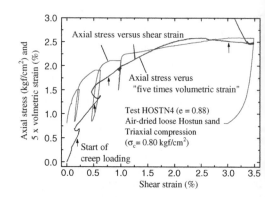

Figure 2-54. Relationships between axial stress and shear and volumetric strains from a drained TC on Hostun sand (Hoque, 1996).

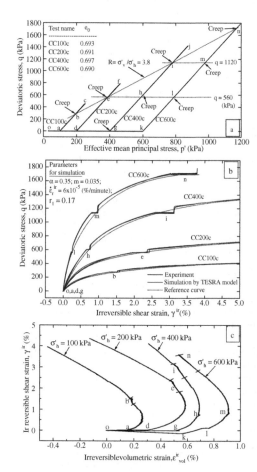

Figure 2-55. Drained CT tests on saturated Toyoura sand at different confining pressures (from 100 to 600 kPa) including creep periods (Nawir *et al.*, 2003a) (cf. section 3 for simulations in figure b).

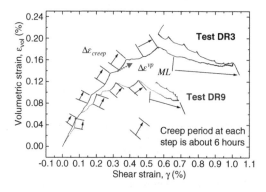

Figure 2-53. Shear strain versus volumetric strain for the 2 drained TC tests DR3 and DR9 on Chiba gravel presented in Figure 2-28, from Ahn Dan *et al.*, 2001.

dilates, the dilation is less pronounced during creep loading periods.

This conclusion is also valid at low stress level. Kuwano and Jardine (2002) observed considerable creep deformation on sand and glass beads, even for isotropic stress condition at relatively low pressure (where breakage of particles is unlikely to be significant). The strain increment direction changes during creep at non isotropic states.

In addition, it can also be seen from Figure 2-54 and Figure 2-55 that, when monotonic loading is restarted after a creep loading period, the strain path direction changes again tending to rejoin the original curve during continuous monotonic loading after some strain increment.

The analysis of these stress-strain behaviours immediately after the restart of loading after a creep loading period shows, as it is stated previously (cf. section 2.1.1), that the behaviour becomes essentially linear (cf. Figure 2-23) if the creep period is long enough for the viscous strain to sufficiently develop (cf. Figure 2-23). This linearity is confirmed by systematic studies performed on different types of geomaterials and for different stress paths (including or not rotation of axes) at the University of Tokyo and the "Ecole des TPE".

In undrained loading tests, the volumetric strain is kept constant and at least one of the stain component belongs to the loading space (cf. section 1.1). For more general information on the soil skeleton behaviour, the stress component evolution (for example the mean stress p') becomes necessary in addition to the strain evolution in the strain space. The TC test described in Figure 2-30 included alternative drained and undrained creep loading periods at different shear stress levels. The relationships between the shear strain and the volumetric strain from the start of TC

loading is presented in Figure 2-56. In this figure, the result from another test in which the specimen was subjected to drained continuous monotonic loading is also presented for the reference. During the drained creep periods, the strain increment path twists toward the positive volumetric strain axis, showing that the behaviour becomes more contractive than during monotonic loading. During the undrained creep loading, the effective mean pressure decrease, created by a tendency that the aggregate skeleton contract, which is compensated by a decrease in the effective mean principal stress under the imposed constant volume condition. Then, this decrease in the effective mean principal stress is in agreement with the contractive behaviour during drained creep.

Figure 2-57 shows the result from three undrained CT tests on Metramo silty sand with several undrained creep loading periods and a number of stepwise changes in the strain rate (Santucci de Magistris et al., 2004). The stress-strain curves presented in Figure 2-57a reveal that the viscous aspect of the behaviour is of "isotach" type at low strain levels and progressively becomes

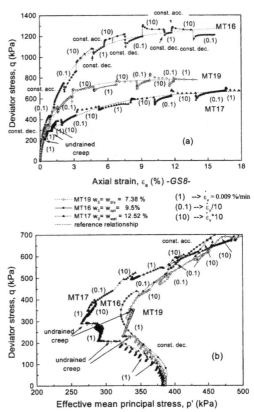

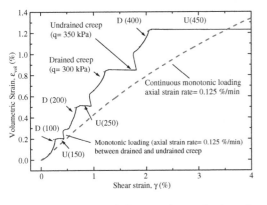

Figure 2-56. Relationship between shear and volumetric strains from a TC test on saturated silica sand with a series of drained and undrained creep loadings, presented in Figure 2-30 (Kiyota, 2004).

Figure 2-57. Undrained TC tests on saturated Metramo silty sand with stepwise change rates and undrained creep periods, from Santucci de Magistris et al., 2004.

91

clearly "viscous evanescent type" with an increase in the shear strain (cf. section 2.1.1.2 and Figure 2-57). The result presented in Figure 2-57b is in agreement with the observation made previously on other types of materials relatively to the volume variation.

2.1.2.3 Cyclic loadings

Figure 2-58a & b present the relationships between the deviator stress and the volumetric strain and between the shear and volumetric strains from a drained TC test on Toyoura sand (Hirakawa, 2003). The relationships between the deviator stress and the vertical strain is presented in Figure 2-42. In this test, creep loading periods and small stress amplitude cyclic loadings were alternatively applied at different shear stress levels during otherwise monotonic loading at a constant rate of deviator stress. Figure 2-59 shows the results from three drained TC tests on Toyoura sand (Ko *et al.*, 2003). In the first test (test S1), the specimen was subjected to continuous drained monotonic loading, while, in the second and third tests (tests 1 & 2), creep and cyclic loading was applied alternatively in the two different ways. In these two sets of tests presented in Figure 2-58 and Figure 2-59, at different shear stress levels, creep load is applied at the maximum stress during cyclic loading in the other corresponding test, which allows

a direct comparison of the respective effects of small stress amplitude cycle loading and creep loading. Figure 2-60(a) shows the relationships between the increment of shear strain that took place during respective creep loading and the deviator stress during creep loading and between the shear strain increment that took placed during respective cyclic loading and the maximum deviator stress during cyclic loading, obtained from the test described in Figure 2-42 and another similar test performed under the same test condition except that the creep and cyclic loading periods were applied in the inversed order. The period of respective creep and cyclic loading was two hours. Figure 2-60(b) shows the similar plot for the increment of volumetric strain from the creep and cyclic loading, while Figure 2-60(c) shows the ratio of shear and volumetric strain increments presented in figures (a) and (b). In this figure,

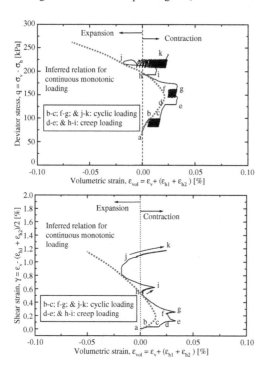

Figure 2-58. Cyclic loadings and creep stages during drained CT test on Toyoura sand (cf Figure 2-42 for stress-strain evolution) Dashed line is a deduced path for monotonic loading (ML) at constant loading rate (Hirakawa, 2003).

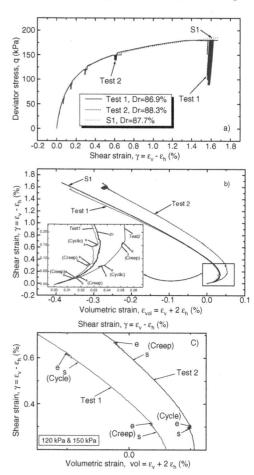

Figure 2-59. Drained TC tests on air-dried Toyoura sand with alternative cyclic and creep periods for tests 1 and 2. Test S1 is a reference monotonic loading test at constant strain rate. (Ko *et al.*, 2003). Partial results are presented in Figure 2-43.

the ratio during monotonic loading between the successive cyclic and creep loading periods is also plotted.

The following trends of behaviour may be seen from these three figures (Figure 2-58 to Figure 2-60):

1) The path direction in the strain space is similar for both types of loading (cyclic and creep loadings), while it is more contractive in terms of the absolute value of volumetric strain increment and the ratio of volumetric to shear strain increments than during continuous monotonic loading.

2) It can also be confirmed from Figure 2-59 and Figure 2-60 that creep deformation is predominant in the residual strain developed during cyclic loading with relatively small stress amplitude. These results are consistent with those described in sections 2.1.1.2(2) and 2.1.1.2(5).

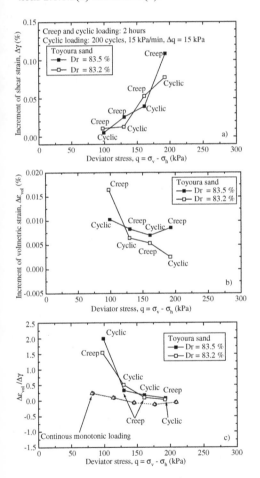

Figure 2-60. Volumetric strain increase (a) and ratio of the volumetric strain by the shear strain increases (b) versus deviator, for the test of Figure 2-42 together with another test, performed in identical condition except that the creep and cyclic periods are inversed (Hirakawa, 2003).

Figure 2-61 concerns 3 TC tests on dry Hostun sand including a large loading-unloading cycle up to negative values of the deviator stress (q). Two radial pressures (200 kPa and 400 kPa) for the same void ratio (0.95) and two void ratios (loose and more dense) for the same confining pressure (200 kPa) were considered. At different stress levels creep periods, of about 3 hours, are applied. The zoom-up plot of the large cycles in the axes axial strain versus stress ratio illustrates again the previously announced result concerning the direction of strain path during creep. It is not only controlled by the actual stress but seems also very influenced by the stress history (last stress reversal among others). The plot in the axial strain versus volumetric strain shows that the strain increment turns slightly during creep period. The strain direction is then not the same as the one for monotonic loading and seems to tend toward more contractive behaviour. This last affirmation needs to be verified on a wider set of experimental data.

2.1.2.4 Ageing effects

The aged clay exhibits generally less contractive behaviour than less aged clays, as typically seen from Figure 2-49. There is no conclusive data showing the

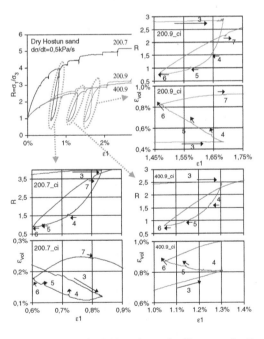

Figure 2-61. Drained TC cycles on dry Hostun sand with creep periods of about 3 hours at different stress levels. Axial strain-stress ratio general plot and zoom up for the large cycles, in the axes axial strain versus stress ratio and axial strain versus volumetric strain. Radial pressure and void ratio are respectively for tests 200.7, 200.9, 400.9: 200 kPa, 200 kPa, 400 kPa and 0.72, 0.95, 0.95 (Pham Van Bang, 2004).

effects of ageing process on the strain path during loading. The results from TC tests with creep loading stages on the cement-mixed soil are not useful to discuss on this issue, as the effects of cement hydration masks the ageing effect on the strain path, as typically seen from Figure 2-50.

2.2 Physical boundary-value model tests and full-scale behaviour

Among a number of different types of physical model tests that were performed at the University of Tokyo to evaluate the effects of material viscous properties on the overall behaviour of structure and ground models for specific practical engineering applications and full-scale field cases in which the residual deformation of structures were of great concern, results from two typical model tests and one field full-scale case are reported below. The model and full-scale behaviours in terms of loaddisplacement-time relation are first described while referring to qualitative as well as quantitative information of the constitutive behaviour of the materials constituting the model or full-scale structure.

2.2.1 Strip footing on level ground of unreinforced Toyoura sand under plane strain conditions

Figure 2-62 shows the set up of model footing loading test using a rough and rigid strip footing placed on the model level ground of air-dried Toyoura sand. A series of model tests were performed under plane strain conditions using a sand box having rigid and well-lubricated side walls (Tatsuoka et al., 1991b; Hirakawa et al., 2003a). Central and vertical load was applied to the footing controlling the footing settlement rate while changing many times the footing settlement rate suddenly and performing creep loading and relaxation loading during otherwise monotonic loading at a constant footing settlement rate. Figure 2-63(a) shows the overall relationship between the footing average contact pressure and the footing settlement from a typical test (test 1). In this figure, the slope k_{eq} means the elastic behaviour estimated by applying unload/reload cycles with a small amplitude of footing load performed in another test. Figure 2-63(b) shows the zoom up figure of part of the relations shown in Figure 2-63(a). Figure 2-64 shows the overall relationship between the footing average contact pressure and the footing settlement from another similar test (test 2) in which the footing settlement rate was stepwise changed not only in the pre-peak regime but also in the post-peak regime. The following trends of behaviour may be seen from these figures:

1) The average footing contact pressure changes suddenly upon a step change in the footing settlement rate, the noticeable creep settlement takes place and the footing load relaxes with time when keeping

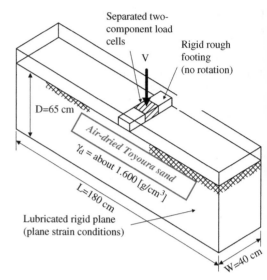

Figure 2-62.   Set up of model footing test using a rough and rigid strip footing on air-dried Toyoura sand; (Hirakawa et al., 2003a).

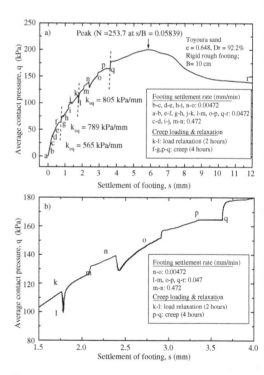

Figure 2-63.   (a) Overall relationship between average contact pressure of footing and footing settlement; and (b) zoomed up figure of part of figure (a), test 1 (Hirakawa et al., 2003a).

94

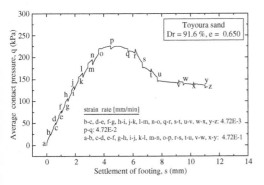

Figure 2-64.   Overall relationship between average contact pressure of footing and footing settlement, test 2 (Hirakawa *et al.*, 2003a).

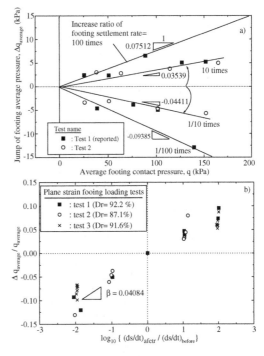

Figure 2-65.   a) Relationship between contact pressure jump and instantaneous contact pressure for different ratios of footing settlement rates after and before a step change tests 1 & 2; and b) viscous properties from model tests 1, 2 & 3 (Hirakawa *et al.*, 2003a).

the footing settlement constant. These phenomena confirm the important effects of the viscous properties of air-dried Toyoura sand in such a boundary value problem.

2) The viscous effects on the average footing contact pressure decays with an increase in the footing settlement when monotonic loading at a constant footing settlement rate continues following a step change in the footing settlement rate (i.e., the viscous evanescent or TESRA viscous properties). The decay rate of footing pressure in the post-peak regime is much larger than the one in the pre-peak regime. This difference could be explained by the fact that the shear bands start developing in the ground before the peak footing load and becomes significant and dominant in the post-peak regime, as reported by Tatsuoka *et al.* (1991b) and Siddiquee *et al.* (1999). That is, the ratio of the shear strain increment in the shear bands to the footing settlement increment in the post-peak regime is much larger than the ratio of the maximum shear strain increment in the ground to the footing settlement increment in the pre-peak regime, which leads to a larger decay rate of the viscous component of footing pressure per footing settlement increment. These trends of behaviour are well in accordance with the viscous properties of Toyoura sand observed in TC and PSC tests as described in the previous paragraphs.

Figure 2-65(a) shows the relationship between the jump in the average footing contact pressure upon respective jump in the footing settlement rate and the instantaneous average footing contact pressure for different ratios of footing settlement rates before and after a step change, produced by using the data presented in Figure 2-63 and Figure 2-64. It may be seen that the jump in the average footing pressure is always rather proportional to the instantaneous average footing pressure for the respective ratio of footing settlement rates after and before a step change. This trend of behaviour

is consistent with the one observed in the TC and PSC tests on Toyoura sand (and other sands and gravels). Figure 2-65(b) shows the relationships between the ratio of the footing pressure jump to the instantaneous footing pressure and the logarithm of the ratio of footing settlement rates after and before respective step change from the two tests described above (tests 1 & 2 presented in Figure 2-63 and Figure 2-64) and another one (test 3). It may be seen that the relation is rather linear. The slope is about 0.0408, which is surprisingly similar to the values from oedometer, TC and PSC tests on Toyoura sand (see Figure 2-38).

In the rigorous sense, the average footing contact pressure and the footing settlement in the model footing tests are not equivalent to, respectively, the stress ratio, $R = \sigma_v/\sigma_h$, and the shear strain, $\gamma = \varepsilon_v - \varepsilon_h$ (or the vertical strain, ε_v) in the TC and PSC tests at constant confining pressure and the vertical stress, σ_v, and the vertical strain, ε_v, in the 1D compression tests. Therefore, a comprehensive numerical analysis, such as FEM analysis, is necessary to understand this apparent resemblance in the viscous property between the model footing tests and the element tests described above. In fact, it will be shown later in this paper that the viscous properties observed in the model test can be simulated by FEM analysis incorporating the material viscous properties of Toyoura sand evaluated by element tests. It can be concluded based on the above that the material viscous properties of soil constituting the ground control the viscous aspects observed in the overall load-displacement behaviour of footing load, which is quite natural.

2.2.2 Axial loading of 3D grid-reinforced gravel structure

Another example showing similar viscous properties as the element stress-strain tests is presented in this section. Figure 2-66 shows the set up of the model test (Hirakawa *et al.*, 2003b). Models of reinforced-soil structure simulating the geogrid reinforced-soil bridge pier (Uchimura *et al.*, 2003 & 2004), described in the next section, were prepared by compacting air-dried well-graded crushed quarry sandstone gravel (called model Chiba gravel with $D_{max} = 5$ mm; $D_{50} = 0.8$ mm, $U_c = 2.1$ and Gs = 2.74). The models were 60 cm high and 35 cm by 35 cm square in cross-section with the backfill compacted to a dry density ρ_d of 1.789 g/cm^3. The backfill consisted of twelve tamped 5 cm-thick sub-layers reinforced with grid reinforcement layers with an aperture of 8 mm, each consisting of 34 phosphor bronze strips (3.5 mm-wide, 0.2 mm-thick and 350 mm-long). The periphery of each sub-layer was protected with about 3.5 cm-diameter bags filled with Chiba gravel, wrapped-around with a polymer geogrid made of polyvinyl alcohol fiber. As the effects of the viscous properties of reinforcement can be deemed negligible, any viscous behaviour of the model, if any, should be considered to be due to the viscous properties of the backfill (i.e., model Chiba gravel). The viscous properties of the model Chiba gravel were evaluated by TC tests and 1D compression tests (Hirakawa *et al.*, 2003b) and part of the test results are presented in Figure 2-38. The axial compression of the models, which were free from the effects of bedding error, were measured by a pair of Local Deformation Transducers: LDTs (Goto *et al.*, 1991).

Figure 2-67 shows the relationship between the vertical stress, σ_v, and the vertical strain, ε_v, both averaged for the reinforced backfill, from a test in which the strain rate, $\dot{\varepsilon}_v$, was stepwise changed many times, two creep loading tests and one load relaxation test were performed during otherwise monotonic loading at a constant $\dot{\varepsilon}_v$. At the peak load, a series of cyclic loading histories increasing stepwise the stress amplitude was applied, followed by global unloading. At several stress levels, multiple unload/reload cycles with a small amplitude of σ_v were applied to evaluate the elastic properties of the model. The relationship between σ_v and the average elastic strain of the backfill estimated based on the elastic modulus evaluated as above is indicated in this figure. It may be seen that

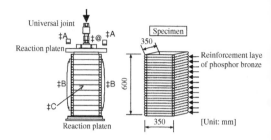

Figure 2-66.   Set up of model loading test on reinforced gravel backfill pier (Hirakawa *et al.*, 2003b).

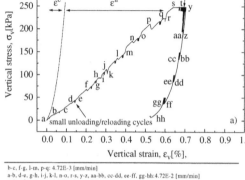

b-c, f-g, l-m, p-q: 4.72E-3 [mm/min]
a-b, d-e, g-h, i-j, k-l, n-o, r-s, y-z, aa-bb, cc-dd, ee-ff, gg-hh:4.72E-2 [mm/min]
c-d, e-f, m-n, o-p: 4.72E-1 [mm/min]
j-k: stress relaxation (for 1 hour)
h-i, q-r: creep (for 1 hour)
s-t, x-y: creep (for 6 hours)
t-u: cyclic loading, Amp.=6.2 kPa, 4.72E-2 [mm/min], 100 cycles
u-v, z-aa, bb-cc, dd-ee, ff-gg: cyclic loading, Amp.=12.4 kPa, 4.72E-2 [mm/min], 100 cycles
v-w: cyclic loading, Amp.=24.8 kPa, 4.72E-2 [mm/min], 100 cycles
w-x: cyclic loading, Amp.=49.6 kPa, 4.72E-2 [mm/min], 100 cycles

Figure 2-67.   Overall relationship between average vertical stress and strain of the backfill of the model pier of reinforced gravel (Hirakawa *et al.*, 2003b).

the model exhibits noticeable viscous properties as seen in the element tests on the backfill materials. Figure 2-68 shows the relationships between the ratio of the jump in σ_v observed upon a step change in $\dot{\varepsilon}_v$ to the instantaneous σ_v and the logarithm of the ratio of the vertical strain rates after and before a step change obtained from this model test. A linear relation as seen in the element tests can be seen, while the slope is similar to the β values from the element tests, presented in Figure 2-38. By using the same type of models as described in Figure 2-66, load-controlled tests were performed as shown in Figure 2-69 (Hirakawa *et al.*, 2003b). As shown in Figure 2-69, the models were first monotonically loaded to $\sigma_v = 250$ kPa at a constant stress rate of 0.2 kPa/sec. Subsequently, the models were subjected to either; a) immediate unloading (only in test *no creep/cyclic*), or b) sustained loading at $\sigma_v = 250$ kPa for 48 hours (only test *creep/cyclic*);

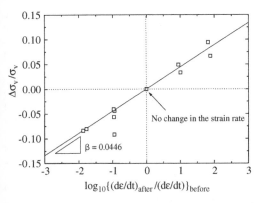

Figure 2-68. Relationship between the ratio of the stress jump and instantaneous stress and the logarithm of the ratio of the strain rates after and before a step change from the test results presented in Figure 2-67 (Hirakawa *et al.*, 2003b).

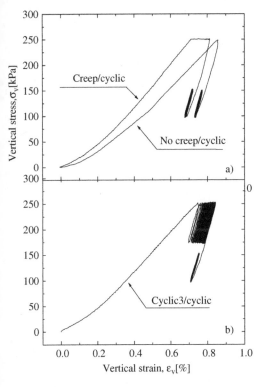

Figure 2-69. Overall relationship between average vertical stress and strain of the backfill of the model pier of reinforced gravel for different loading histories (Hirakawa *et al.*, 2003b).

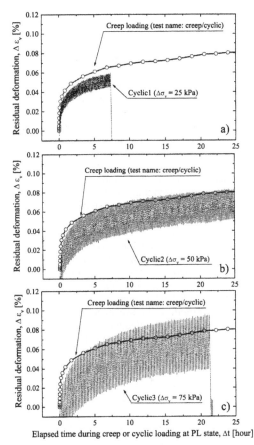

Figure 2-70. Time histories of residual strain during creep loading and cyclic loading with different stress amplitudes (Hirakawa, 2003).

or c) cyclic loading at a stress rate of $\dot{\sigma}_v = 0.2$ kPa/min with a stress amplitude equal to either 25 or 50 or 75 kPa by keeping the peak stress to 250 kPa. Figure 2-70 compares the time histories of average vertical

strain, ε_v, observed during creep and cyclic loading applied at the preload state. The following trends of behaviour may be seen from these figures:

1) The models exhibited significant residual axial strains by sustained and cyclic loading at the preloading state at $\sigma_v = 250$ kPa.

2) The residual vertical stain $\Delta\varepsilon_v$ developed by cyclic loading with a stress amplitude of $\Delta\sigma_v = 25$ kPa was noticeably smaller than the one developed by creep loading for the same loading period.

3) However, the value of $\Delta\varepsilon_v$ developed by cyclic loading with $\Delta\sigma_v = 50$ kPa was very similar to the one developed by creep loading for the same loading period.

4) The decreasing rate of residual strain rate was noticeably smaller in the cyclic loading test with $\Delta\sigma_v = 75$ kPa than in the creep loading test. Therefore, the value of $\Delta\varepsilon_v$ after some number of cycles developed in this cyclic loading test became

noticeably larger than the one developed in the sustained loading test.

These test results show again that the residual strain due to the material viscous properties could be dominant in the residual strain developing during these cyclic loading tests. It is also true that the rate-independent effects of cyclic loading increase with an increase in the stress amplitude, although the residual strain by this factor is not independent of the residual strain taking place by the viscous properties, but these two factors are coupled to each other.

Another important point seen from these test results is that the residual strain by cyclic loading could be made substantially small by applying a preloading history, which has an significant practical implication as shown below.

2.2.3 Prototype reinforced gravel

The bridge pier described in Figure 2-71 was constructed to support a pair of temporary railway girders, each 16.5 m in length, in the summer of 1996 in Fukuoka City, Japan (Uchimura et al., 2003). The pier consisted of reinforced gravel backfill with four steel tie rods, vertically installed through the backfill and anchored to the foundation ground at their bottom ends. First, the backfill was vertically preloaded by

using hydraulic jacks set at the top ends of the tie rods. Then the preload was partially released from the preload level to a prescribed non-zero prestress level. Finally, the top ends of the tie rods were fixed to the top reaction block by using nuts before removing the jacks. A certain level of compressive stress remained in the backfill as prestress during the service period. Figure 221.11 shows the relationship between the tie rod tension and the settlement of the pier during the preloading and prestressing procedure (until point 18), service for about three and a half year (until point 19) and full-scale loading tests (until point 22). The average vertical stress at the top of the backfill, when no external load was applied, was equal to the tie rod tension divided by cross sectional area of the backfill. During the period of service, every day more than 120 trains passed over the bridge supported by the pier. As seen from Figure 2-72, the most important factors among the effects of the preloading and prestressing procedures are: 1) the backfill becomes nearly elastic and stiffer by applying and partially unloading the preload; 2) the backfill shows higher elastic stiffness because the compressive prestress always remains in it during the service period; and 3) the backfill shows a minimum residual strain caused by the effects of the viscous properties of gravel and the rate-independent effects of cyclic loading for a long period of survice. It may be seen from Figure 2-72 that the residual compression of the backfill and the associated drop in the prestress in the tie rods for the period of service was surprising small.

The curve denoted by solid circles in Figure 2-72 in is the relationship between the applied load and the integrated value of the instantaneous compressive strain increments that occurred during the primary loading process. The difference between the total compressive strains and the instantaneous compressive strains consists of: 1) viscous compression caused by creep deformation; and 2) residual compression caused by

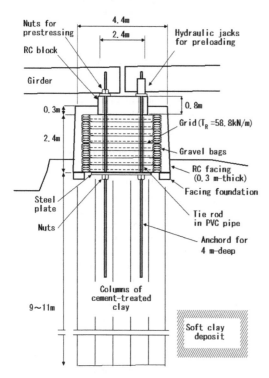

Figure 2-71. PLPS reinforced soil pier (Uchimura et al., 2003).

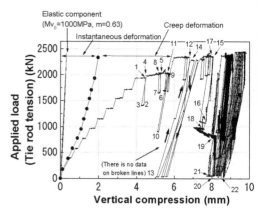

Figure 2-72. Relationship between tie rod tension and settlement of the pier (Uchimura et al., 2003).

cyclic loading. The difference increases with the load level, and 85% of the largest total compression is the irreversible component. This fact shows that a significant creep deformation and residual deformation by cyclic loading took place when creep and cyclic loads were applied despite that the backfill consisted of a well-graded gravel and was well compacted, showing the importance a preloading history to minimise the residual deformation of the backfill.

3 SIMULATION (BY THREE-COMPONENTS MODELS) AND ASSOCIATED ISSUES

The experimental results presented in this paper first focus on the general stress-strain response in terms of the relationship between one stress component and one strain component before investigating the general behaviour under general stress and strain conditions. This scheme is also followed in the next paragraphs where the one-dimension case is considered before presenting some extension for the general 3-dimension behaviour. The previous paragraphs reveal that, even in the case of one dimension consideration, viscous effects have specific features that are not simple to capture or obvious to understand.

The whole modelling analysis is made inside the framework of the three-component model (Figure 1-5), which has been found to be quite adapted and powerful. This general framework is presented in section 1.3.

As shown in Figure 1-5, a large variety of choices are possible for the three bodies, "EP1", "EP2" and "V". Following the respective choice, the behaviour described by the model will catch more or less sophisticated aspects in relation with experimental observations.

If no viscous effects are considered, no viscous body "V" is needed and the two non-viscous (or elastoplastic) bodies can be replaced by a single "EP" body. From the previously presented experimental results, this non-viscous behaviour for geomaterials is always an approximation, which could be justified depending on given conditions in the loading and time domains. A great number of "EP" type laws have been proposed in the literature. Some represented ones are categorized in Figure 1-5. This case of purely non-viscous behaviour is not considered further.

The form of the EP1 body can reasonably be approximated as hypoelastic eventually including ageing effects (in that case, the parameters t_c, is needed, cf. section 1.2). One version of the laws proposed by the authors, not considering this hypothesis of hypoelastic behaviour, was developed for bituminous mixes. The version is called "DBN" law. Figure 2-19 gives examples of modelling with such a version of the law. Other examples of simulation of thermo-mechanical loadings are proposed in Di Benedetto et al. (2000)

and Neifar et al. (2001). In the next sections, the EP1 body will be considered as hypoelastic. Then the non viscous strain can be replaced by an elastic one (possibly including ageing effects) noted as: ε^e. The equation 1-11 will then be preferred to equation 1-2.

$$\dot{\varepsilon} = \dot{\varepsilon}^e + \dot{\varepsilon}^{vp} \qquad \text{(equation 1-11 repeated)}$$

The hypoelastic model of the Universiry of Tokyo (Hoque and Tatsuoka, 1998; Tatsuoka et al., 2001a) and the hypoelastic model "DBGS" of ENTPE (Di Benedetto et al., 2001a & 2001b; Geoffroy et al., 2003), which are both able to describe anisotropic behaviour, were developed to simulate the non viscous component "ε^e". This part of the simulation is out of the scope of the paper and is not presented.

3.1 *Simplified ID case*

In the one-dimensional case consideration, equations 1-2 and 1-6 to 1-9 are no more tensorial and become scalar equations. In these equations the stress and stress variable (σ and ε) are generic and can take different values following the considered loading and response axes. For example, a principal stress "σ_i", the deviatoric stress "q", the stress ratio "R", the mobilized friction angle "ϕ_{mob}", ... (respectively, a principal strain "ε_i", the shear strain "γ", the volumetric strain "ε_{vol}",..) can be chosen as the stress (respectively, the strain) variable.

The next paragraphs present different choices of EP2 and V bodies and the main behaviour tendency that can be modelled following the considered choice. Simulations for very different kinds of geomaterials are also given. It is to be reminded that EP1 body is considered as hypoelastic so that it can be written as:

$$\dot{\varepsilon}^e = \frac{\dot{\sigma}}{E^{nv}} \qquad (3\text{-}1)$$

where E^{nv} is the tangent modulus of the non-viscous body, which depends only on the stain "ε^{nv}" = "ε^e" (or of the stress σ) because of the postulated hypothesis. From equations 1-6 and 3-1 it follows that: $M^{nv} = 1/E^{nv}$

3.1.1 *Linear case*

If the behaviour is linear, the EP2 body becomes a linear spring and the "V" body is linear viscoelastic. An extreme case consists of the body having only two springs (with stiffness values of K_0 and K) and one linear dashpot (with a viscosity η) (cf. Figure 3-1), which is called "Simple Asymptotic Body" (SAB) by Di Benedetto and Tatsuoka (1997). The analyses proposed in Di Benedetto and Tatsuoka (1997) show that this body can capture rather well the behaviour of very different kinds of geomaterials in the small strain domain. Not only the stress-strain loop but also

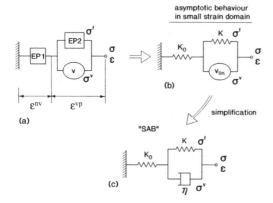

(a)

(b)

"SAB"

(c)

Figure 3-1. Three-component model and asymptotic models (SAB: Simple Asymptotic Body) (Di Benedetto and Tatsuoka, 1997).

the damping can be evaluated. The good ability of such a model for geomaterials is due to the existence of a linear domain for small strain amplitude cycles and also after creep periods, as described and shown in the section 2.1.1.1. An example of simulation with the SAB model compared with experimental data is given in Figure 3-2.

As a general trend, all the non-linear formalisms should tend towards the linear case in the small amplitude domain. The linear behaviour corresponds to an asymptotic one, which explains the choice of the terminology for "asymptotic body" and "simple asymptotic body".

3.1.2 "EP2" body
The EP2 body gives the link between the viscoplastic strain "ε^{vp}" and the inviscid stress "σ^f" (cf. equation 1-7).

3.1.2.1 Without ageing
For a given one-dimension monotonic loading condition, the following general form of the stress strain curve can be postulated:

$$\sigma^f = l_{1Dmon}(\varepsilon^{vp}) \qquad (3\text{-}2)$$

or when considering stress and strain rates or increments:

$$\dot{\varepsilon}^{vp} = \frac{\dot{\sigma}^f}{E^f} \text{ or } d\varepsilon^{vp} = \frac{d\sigma^f}{E^f} \qquad (3\text{-}3)$$

where l_{1Dmon} is a function of ε^{vp} only and E^f is the tangent modulus of the EP2 body, equal to the derivative of the stress-strain curve (monotonic loading case only). E^f depends only on the strain value: $E^f(\varepsilon^{vp})$. If

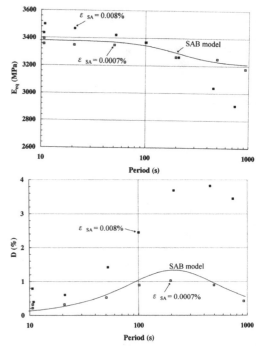

Figure 3-2. Modelling with the SAB model (cf. Figure 3-1) of sinusoidal strain shape tests on soft rocks ($K_0 = 3380$ MPa, $K = 61455$ MPa, $\eta = 2100000$ Mpa.s) tested by Suzuki (1994) (figure 12 of Di Benedetto and Tatsuoka, 1997).

no softening is considered, it can be also expressed as a function of the inviscid stress only: $E^f(\sigma^f)$. This means that an independent set of memory parameter of EP2 (h^f) (cf. equation 1-7) is given by σ^f only or ε^{vp} only ($h^f \equiv \sigma^f$ or $h^f \equiv \varepsilon^{vp}$).

If cyclic path is considered, equation 20 is no more valid and the stress strain curve is given by:

$$\sigma^f = l_{1D}(\varepsilon^{vp}, h^f_{cycle}) \qquad (3\text{-}4)$$

where "h^f_{cycle}" are the parameters to take into account the irreversible and non linear behaviour in particular during reversal of the loading. The equation 3-3 remains valid, but the tangent modulus becomes also a function of "h^f_{cycle}" and of the direction of the strain rate:

$$E^f(\varepsilon^{vp}, h^f_{cycle}, \text{dir } \dot{\varepsilon}^{vp}) \qquad (3\text{-}5)$$

It can be noted in the considered one – dimension case that: $E^f = N^f = 1/M^f$ (cf. equations 1-7, 1-10 and 3-3) and the direction "dir ε^{vp}" is the sign of $\dot{\varepsilon}^{vp}$ (cf Figure 3-3).

Many possible formulations of the function l_{1Dmon} have been proposed in the literature (Kondner, 1963;

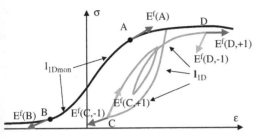

Figure 3-3. General illustration of stress-strain curve for monotonic loading (l_{1Dmon} equation 3-2) and cyclic loading (l_{1Dmon} equation 3-3). The tangent modulus E^f is also indicated. For cyclic loading it depends on the direction of the strain increment (equation 3-5).

Di Benedetto and Darve, 1983; Mohkam, 1983; Tatsuoka and Shibuya, 1991a; Hachi-Benmebarek, 2000; Balakrishnaier and Koseki, 2001; ...). The following one consisting in a sum of exponentials is proposed in Di Benedetto *et al.* (2002):

$$\sigma^f = R_0 + A_1\left\{1-\exp\left(-\frac{\varepsilon^{vp}}{c_1}\right)\right\} + A_2\left\{1-\exp\left(-\frac{\varepsilon^{vp}}{c_2}\right)\right\}$$

(3-6)

where R_0, A_1, A_2, c_1, c_2 are constants.

The function, l_{1D}, for cyclic loading histories could be expressed as a function of the monotonic loading function, l_{1Dmon}. Tatsuoka *et al.* (1999a & 2001a) proposed the following "hysteresis rule", which is applied to derive the function for each loading or unloading branch:

$$l_{1D}(x/n) = c_3\{l_{1Dmon}(x/c_3)\}$$

(3-7)

where c_3 is a constant, which could take a value different from two, corresponding to the classical "Masing" rule case. Then the parameters "h^f_{cycle}" can be chosen equal to the set of strain values at the reversal of loadings. A similar expression is proposed in Di Benedetto *et al.* (2000).

It is not the objective of this paper to discuss more in details the possible forms of functions l_{1Dmon} and l_{1D}. Although more powerful expressions than equations 3-6 and 3-7 could probably be found, these two equations are used herein for the simulation by the new isotach, TESRA and general TESRA models, proposed in this paper. The function l_{1Dmon} (or l_{1D}) gives the so-called reference curve, which is plotted in the figures showing simulation with these models.

3.1.2.2   Introduction of ageing

The case of ageing behaviour affecting the body EP2 cannot be treated by equation 3-3 because at constant stress the strain can change due to ageing of the material. For the same reason, at constant strain, stress can develop with time (due to ageing effects). Then the following equation can be considered:

$$d\sigma^f = E^f(h^f, t_c, dir\,\dot{\varepsilon}^{vp})\cdot d\varepsilon^{vp} + H(h^f, t_c)\cdot dt \quad (3\text{-}8)$$

The first term on the right side of equation 23-1 is the stress (inviscid) increment created by the (visco-plastic) strain increment $d\varepsilon^{vp}$. The second term is the (inviscid) stress increment created by ageing effects. This last term (H dt) is supposed to be not affected by the rate of loading, then, it is not due to viscous effect (cf. section 1.2). An example of physical phenomenon where the value of H is not nil is the contraction observed during the hydration of concrete: at constant fixed strain, a tension stress develops and increases with time.

It is classical to treat the first term "$E^f d\varepsilon^{vp}$" within the framework of the classical elasto-plastic theory. A yield stress σ^f_y is then introduced (for the inviscid stress σ^f). If σ^f is different from σ^f_y, the behaviour of EP2 is rigid (only EP1 is active in the three-component model) or elastic. σ^f equal σ^f_y when yield loading condition is respected ($\sigma = \sigma^f_y$) (cf. also section 3.2.2 and Figure 3-15). The evolution of σ^f_y is given by:

$$d\sigma^f_y = E^f_\infty \cdot d\varepsilon^{vp} + F^f_0 \cdot dt \quad\quad (3\text{-}9)$$

where E^f_∞ is the tangent modulus obtained at "infinitely high" strain rate or stress rate ($\dot{\varepsilon}^{vp} = \infty$ or $\dot{\sigma}^f = \infty$). In that case ageing has "no time" to have any effect. And F^f_0 is the yield stress rate that would be observed if the strain was kept constant ($d\varepsilon^{vp} = 0$ or $\dot{\varepsilon}^{vp} = 0$).

The tangent modulus E^f_∞ and the stress rate F^f_0 depend on the specific time "t_c" (cf. paragraph 1.2). The time t_c is then included in the set of memory parameters h^f. The expression taken for the proposed simulation considers that the modulus (equation 3-9) is written as:

$$E^f_\infty(\varepsilon^{vp}, t_c) \quad \text{for monotonic loading case} \quad (3\text{-}10)$$

In the same way:

$$F^f_0(\varepsilon^{vp}, t_c) \quad\quad\quad\quad (3\text{-}11)$$

The integration of equation 3-9 gives the yield stress at time t_c, where the strain is ε^{vp} :

$$\sigma^f_{y\,(\varepsilon^{vp},t_c)} = \int_0^{\varepsilon^{vp}} E^f_\infty(\tau, t_{(\tau)})\cdot d\tau + \int_0^{t_c} F^f_0(\tau_{(t)}, t)\cdot dt \quad (3\text{-}12)$$

101

At the origin, time and strain are defined as zero, which means that the origin for time is the starting moment of ageing. For the integration of equation 3-12, the current time is denoted as t and the current strain as τ. Equation 3-12 is in general not totally differential, which means that two different yield stress values can be obtained at the same t_c and ε^{vp} values, for two different loading paths.

Tatsuoka et al. (2003) assumed H = 0 (in equation 3-8) for the first approximation and the following equations as the expressions of E_∞^f and F_0^f:

$$E_\infty^f(\tau, t) = E_0^f(\tau) \cdot A^f\{t_{(\tau)}\} \tag{3-13}$$

$$F_0^f(\tau, t) = \{a^f \cdot \sigma_0^f(\tau_{(t)}) + b^f\} \cdot \alpha^f(t) \tag{3-14}$$

where $A^f(t)$ is the ageing function for σ^f, which increases (respectively decreases) with time from $A^f(0) = 1.0$ in the case of positive (respectively negative) ageing. $E_0^f(\tau)$ is the basic stiffness, not affected by ageing. This stiffness is obtained at the strain "τ" for infinitely high loading rate, which results in the stress $\sigma_0^f(\tau)$. a^f and b^f are constants; and $\alpha^f(t)$ is a function of time t.

When $a^f = 1.0$ and $b^f = 0.0$ and $\alpha^f(t) = d\{A^f(t)\}/dt$, σ_y^f becomes independent of loading history, obtained as:

$$\sigma_y^f(\varepsilon^{vp}, t_c) = \sigma_0^f(\varepsilon^{vp}) \cdot A^f(t_c) \tag{3-15}$$

Equation 3-15 indicates that ageing effects are the same whatever the strain or stress values. The effects of ageing are then decoupled from the other effects. For the data from test JA003 in Figure 2-50, the stress-strain behaviour upon the restart of ML after drained creep tends to ultimately become similar as the one for continuous ML after the same total ageing period. In this case, equation 3-15 is rather relevant. However, Kongsukprasert and Tatsuoka (2003a & 2003b) showed a set of data for which equation 3-15 is only the first approximation.

Examples of simulation including viscous and ageing effects using equation 3-15 and H = 0 in equation 3-8 are presented in the section 3135.

3.1.3 "V" body and associated behaviours
The "V" body represents the link between the viscous stress "σ^v" and the viscoplastic strain evolution "ε^{vp}" (as resulting from the general equation 1-8 it is the strain rate "$\dot{\varepsilon}^{vp}$" which is predominant).

For the 1D case equation 1-8, which is considered for the "V" body, is also 1D (scalar).

3.1.3.1 Isotach case without ageing
The isotach type behaviour is illustrated in Figure 2-22; i.e., each stress-strain curve is uniquely associated with a strain rate (more rigorously with a viscoplastic strain rate $\dot{\varepsilon}^{vp}$, which is in many cases very close to the total strain rate, $\dot{\varepsilon}$). For the modelling it means that the memory parameters of the body V (h^v) are only coming from the set of the memory parameters of EP2 (h^f): $h^v \epsilon h^f$. It is reminded that the current time must not appears in the law for objectivity reason.

(1) Simple isotach case
The simple isotach case corresponds to the case where no h^v parameters are considered. It comes:

$$\sigma^v = f(\dot{\varepsilon}^{vp}) \tag{3-16}$$

In the following, only positive values of the strain rate $\dot{\varepsilon}^{vp}$ will be considered. In case of negative value the general following expression should be taken:

$$\sigma^v = \text{sgn}(\dot{\varepsilon}^{vp}) f(|\dot{\varepsilon}^{vp}|) \tag{3-17}$$

where sgn (x) and |x| are the sign and norm of x.

A commonly used expression is derived from the "Norton-Hoff" formulation:

$$\sigma^v = f(\dot{\varepsilon}^{vp}) = \eta_0 \left(\frac{\dot{\varepsilon}^{vp}}{\dot{\varepsilon}_0^{vp}}\right)^{1-b} \tag{3-18}$$

where η_0 and b are constants and $\dot{\varepsilon}^{vp}_0$ stands for the unity respect. The case where b = 0 is the classical linear behaviour represented by the linear dashpot.

For air dried Hostun sand, Di Benedetto et al. (1999b) and Sauzéat (2003) applied equation 3-18. It was found that the coefficient b seems independent of the previous loadings and equal to −0.95 (b = −0.95). Meanwhile it was found that the coefficient η_0 can't be considered as constant (Pham Van Bang et al., 2003) (cf. section 3.1.3.2). Figure 3-4 from Di Benedetto et al. (1999b) gives a confirmation of the above statements.

As the inviscid stress level seems to have an influence on the viscous stress evolution. This stress should then be added to the parameters h^v. In that way a coupling is introduced between the two bodies EP2 and V.

(2) Isotach case with simple coupling
Tatsuoka et al. (1999b, 2002) and Di Benedetto et al. (2002) showed that the following decoupled form is relevant to geomaterials:

$$\sigma^v = f(h^v, \dot{\varepsilon}^{vp}) = H_v(\sigma^f) \cdot g_v(\dot{\varepsilon}^{vp}). \tag{3-19}$$

where $H_v(\sigma^f)$ means a function of σ^f; and $g_v(\dot{\varepsilon}^{vp})$ is a function of $\dot{\varepsilon}^{vp}$, called the viscosity function. In that case, the unique memory parameter h^v (equation 1-8) is the inviscid stress: σ^f.

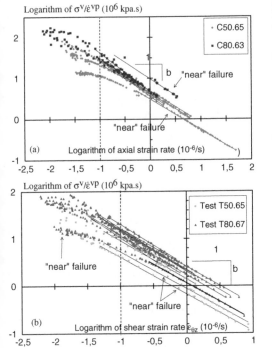

Logarithm of $\sigma^V/\dot{\varepsilon}^{vp}$ (10^6 kpa.s)

- C50.65
- C80.63

b "near" failure

"near" failure

(a) Logarithm of axial strain rate (10^{-6}/s)

Logarithm of $\sigma^V/\dot{\varepsilon}^{vp}$ (10^6 kpa.s)

- Test T50.65
- Test T80.67

"near" failure

"near" failure

b

(b) Logarithm of shear strain rate $\dot{\varepsilon}_{\theta z}$ (10^{-6}/s)

Figure 3-4. Creep periods represented in the axes: Log of viscosity (= $\sigma/\dot{\varepsilon}$) – Log of strain rate, for CT tests (a) and pure shear tests (b) performed on hollow cylinder samples of dry Hostun sand (Di Benedetto et al., 1999b). "b" is close to −0.95.

The "new isotach" model proposed at the University of Tokyo (Tatsuoka et al., 2001a) is a simple form of equation 3-19. It assumes that σ^v is proportional to σ^f (in the monotonic loading case). This assumption results in the analysis of drained PSC tests on sand and undrained TC tests on soft clay (Tatsuoka et al. 1999b). Using the principal stress ratio R (= σ_1/σ_2) as σ, it comes:

$$\sigma^v = \sigma^f \cdot g_v(\dot{\varepsilon}^{vp}) \qquad (3\text{-}20)$$

Komoto et al. (2003) showed that equation 3-20 (with $\sigma = R = \sigma_1/\sigma_2$) is relevant to undisturbed samples of Pleistocene clay and reconstituted specimens in CD TC. Tatsuoka et al. (1999b, 2002) and Di Benedetto et al. (2002) showed that the following non-linear function is relevant to $g_v(\dot{\varepsilon}^{vp})$:

$$g_v(\dot{\varepsilon}^{vp}) = \alpha \cdot [1 - \exp\{1 - (\frac{|\dot{\varepsilon}^{vp}|}{\dot{\varepsilon}_r} + 1)^m\}] \quad (\geq 0) \qquad (3\text{-}21)$$

where α, $\dot{\varepsilon}_r^{vp}$ and m are the positive constants; and $|\dot{\varepsilon}^{vp}|$ is the absolute value of $\dot{\varepsilon}^{vp}$ so that Eq. 34a is valid for

both cases with positive and negative values of $\dot{\varepsilon}^{vp}$. Note that $g_v(\dot{\varepsilon}^{vp}) = 0$ when $\dot{\varepsilon}^{vp} = 0$ and $g_v(\dot{\varepsilon}^{vp}) = \alpha$ when $\dot{\varepsilon}^{vp} = \infty$.

On the other hand, Di Benedetto et al. (1999b) chose the following g_v function deduced from equation 3-18:

$$g_{v2}(\dot{\varepsilon}^{vp}) = \alpha * \left(\frac{\dot{\varepsilon}^{vp}}{\dot{\varepsilon}_0^{vp}}\right)^{1+b} \qquad (3\text{-}22)$$

where $\alpha*$ is a constant.

It can be shown that the two expressions of equations 3-21 & 3-22 give similar evolution on a wide range of values for $\dot{\varepsilon}^{vp}$. Differences are obtained only for very small and very large values of $\dot{\varepsilon}^{vp}$. Then, the use of any of these two equations for the usual experimental and practical values of strain rate gives similar results.

From equations 3-20 and 3-2, it can be seen that the viscous stress increment consists of two terms as follows:

$$d\sigma^v = \left(\frac{\partial \sigma^f}{\partial \varepsilon^{vp}}\right) \cdot g_v(\dot{\varepsilon}^{vp}) \cdot d\varepsilon^{vp} + \sigma^f \cdot \left(\frac{\partial [g_v(\dot{\varepsilon}^{vp})]}{\partial \dot{\varepsilon}^{vp}}\right) \cdot d\dot{\varepsilon}^{vp}$$
$$(3\text{-}23)$$

Examples of simulation with the "new isotach" model are presented in Figure 2-20 and Figure 2-27 respectively for saturated kaolin and reconstituted Ohimachi clay. Figure 3-5 and Figure 3-6 show that this version of the law is also relevant for sedimentary soft rocks and undisturbed Ohimachi clay. These simulations use analytical expressions of equations 3-6, 3-20 and 3-21.

Pham Van Bang et al. (2003) and Sauzéat et al. (2003) proposed for sand the following equation for σ^v:

$$\sigma^v = \sigma_{1max} \cdot g_{v2}(\dot{\varepsilon}^{vp}) \qquad (3\text{-}24)$$

where g_{v2} is given by equation 3-22 and σ_{1max} is the major principal stress (σ in the 1D case). As the inviscid stress and total stress are usually rather "close" to each other for sand, σ^f_{1max} could also be chosen. A comparison of equations 3-18 and 3-24 shows that the coefficient η_0 (equation 3-18) depends on the stress value.

$$\eta_0 = \eta = \alpha * \sigma_{1max} \qquad (3\text{-}25)$$

The expression (equation 3-24) presents the advantage to be easy to generalise for the three-dimension case (cf. section 3.2). In addition, it seems to remain valid on large loading unloading cycles. An example of simulation of relaxation periods during TC test on dry Hostun sand with equation 3-24 is presented in Figure 2-46. The value of the parameter η obtained by optimisation of the simulations for each relaxation

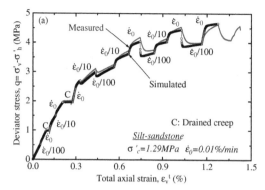

Figure 3-5. Simulation by the New Isotach model of the stress-strain behaviour in CD triaxial compression test on sedimentary soft rock (Tatsuoka et al., 2000; Hayano et al., 2001).

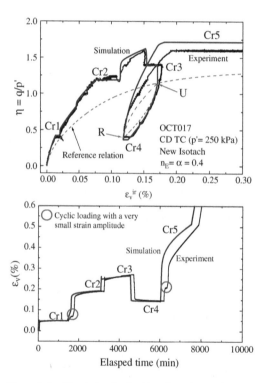

Figure 3-6. Simulation by the New Isotach model of the stress-strain behaviour in CD TC, undisturbed Pleistocene clay from Ohimachi (Tatsuoka et al., 2002).

periods is also indicated in in Figure 2-46b. The value of α^* for this sand is equal to 0.15 ($\alpha^* = 0.15$). The simulations presented in in Figure 2-46 are performed with the viscous evanescent formulation (cf. section 3.1.3.3). As the strain remains constant during relaxation periods, the evanescent properties, which are

proportional to the viscoplastic strain, are negligible. Then, equation 3-24 and the viscous evanescent model give very close results.

3.1.3.2 Possible evolution for isotach case – and ageing

As can be seen from Figure 2-38 for Kitan clay, the β coefficient (cf. Figure 2-36 and equation 2-15) is different among the undisturbed samples retrieved from different depths (Komoto et al., 2003). As the samples at the depth of 63 to 82 meters have a geological age of 0.27 to 0.87 million years, and those at the depth of 82 to 120 meters have a geological age of 1.6 to more than 2.6 million years, the difference among these β values can be explained by different degrees of ageing. This coefficient β has a direct link to the function g_v (cf. equation 3-19). As it could be expected to be the case in more general cases, these results reveal that the viscous properties could be significantly affected by ageing. To model this tendency, a possible formulation in the isotach case may introduce the parameter t_c as a variable of the function g_v. That is, equation 3-19 can be generalised in the following way:

$$\sigma^v = H_v(\sigma^f) \cdot g_v(\dot{\varepsilon}^{vp}, t_c).$$

$$(3-26)$$

The behaviour described by equation 3-26 remains isotach if the loading is applied during a period, which is small enough to create negligible variation of g_v. For the considered Kitan clay the β coefficient is affected over more than million years, while the loading is less than one month. The two times are at utterly different scales and t_c in equation 38 can be considered as constant for any practical purpose, which results in an isotach behaviour (as represented by equation 3-19).

3.1.3.3 Case with viscous evanescent properties – decay function

The "pure viscous evanescent" type behaviour is described in the section 2.1.1.2(1), and illustrated in Figure 2-22. This trend of viscous behaviour can be observed for different types of geomaterials, typically clean sands (cf. section 2.1.1.2(1)). Considering the lack for such description in the literature, the authors developed a specific expression for this trend of behaviour (Di Benedetto et al., 2002). The decay function "g_{decay}" is then introduced whose definition is obtained based on the assumed behaviour during monotonic loading at a constant visco-plastic strain rate abruptly starting from a stable state (i.e., with nil stress, strain and strain rate). The viscous stress is then given by:

$$\sigma^v = f_{isot}(h^v_{isot}, \dot{\varepsilon}^{vp}_0) \cdot g_{decay}(\varepsilon^{vp}) \; ; \; \text{with } \dot{\varepsilon}^{vp}_0 \text{ constant}$$

$$(40)$$

where f_{isot} is the same function as function f in the isotach case (see for example equation 3-19). g_{decay} is a

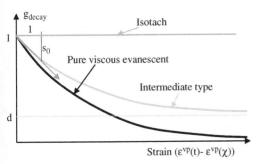

Figure 3-7. Shape of the g_{decay} function. The case d = 0 and s_0 function of the actual strain (ε^{vp}(t)) corresponds to the general TESRA model. Very general intermediate type behaviour (cf figure 2112.6) can be obtained if d and s_0 are function of the actual strain (ε^{vp}(t)).

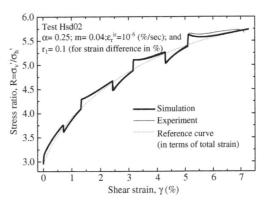

Figure 3-8. Result from a CD PSC at $\sigma_3 = 396\,kPa$ on Hostun sand and its simulation by the TESRA model (Di Benedetto et al., 2002).

monotonously decreasing function (cf. Figure 3-7) equal to the unity in zero ($g_{decay}(0) = 1$). In the case of constant g_{decay} function ($g_{decay} = 1$), no evanescent property exists and the behaviour is isotach.

If g_{decay} decreases toward 0 the behaviour is (purely) viscous evanescent as described in Figure 2-22. If g_{decay} tends toward a non-nil constant ($g_{decay}(\infty) = d > 1$) the behaviour is of the intermediate type (cf. Figure 2-22).

The choice of the viscous strain (ε^{vp}) as the variable for g_{decay}, and not the time (t), rises some questions concerning the micro-structural origin of the viscous component for geomaterials exhibiting this behaviour (**such as clean sands, ...**). A classical viscous behaviour, as observed for polymers or bitumen, would need the time. At present, we have no more explanation on this conceptual difference, which consists in using the strain instead of the time.

If the superposition of the stress path increments applied during the loading history is weighted by the function g_{decay}, the viscous stress at the present time "t" for any strain (or stain rate) evolution is given by:

$$\sigma^{v}_{(t)} = f(h^{v}, \dot{\varepsilon}^{vp})$$
$$= \int_{\chi=0}^{t} d\{f_{isot}(h^{v}_{isot(\chi)}, \dot{\varepsilon}^{vp}_{(\chi)})\} \cdot g_{decay}(\varepsilon^{vp}_{(t)} - \varepsilon^{vp}_{(\chi)}) \qquad (3\text{-}27)$$

where "χ" is the intermediate time needed for the integration. $d\{f_{isot}(h^{v}_{(\chi)}, \dot{\varepsilon}^{vp}_{(\chi)})\}$ is the increment (or variation) of f that is due to the variation of h^{v} and $\dot{\varepsilon}^{vp}$ between the moments χ and $\chi + d\chi$. Equation (41) can be rewritten if $\dot{\varepsilon}^{vp}$ is continuous:

$$\sigma^{v}_{(t)} = f_{isot}(h^{v}_{isot(t)}, \dot{\varepsilon}^{vp}_{(t)})$$
$$+ \int_{\chi=0}^{t} f_{isot}(h^{v}_{isot(\chi)}, \dot{\varepsilon}^{vp}_{(\chi)}) \cdot g'_{decay}(\varepsilon^{vp}_{(t)} - \varepsilon^{vp}_{(\chi)}) \cdot d\varepsilon^{vp}_{(\chi)} \qquad (3\text{-}28)$$

where g'_{decay} is the derivative of the function g_{decay}. More details and explanations on the equation 3-27 and 3-28 are given in Di Benedetto et al. (2002).

These last two equations are the basis for the TESRA (Temporary Effects of Strain Rate Acceleration) and the ENTPE VE (Viscous Evanescent) models. The memory parameters h^{v} (equation 1-8 and 3-27) include the whole history of each of the h^{v}_{isot} parameters and the history of the viscoplastic strain rate $\dot{\varepsilon}^{yp}$. The function g_{decay} is chosen as follow:

$$g_{decay}(x) = (r_1)^x \qquad (0 \leq r_1 \leq 1) \qquad (3\text{-}29)$$

where r_1 is a positive constant lower or equal to unity. If $r_1 = 1$ the decay is nil and the modelled behaviour becomes isotach. The TESRA model uses the analytical expressions given in equations 3-20, 3-21 and 3-29 to express the "V" body behaviour. The VE model considers equations 3-22, 3-24 and 3-29. These two models give rather similar behaviours as can be deduced from the previous explanations.

Examples of simulation for Toyoura sand with the TESRA model are given in Figure 2-17, Figure 2-33, Figure 2-34, Figure 2-35, Figure 2-55. Figure 3-8 shows simulation and experimental results for a drained PSC test, with stepwise changes of the strain rate, for Hostun sand. The particular behaviour obtained for creep periods with different previous strain rates, as illustrated in Figure 2-18 (cf. section 2.1.1.2-1a) on Toyoura sand, is well simulated as can be observed in Figure 3-9. It is reminded that this viscous evanescent property is particularly pronounced for clean sand as can be seen on the previously cited figures.

As mentioned in the preceding section, an intermediate type of behaviour (cf. Figure 2-22) can be obtained if g_{decay} decreases toward a strictly positive

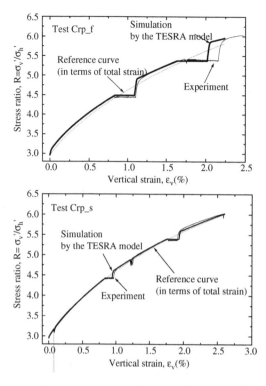

Figure 3-9. Comparison between simulated and measured R versus axial strain, for PSC tests Crp_f and Crp_s on saturated Toyoura sand (cf. Figure 2-18 for experimental specific feature) (Tatsuoka *et al.*, 2002).

value at high strain rate. In that case the shape of the stress-strain curve, during monotonic loading at a constant viscoplastic strain rate, remains the same, but exhibits a sudden change in the stress when changing stepwise the strain rate irrespective of the stress level. Another type of intermediate behaviour is sometime observed showing a transition from purely isotach properties (at low stress levels) to more viscous evanescent behaviour (at higher stress levels). For example, Figure 2-21 shows such an intermediate type of behaviour for Fujinomori clay. To describe this other kind of intermediate behaviour, the "general TESRA" model was proposed (Tatsuoka *et al.*, 2002). The difference from the TESRA model is in the expression of the g_{decay} function. In equation 3-29, the parameter r_1 becomes dependent of the strain value "ε^{vp}" as:

$$r_1(\varepsilon^{vp}_{(t)}) \quad \text{in equation 3-29} \qquad (3\text{-}30)$$

The chosen function for r1 is given in Tatsuoka *et al.* (2002). An example of simulation with the general TESRA is presented in Figure 2-21 for the Fujinomori clay.

A more general way to treat the intermediate type behaviour (cf. Figure 2-22) is to consider that the g_{decay} function depends on both the instantaneous viscoplastic strain and the difference between the instantaneous and the previous intermediate visco-plastic strain as:

$$g_{decay}(\varepsilon^{vp}_{(t)}, \varepsilon^{vp}_{(t)} - \varepsilon^{vp}_{(\chi)}) \qquad (3\text{-}31)$$

To take into account ageing effects in the decay properties, it is necessary to introduce the time t_c as variable in the decay function as:

$$g_{decay}(\varepsilon^{vp}_{(t)}, t_c, \varepsilon^{vp}_{(t)} - \varepsilon^{vp}_{(\chi)}) \qquad (3\text{-}32)$$

Experiments on this aspect are lacking and the authors decided not to touch on the type of evolution in this note any more. This topic remains then totally open.

3.1.3.4 Specific analysis of the "jump" when stepwise change in the strain rate

The stress jump "ΔR" actually determined (cf. Figure 2-36 to Figure 2-40) is the stress change caused solely by a change in the strain rate. On the other hand, the (hypo-)elastic strain rate ($\dot{\varepsilon}^e$) is generally very small when compared to the viscoplastic strain rate($\dot{\varepsilon}^{vp}$). So we obtain:

$$\Delta R \approx \Delta R^v$$
$$\approx \left[f(h^v_{after}, \dot{\varepsilon}^{vp}_{after}) - f(h^v_{before}, \dot{\varepsilon}^{vp}_{before}) \right]$$
$$\approx \left[f(h^v_{after}, \dot{\varepsilon}_{after}) - f(h^v_{before}, \dot{\varepsilon}_{before}) \right] \qquad (3\text{-}33)$$

where the label "after" (and "before") means immediately after (and before) a stepwise change.

The strain difference before the stepwise change and at the "apparent yielding point" just after the stepwise change is rather small. Then, the decay function has negligible effect on the stress-strain response and the jump can be analysed only in the isotach case. In the formulation proposed for the TESRA model (equations 3-19 and 3-21) and the one for the ENTPE VE model (equations 3-29 and 3-24) the simulated jump ratio (DR/R) becomes only connected with the jump in the g_v and g_{v2} functions. It comes:

$$\Delta R / R \approx \Delta R^v / R$$
$$\approx g_v(\dot{\varepsilon}_{after}) - g_v(\dot{\varepsilon}_{before}) \quad \text{;TESRA model}$$
$$\approx g_{v2}(\dot{\varepsilon}_{after}) - g_{v2}(\dot{\varepsilon}_{before}) \quad \text{;ENTPE VE model}$$
$$(3\text{-}34)$$

These two equations show the interest of the analysis of the stress jump during the stepwise strain rate

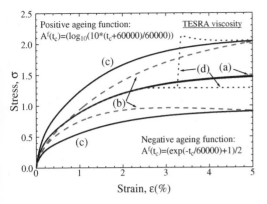

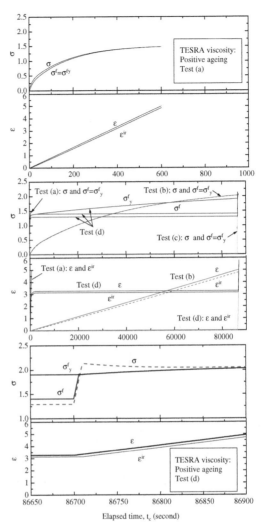

Figure 3-10. Effects of ageing and loading rate in numerical experiments (Tatsuoka *et al.*, 2003).

changes. A high linearity between the $\Delta R/R$ and $\log(\varepsilon^{vp}_{after}/\varepsilon^{vp}_{before})$ of the experimental data led to the introduction of the "rate-sensitivity coefficient" β (section 2.1.1.2-4 (equation 2-15)), whose values are reported in Figure 2-36 to Figure 2-40 and in appendix 2. These results validate equation 3-20 or 3-24. The coefficient β is also very useful to express the characteristics of the g_v functions (Di Benedetto *et al.*, 2002; Pham Van Bang *et al.*, 2003). On the other hand, data are still missing for large stress or strain amplitude cyclic loading path histories to validate the proposed expressions.

3.1.3.5 Illustration of ageing

To illustrate positive and negative ageing effects, Tatsuoka *et al.* (2003) performed the following numerical experiments based on the TESRA type "V" body (Figure 3-10 to Figure 3-12). Figure 3-11 and Figure 3-12 show the time histories of stress and strain in the case of, respectively, positive and negative ageing.

Test a: the specimen is not aged at the origin ($\sigma = 0$) before the start of ML at a high strain rate $\dot{\varepsilon}$. The total elapsed time t_c (from the start of monotonous loading) when the strain ε becomes 5% is 600 seconds.

Test b: the specimen is not aged at the origin before the start of ML at a low $\dot{\varepsilon}$. The total elapsed time t_c when $\varepsilon = 5\%$ is 87,000 seconds.

Test c: the specimen is aged at the origin for 86,400 seconds before the start of ML at a high $\dot{\varepsilon}$. As test b, the total elapsed time t_c when $\varepsilon = 5\%$ is 87,000 seconds.

Test d: the specimen is not aged at the origin before the start of ML at a high $\dot{\varepsilon}$ as test a. The specimen is aged at an intermediate stage for 86,400 seconds. The total elapsed time t_c when $\varepsilon = 5\%$ is 87,000 seconds.

The following assumptions were used for the simulation (first approximation): 1) no ageing effects on

Figure 3-11. Time histories of strain and stress in the case of positive ageing in numerical experiments (cf. Figure 3-10). (Tatsuoka *et al.*, 2003).

the elastic property; 2) $r_1 = 0.0001$ (equation 3-29) and no ageing effects on the function g_v (equation 3-21); 3) the use of ageing functions $A^f(t_c)$ (equation 3-15) for positive and negative ageing as shown in Figure 3-10; 4) $H = 0$ in equation 3-8 and 5) the behaviour of EP2 is perfectly rigid when unloading and reloading up to the monotonic stress-strain curve (up to σ^v_y) corresponding to the yield limit.

The following trends of behaviour may be seen from Figure 3-10 to Figure 3-12:

Positive ageing: Firstly, the stiffness at small strains (less than about 0.5%) is lower in test b (at a low $\dot{\varepsilon}$) than test a (at a high $\dot{\varepsilon}$), due to loading rate effects that are

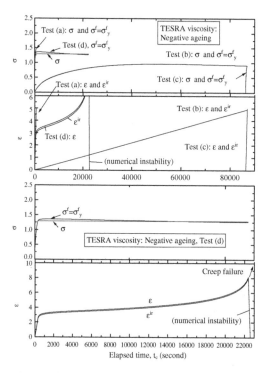

Figure 3-12. Time histories of strain and stress in the case of negative ageing in numerical experiments. (cf. Figure 3-10). (Tatsuoka *et al.*, 2003).

lower than positive ageing effects in test *b*. At larger strains, the stiffness gradually becomes larger and the strength ultimately becomes higher in test *b* than in test *a*. This is due to that positive ageing effects have become more dominant over loading rate effects in test *b* while the viscous stress has decayed in test *a*. Kongsukprasert and Tatsuoka (2003b) report data from two CD TC tests performed at a very low strain rate on cement-mixed gravel showing the trend of behaviour in test *b*. Secondly, the pre-peak stiffness and peak strength in test *c* (at a high $\dot{\varepsilon}$) are consistently larger than in test *a* (at the same $\dot{\varepsilon}$) due to positive ageing effects developed at the origin in test *c*. Lastly, in test *d*, the development of creep strain totally stops soon after the start of creep loading. After that moment, σ_y^f (which is on the ML actual curve) continuously increases by positive ageing effects. Upon the restart of ML at a high $\dot{\varepsilon}$, a very stiff behaviour appears due to the rigid behaviour of EP1 while σ^f is catching up σ_y^f until the actual ML curve is rejoined (restarts of yielding) (Figure 3-11, bottom). Then, an overshooting in σ takes place, σ becoming larger than $\sigma^f = \sigma_y^f$ due to the viscous property. Subsequently, the stress-strain behaviour tends to join the stress-strain relation for continuous ML at the same $\dot{\varepsilon}$ and for the same total ageing time in test *c*.

Negative ageing: Firstly, the pre-peak stress-strain behaviour during ML in test *b* (at a low $\dot{\varepsilon}$) is less stiff than in test *a* (at a high $\dot{\varepsilon}$) due to smaller viscous stress and larger negative ageing in test *b*. Secondly, the pre-peak stiffness and peak strength in test *c* (at a high $\dot{\varepsilon}$) is consistently lower than in test *a* (at the same $\dot{\varepsilon}$), due to negative ageing effects developed before the start of ML in test *c*. Thirdly, in test *b*, the stress σ exhibits the peak value around $\varepsilon = 2.5\%$, much before the peak strength is reached in the inviscid $\sigma^f - \varepsilon^{ir}$ relation. This is due to that, because of a low strain rate, the decreasing rate of σ by negative ageing effects becomes larger than the increasing rate of σ by an increase in ε^{ir} (i.e., the strain-hardening effects).

Lastly, in test *d*, the development of creep strain does not stop but it is accelerated from some moment after the start of creep loading, ultimately resulting into failure. This is due to that σ_y^f decreases with time due to negative ageing effects during creep loading and, sometime after the start of creep loading, it becomes the same with and then smaller than the applied constant total load σ. This phenomenon is the same as the one that takes place during creep loading at a constant σ that is larger than the inviscid strength in the case of no ageing effects as tests 1-3 illustrated in Figure 3-13. In this illustration, the isotach type viscosity is assumed. In any creep test, $\dot{\varepsilon}$ first becomes smaller with time immediately after the start of creep loading due to a decrease in σ^v with time. In tests 1-3, as σ at the creep loading stage is higher than the inviscid strength, σ_{peak}^f, $\dot{\varepsilon}$ starts increasing after having exhibited the minimum when σ^v is the minimum. Then, $\dot{\varepsilon}$ increases with time due to an increase in σ^v with time. Finally, creep failure takes place in the post-peak regime in the inviscid stress-strain relation. When the creep stress is lower than σ_{peak}^f, creep straining ultimately stops (test *4*). The failure would also be obtained if the stress-strain curve has no peak (no softening behaviour) but with an always decreasing creep strain rate (cf. Figure 3-13).

Another example of positive ageing effect is reported in Figure 3-14a, which shows the results from three CD TC tests on cement-mixed sand (Kongsukprasert *et al.*, 2001). Natural sand from Aomori ($G_s = 2.80$, $U_c = 3.0$ & $D_{max} = 2$ mm) was mixed with ordinary Portland cement in a cement/sand ratio of 4.36% by weight. At water content of about 22.5%, close to the optimum water content, the specimens were compacted to a dry density of 1.23–1.24 g/cm³ and cured under the atmospheric pressure at constant water content for 11 days. The specimens, moist as prepared, were isotropically consolidated at 200 kPa for 20 hours (tests A11APSC and C11APSC) and 92 hours (test Cc11APSC) before the start of ML at $\sigma_3 = 200$ kPa and at $\dot{\varepsilon}_a = 0.03\%$/min. In test A11APSC, the specimen was cured at an anisotropic stress state for 72 hours during otherwise ML drained

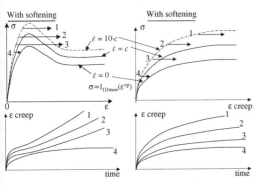

Figure 3-13. Creep tests at different stress levels. Shape of the creep stain versus time for isotach behaviour (cf. Figure 2-22), in case of softening (left) and without softening (right). Creep periods 1, 2 and 3 will result in failure.

triaxial compression at constant $\dot{\varepsilon}_a$. Before the start of creep loading, the measured strength in test A11APSC was slightly smaller than the one in test C11APSC due likely to a slightly smaller dry density. Therefore, in Figure 3-14a, the measured stress values in test A11APSC have been proportionally increased so that the stress values before the start of creep loading in test A11APSC become the same as test C11APSC. The effects of curing at an anisotropic stress state seen in this figure are essentially the same as the one seen in Figure 2-50. Figure 3-14b shows the result from the simulation. For the ageing function shown in Figure 3-14b, it was assumed that $t_c = 0$ when the elapsed curing period became 20 hours at the isotropic consolidation stress state. The TESRA viscosity with $r_1 = 0.001$ (equation 3-29) was assumed. The major trends of observed behaviour are well simulated.

3.2 3D formalism

Formalism of the three-component model in the general three-dimensional case is proposed in the next sections. It is not the intention of the authors to go inside the details in this note. Only the basic equations are given. Thus, they are described only enough to give the ingredients allowing to propose a viscous formulation in this framework.

3.2.1 "EP1" body

As already mentioned in the beginning of chapter 3, the EP1 body of the three-component model (Figure 1-5) can reasonably be considered as hypoelastic for the geomaterials. The reader could refer to previous publication of the authors for more details on the hypoelastic model of the University of Tokyo (Hoque and Tatsuoka, 1998; Tatsuoka et al., 2001a) and the hypoelastic model "DBGS" of ENTPE (Di Benedetto et al.,

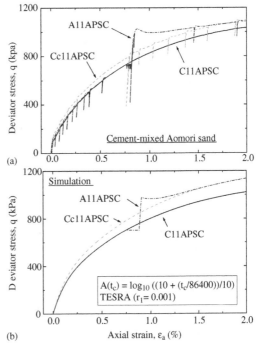

Figure 3-14. a) Result from CD TC tests of cement-mixed sand with drained ageing (Kongsukprasert et al., 2001); b) their simulation (Tatsuoka et al., 2003).

2001a & 2001b; Geoffroy et al., 2003), which are both able to simulate the actual anisotropic elastic behaviour of geomaterial. This part of the simulation is not touched upon in this paper.

3.2.2 "EP2" body

The general behaviour of the EP2 body is described by equation 1-7 (or equation 1-10) if ageing is not considered. Any of the many traditional or more sophisticated non-viscous models, proposed in the literature, can be used to express the M^f tensor. For example the following formalisms, which are also listed in Figure 1-5, are among possible candidates: elasticity, pure plasticity, elastoplasticity, hypoelasticity, hypoplasticity, interpolation type and so on. It is clear that irreversibility and non-linear behaviour must be integrated as these characteristics are strongly present for geomaterials in the common loading domains. For example, Di Benedetto (1987) and Di Benedetto et al. (1999b) proposed an EP2 law of the interpolation type and Tatsuoka and Molenkamp (1983) proposed a constitutive elastoplastic model with 2 plastic mechanisms.

As the viscous properties are focused in this paper, this EP2 formalism is not more described. One important remark is that analysis of the EP2 response

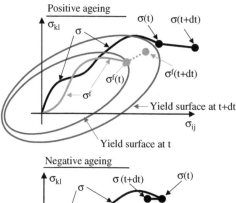

Positive ageing

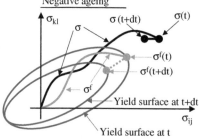

Negative ageing

Figure 3-15. Illustration in the case of elastoplastic model of positive ageing and negative ageing. Due to ageing the yield surface expands (positive ageing) or shrinks (negative ageing), which can result in no plastic strain increment even at increasing stress level (positive ageing) or existence of a plastic strain increment even at decreasing stress level (negative ageing).

needs the determination of the (objective) inviscid stress rate $\dot{\sigma}^f$, which could be very different from the stress rate. This determination can be made by the "mapping rule", which gives the relative evolution of σ^f and σ^v in the stress plane. Di Benedetto (1987) proposed in the case of isotach behaviour that the inviscid stress rate always remains collinear and of the same direction of the viscous stress σ^v (cf. Figure 3-16). With respect to the viscous body V behaviour, the mapping rule is described more in detail in the next section.

One of the specific features of the three-component model formalism with a mapping rule is that the viscoplastic strain evolution is controlled by the instantaneous inviscid stress and not by the instantaneous total stress component as it is the case with the Perzyna over-stress model (Perzyna, 1963) and the other classical visco-plastic models. Consequences of this difference could be rather subtle while this point is delicate to validate experimentally.

As explained before, the introduction of the ageing properties in the EP2 body needs the time t_c as memory parameter. In the particular case of elastoplasticity theory, a classical way to proceed is to make the yield surface changing not only with the stress path

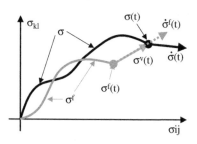

Figure 3-16. Illustration of the mapping rule in the case of isotach behaviour (cf. Figure 2-22). The equality: $\sigma = \sigma^f + \sigma^v$, is always verified.

but also with the time ($t = t_c$). In case of positive aging, the size of the yield surface increases with time and it decreases in case of negative ageing. Then, if a creep period is applied for a stress state initially on the yield surface, the consequences are as follows:

– In case of *positive ageing*, as the yield surface expands, the material remains at a constant irreversible (or viscoplastic) strain but enters progressively inside the elastic domain. Then, a large elastic zone can exist during reloading.

– In case of *negative ageing*, the shrinkage of the yield surface, on which must remain the inviscid stress, creates an irreversible strain and the material shows strain evolution at constant stress value. This evolution is independent of any viscous effect and the "creep" phenomenon is due to ageing.

Explanation on the previous behaviour is presented in Figure 3-15.

Such a formulation in the one-dimensional case is given in section 3.1.2.2 (equation 3-9 with $F^f_0 = 0$) and examples of simulations for positive and negative ageing are given in Figure 3-11 and Figure 3-12.

3.2.3 "V" body

3.2.3.1 Isotach case

The mapping rule proposed by Di Benedetto (1987) is presented in Figure 3-16 and expressed by:

$$\overline{\dot{\sigma}^f} = \lambda \overline{\sigma^v} \qquad (3\text{-}35)$$

where λ is a positive constant ($\lambda > 0$) and $\overline{A}$ denotes that A is a tensor. This new notation is used from this section to facilitate the understanding of the equations.

When considering equations 3-35 and 1-7, only one extra scalar equation is needed to obtain the viscoplastic strain rate and the coefficient λ. This equation is given by the V body properties by expressing a relation between the norm of the inviscid stress $\|\overline{\sigma^v}\|$ and the norm of the viscoplastic strain rate $\|\overline{\dot{\varepsilon}^{vp}}\|$. As this equation is scalar, all the developments presented

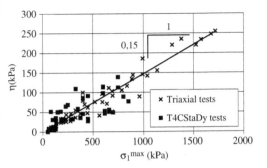

Figure 3-17. Values of η (equation 3-38) versus σ_{1max} for creep periods, applied with Hollow cylinder T4CstaDy tests (Sauzéat et al., 2003; Sauzéat, 2003) and triaxial tests (Pham Van Bang et al., 2003) on dry Hostun sand. More than 100 periods are considered at different stress levels with and without rotation of axes.

in the one-dimension case can be directly considered when replacing the tensor variables by their norm (or other chosen invariants). With these considerations, equation 1-8 can be rewritten as:

$$\left\|\overline{\sigma}^v\right\| = f_{isot}(h^v, \left\|\overline{\varepsilon}^{vp}\right\|) \tag{3-36}$$

where the memory parameters h^v (which could be tensor) are taken from the set of parameter h^f (of the EP2 body) to respect the isotach property, as explained in the one-dimensional case.

As the general case is treated by a scalar equation (equation 3-36) the expressions proposed in the one dimension case (cf. section 3.1.3) can be directly used. In particular equations 3-22 and 3-24 can be rewritten as:

$$\left\|\overline{\sigma}^v\right\| = \alpha * \sigma_{1max}\left(\frac{\left\|\dot{\overline{\varepsilon}}^{vp}\right\|}{\dot{\varepsilon}_0^{vp}}\right)^{1+b} \tag{3-37}$$

where only two constants $\alpha*$ and b have to be determined. In the one dimension case, the parameter b was found equal to -0.95 on a large series of creep periods performed on CT tests and hollow cylinder tests (T4Cstady) on air dried Hostun sand at different stress levels (cf. for example Figure 3-4). Figure 3-17 represents a plot of experimental values of the ratio η (equation 3-38) obtained with $b = -0.95$ by the best fitting of these creep periods with and without rotation of axes. The linearity between η and σ_{1max} seems well respected for these more than 100 creep periods, as can be also deduced from the plot in the axes η versus σ_{1max}, and the coefficient $\alpha*$ is found equal to 0,15. It has to be underlined that the equation 3-37

has been validated on Hostun sand, with the same values of the constant b and η for relaxation periods (cf. Figure 2-46) and for loadings with stepwise strain rate changes (cf section 21124). This validation on rather different paths involving viscous effect is encouraging. Meanwhile the introduction of variable σ_{1max} is probably a first step, which could be improved or corrected with more experimental data, in particular on large cycle loadings and on general p–q paths during triaxial tests.

$$\eta = \left\|\overline{\sigma}^v\right\| / \left(\frac{\left\|\dot{\overline{\varepsilon}}^{vp}\right\|}{\dot{\varepsilon}_0^{vp}}\right)^{1+b} \tag{3-38}$$

As can be easily concluded, set of experimental data focusing on viscous effects are still missing to improve modelling.

3.2.3.2 Case with viscous evanescent properties

Considering the statements of the previous paragraph, a formalism including viscous evanescent properties (cf. Figure 2-22) can be deduced from the one dimensional case developments. Equation 3-28 is generalised as:

$$\overline{\sigma}^v_{(t)} = f_{isot}(\left\|\overline{\varepsilon}^{vp}_{(t)}\right\|) \cdot \overline{u}^{fd}_{(t)}$$

$$+ \int_{\chi=0}^{t} f_{isot}(\left\|\overline{\varepsilon}^{vp}_{(\chi)}\right\|) \cdot g'_{decay}(\int_{\xi=\chi}^{t} d\left\|\overline{\varepsilon}^{vp}_{(\xi)}\right\|) \cdot \overline{u}^{fd}_{(\chi)} \cdot d\left\|\overline{\varepsilon}^{vp}_{(\chi)}\right\| \tag{3-39}$$

where the parameters h^v have been omitted in the function f_{isot} (cf. equation 3-36) to clarify the writing. χ is an intermediate time for the integration. The g_{decay} function is the same as the one introduced in the one dimensional case. Equation 3-29 has been chosen for different geomaterials. The integral term in the g_{decay} function corresponds to the length of the viscoplastic strain path. $\overline{u}^{ft}$ and $\overline{u}^v$ are the direction of the inviscid stress (objective) rate and of the viscous stress (equations 3-40 and 3-41):

$$\overline{u}^{fd}_{(t)} = dir(\dot{\overline{\sigma}}^f_{(t)}) = \dot{\overline{\sigma}}^f_{(t)} / \left\|\dot{\overline{\sigma}}^f_{(t)}\right\| \tag{3-40}$$

$$\overline{u}^v_{(t)} = dir(\overline{\sigma}^v_{(t)}) = \overline{\sigma}^v_{(t)} / \left\|\overline{\sigma}^v_{(t)}\right\| \tag{3-41}$$

It can be seen from equation 3-39 that the direction of the stress rate is given by the following expression

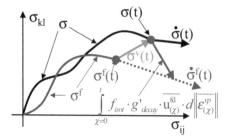

Figure 3-18. Illustration of the mapping rule in the case of pure viscous evanescent and intermediate behaviours (cf. Figure 2-22). The equality: $\sigma = \sigma^f + \sigma^v$, is always verified.

which takes into account the decay property (evanescent behaviour):

$$\overline{u^{fd}_{(t)}} = dir\{\overline{\sigma^v}_{(t)}\}$$

$$-\int_{\chi=0}^{t} f_{isot}\left(\left\|\overline{\varepsilon^{yp}_{(\chi)}}\right\|\right)\cdot g'_{decay}\left(\int_{\xi=\chi}^{t} d\left\|\overline{\varepsilon^{yp}_{(\xi)}}\right\|\right)\cdot\overline{u^{fd}_{(\chi)}}\cdot d\left\|\overline{\varepsilon^{yp}_{(\chi)}}\right\|\} \quad (3\text{-}42)$$

Equation 3-42 expresses the mapping rule in the case of viscous evanescent behaviour. Figure 3-18 illustrates this mapping rule in the viscous evanescent or intermediate cases (cf. Figure 2-22). One cane see that if $g_{decay} = 1$, the isotach case is obtained.

Expression 3-39 can be added to any EP2 formalism and the analytical expressions given in the one dimension case can be adopted. Then there is no great difficulty to make a constitutive model including viscous evanescent behaviour. Meanwhile the integration of this model for soil mechanics construction calculations remains a very open topic. An examples of such simulation with a finite element method code for a two dimension case, is proposed in the section 4.

3.2.3.3 Introduction of aeging

Introduction of ageing effect in the three dimension case for the V body can be made in a similar way that in the one dimension case (cf. sections 3.1.3.2 and 3.1.3.3). The time t_c (cf. section1.2) is, of course, added among the memory parameters.

Considering the great lack of experimental data on that complicated topic involving complex physical phenomena, no more development is given.

4 EXAMPLES OF FEM SIMULATION

Siddiquee et al. (2003a & 2003b) shows that the three-component model described in this paper can be smoothly incorporated into an exiting elasto-plastic FEM code that has been validated by a successful simulation of the footing model test ignoring the viscous effects (Siddiquee et al., 1999, 2001a). For the first

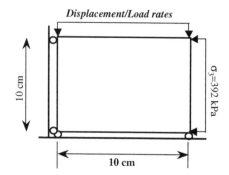

Figure 4-1. Geometric definitions of the problem for the FEM analysis enriched with the TESRA model (Siddiquee et al., 2003a & 2003b).

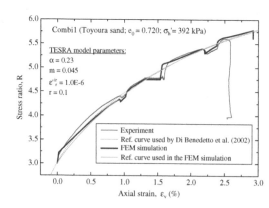

Figure 4-2. Measured stress-strain relation from a drained PSC test (Combi1) on water-saturated Toyoura sand and its FEM simulation by using the TESRA model. (Siddiquee et al., 2003a & 2003b).

step, numerical simulations using a single element shown in Figure 4-1 was performed. In the simulation, the reference stress-strain curve was first simulated by some trial and error by the inviscid FEM analysis to be used in the subsequent FEM analysis that takes into account the viscous effects. The FEM analysis was made at a very low strain rate equal to 10^{-6}%/second. The elasto-plastic model is described in details in Siddiquee et al. (1999, 2001a). It is a single yielding model using shear yield loci, proportional to the failure plane. Compression by an increase in the mean principal stress was assumed to be "elastic" using the equivalent modulus that could simulate the actual isotropic compression properties of sand. It is assumed that the flow properties are controlled by the instantaneous inviscid stress state through a potential function described in terms of inviscid stresses. The strain localisation is modelled by introducing a kind of smeared method, by which the effects of shear band width (i.e., the effects of particle size) can be

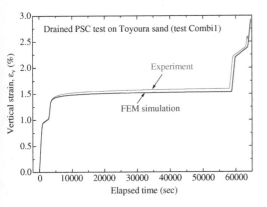

Figure 4-3. Measured time history of axial strain in PSC test "Combi1" and its FEM simulation (Siddiquee *et al.*, 2003a & 2003b).

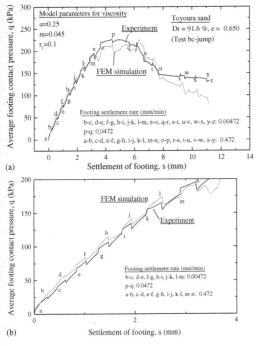

(a)

(b)

Figure 4-4. a) Comparison of model footing test and its FEM simulation incorporating the TESRA viscosity, and b) Zoom-up of the initial load and settlement relation (Siddiquee 2003, unpublished data).

taken into acccount. The viscous properties were modelled to have the TESRA viscosity (cf. section 3.1.3.3) described by Di Benedetto *et al.* (2002) and Tatsuoka *et al.* (2002) for plane strain compression at a fixed confining pressure and Nawir *et al.* (2003a & 2003b) for more general stress conditions (i.e., the viscous

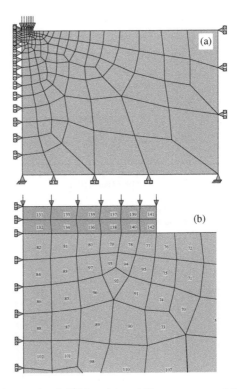

Figure 4-5. a) FEM model; and b) zoom-up around the footing (Siddiquee 2003, unpublished data).

stress component formulated in terms of the principal stress ratio), as described in Siddiquee *et al.* (2001b). Figure 4-2 compares measured stress-strain relation from a drained PSC test (Combi1) on water-saturated Toyoura sand and its FEM simulation by using the TESRA model (Siddiquee *et al.*, 2003a & 2003b). The TESRA model parameters used in the FEM simulation were $\alpha = 0.23$, $m = 0.045$, $\dot{\varepsilon}_r^{ir} = 10^{-6}\%/sec$ and $r_1 = 0.1$, which are the same as those used by the direct TESRA model simulation reported by Di Benedetto et al. (2002) and Tatsuoka et al. (2002a) (shown in Figure 4-2). The time history of axial strain obtained from the FEM simulation is compared with measured one in Figure 4-3. The following trends of behaviour may be seen from these figures:

1) The strain rate increased stepwise at the start of drained PSC loading and gradually increased and decreased at a constant rate two times during otherwise monotonic loading at a constant strain rate. The loading rate effects associated with these strain rate histories are well simulated.
2) The model parameters were determined from the behaviour upon step changes in the strain rate. Despite the above, the time histories of creep strain are well simulated.

113

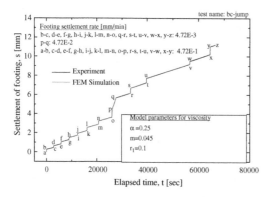

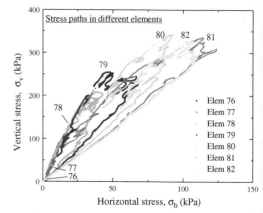

Figure 4-6 Measured and simulated time histories of footing settlement (Siddiquee 2003, unpublished data).

Figure 4-8. Stress paths in the elements immediately below the footing from FEM simulation (Siddiquee 2003, unpublished data).

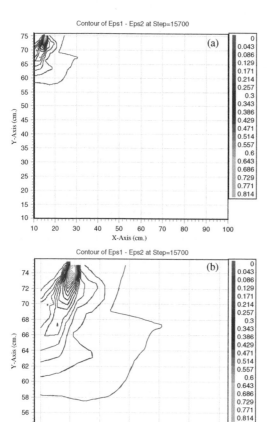

Figure 4-7. a) Distribution of maximum shear strain at the end of loading from FEM simulation; and b) zoom up of figure a) (Siddiquee 2003, unpublished data).

Then, the results from model footing tests on air-dried Toyoura sand, in which the footing settlement rate was stepwise changed many times during otherwise monotonic loading at a constant footing settlement rate, described in Figure 2-64 (Section2.2.1), was simulated by the FEM analysis described above (Siddiquee, 2003; Figure 4-4a) compares the relationship between the footing average pressure and the footing settlement obtained from the model footing test and its FEM simulation, incorporating the TESRA viscosity (Siddiquee, 2003a & 2003b). Figure 4-4b shows the zoom-up of the initial part of the load and settlement relation presented in Figure 4-4a. Figure 4-5a and b show the FEM model.

Figure 4-6 shows the measured time history of footing settlement, which was followed up in the FEM simulation. It may be seen that the measured behaviour, which exhibits obviously viscous properties, is well simulated, in particular the behaviour before the footing load becomes the peak value, when the shear bands have not largely developed in the ground. It can also be seen that the simulation after the peak footing load is less satisfactory, which would be due to that the viscous effects on the shear banding process is not well modelled in this simulation. Figure 4-7 shows the distribution of maximum shear strain at the end of loading obtained from FEM simulation. It may be seen that the failure of ground is highly progressive associated with the development of shear bands. Figure 4-8 shows the stress paths in the elements immediately below the footing from FEM simulation. It may be seen that the stress paths is not smooth due to the viscous effects upon step changes in the footing settlement rate.

These results from the FEM analysis show that the TESRA (or VE) viscosity can be smoothly incorporated to any classical elasto-plastic model to FEM analysis of boundary value problems as described above.

114

5 CONCLUSION

Time effects on a wide range of geomaterials, including sands, reconstituted and undisturbed clays, gravels, natural soils, crushed concretes, soil-cement mixtures, bituminous materials, …, were specifically analysed from extensive experimental campaigns performed on specifically designed devices at University of Tokyo and ENTPE. The great amount of considered data includes results from triaxial tests (two and three dimensional), plane strain compression tests, one-dimensional loading tests and torsional tests. It is completed by results from model and full scale tests. The main conclusions with respect to the time effects are the following:

1) Among time effects, one should clearly distinguish between, i) the viscous or loading rate effect and, ii) the ageing effects. The paper is focused on these two effects, for which clear definition is proposed.
2) All the studied geomaterials exhibit viscous effects. These effects can be observed during stepwise changes of the loading rate or creep or relaxation periods, among others type of path histories. These effects are considered at a macro level. The mechanism, which is at the origin of this behaviour at the particle interfaces for certain types of materials, such as dry sand, still remains to be studied.
3) A consequence of the viscous properties is that a state on the stress-strain monotonic loading curve is not "stable". When loading is stopped, the stress or/and strain does not remains fixed and constant.
4) Experimental evidences show that the viscous effects are not necessarily linked to the presence of water. Then, these effects should be disconnected to the any pore water pressure evolution. In particular the "consolidation" process, which is not due to the skeleton viscous effect, is not treated in the paper.
5) The general influence of the viscous effects on the strain-stress and strain-strain curves, even if very different in magnitude, is rather similar following the considered geomaterials. Two extreme cases could be considered, i) the classical so called "isotach" and ii) a new specifically identified type called "Viscous Evanescent" or "Temporary Strain Rate Acceleration". This second type is clearly obtained for clean sand but is also present for gravels, stiff clays, soft rocks, among others.
6) The newly defined "Rate Sensitivity Coefficient", β, reveals to be a relevant indicator for the characterisation of viscous behaviour sensitivity. Surprisingly this coefficient has relatively small range when considering the wide range of considered geomaterials and testing conditions (see Figure 2-36 to Figure 2-40). It varies from 0.02 for clean dry sand to 0.08 for reconstituted saturated clay.
7) Ageing effects are often coupled with viscous effects. In some cases, corresponding to very different time scales, ageing effects and viscous effects can be decoupled and treated independently. The modelling of ageing effects needs to introduce a time "t_c" having the origin at a physical event of the material.

Concerning modelling, the high ability of the three-component model (see Figure 3-1) to model the behaviour of all the considered geomaterials could be confirmed in the one-dimensional case. The versatility and the potential of such formalism should be underlined. It can be used to simulate very simple behaviour, such as linear viscoelasticity or elasticity, or more sophisticated one including non-linearities and ageing. As any existing non-viscous (or elastoplastic) model can be used for EP1 and EP2 bodies, the input of the results presented herein concerns the body "V", which take into account the viscous effects. In order to model the very specific "Viscous Evanescent" type behaviour, the authors propose a new type of law (VE and TESRA). These new models include a decay of strain rate effects with strain. A three-dimensional expression of the models is also presented.

Some examples of FEM simulation, of model test results exhibiting viscous effects of the evanescent type, are also presented. It appears that the three-component formalism, that has been developed and presented in this paper, can be integrated into any existing elastoplastic FEM code. This last result is of practical importance.

AKNOWLEDGEMENTS

The authors greatly acknowledge the financial support provided by the Japanese Society for the Promotion of Sciences, the "Centre National de la Recherche Scientifique" (CNRS, France) and the Ministry of Education, Science, and Sport from Japan. The funds allow establishing a fruitful cooperation on the topics between the University of Tokyo and ENTPE. All the present and previous colleagues at University of Tokyo and ENTPE helping for this work are also gratefully acknowledged.

APPENDIXES

Appendix 1: Definition of ageing effects
Suppose that a material has some previous strain and stress history until the time t' and it subsequently remains at rest ($\sigma = 0$) until time $t = t''$, which is at a stage sufficiently later than time t' to allow $\dot{\varepsilon} = 0$ condition. Suppose that the response of the material is $R(L)$ for a newly given loading history $L(\Delta t)$ applied for a period from $t = t''$ to $t = t'' + \Delta t$.
For time $t = t^*$, which is larger than time t'';

115

i) in case the response of the material is the same as $R(L)$ for the same loading history $L(\Delta t)$ applied for a period from $t = t^*$ to $t = t^* + \Delta t$, **there is no ageing effects** for the period between t'' and t*: and

ii) in case the response of the material is not the same as $R(L)$ for the same loading history $L(\Delta t)$ applied for a period from $t = t^*$ to $t = t^* + \Delta t$, **there is ageing effects**.

Appendix 2: List of rate sensitivity coefficient βvalues from drained loading tests with step changes in the strain rate

Material	Grading properties	Wet condition	Density [1]	Test method	Range of vertical strain rate $\dot{\varepsilon}_v$ (%/min)	β	Ref.
Original Chiba gravel (moist compacted)	Crushed sandstone: $D_{50} = 7.8$ mm, $D_{max} = 39.6$ mm, $U_c = 11.2$ $G_s = 2.71$	Moist ($w = 5\%$; $S_r = 77$)	Dense, $e_c = 0.19$ (two specimens)	Drained TC at $\sigma'_h = 490$ kPa	0.0006–0.06%/min.	0.0335	Anh Dan et al. (2004)
Model Chiba gravel (air-dried compacted)	Crushed sandstone: $D_{50} = 0.8$ mm, $U_c = 2.1$ $D_{max} = 5.0$ mm,	Air-dried	$e_c = 0.584$ & 0.556 ($\rho_d = 1.760$ & 1.770 g/cm^3)	Drained TC at $\sigma_h' = 40$ kPa	0.008– 0.00008%/m	0.0244	Hirakawa (2003)
	$G_s = 2.74$, $e_{max} = 0.727$, $e_{min} = 0.363$		$e_o = 0.456$ & 0.403 ($\rho_d = 1.820$ &1.889 g/cm^3)	Oedometer; $\sigma'_v = $ up to 7,405 kPa	0.00036– 0.026%/min	0.0380	
Hostun sand (air-pluviated)	Quartz-rich sub-angular: $D_{50} = 0.31$ mm, $U_c = 1.94$, $G_s = 2.65$, $e_{max} = 0.95$, $e_{min} = 0.55$	Air-dried	$e_c = 0.698$ & 0.700: $\rho_d = 1.760$– 1.770 g/cm^3	Drained PSC at $\sigma_h' = 392$ kPa	0.00025– 0.25%/ min	0.0219	Matsushita et al. (1999); Di Benedetto et al. (2002)
		Air-dried	$e_c = 0.72$ & 0.93	Drained TC at $\sigma'_h = 80$, 200% 400 kPa	0.006– 0.6%/min	0.024	Phan Van Bang et al. (2003)
Toyoura sand (air-pluviated)	Quartz-rich sub-angular: $D_{50} = 0.18$ mm, $U_c = 1.64$, $G_s = 2.65$, $e_{max} = 0.99$, $e_{min} = 0.62$	Saturated	$e_c = 0.658$	Drained PSC at $\sigma'_h = 400$ kPa	0.000108– 1.08%/min	0.0219	Matsushita et al. (1999)
		Air-dried	$e_c = 0.674$ & 0.673	Drained PSC at $\sigma'_h = 30$ & 80 kPa	0.004– 0.4%/min	0.0207 & 0.0226	Kongkitkul & Tatsuoka (2003)
		Air-dried	$e_c = 0.633$– 0.649	Drained TC at $\sigma'_h = 40$ kPa	0.00008– 0.008%/min	0.0269	Hirakawa (2003)
		Saturated	$e_c = 0.658$	Drained TC at $\sigma'_h = 200$ kPa	0.00108– 1.077%/min	0.0205	Matsushita et al. (1999)
		Saturated & air-dried	$e_c = 0.605$ –0.925	Drained TC at $\sigma'_h = 100$– 600 kPa	0.0008– 0.12%/min	0.0242	Nawir et al. (2003a)
		Air-dried	$e_c = 0.623$ –0.631	Oedometer, $\sigma'_v = $ up to 7,375 kPa	0.00036– 0.026%/min	0.0292	Hirakawa (2003)

Material	Grading properties	Wet condition	Density [1]	Test method	Range of vertical strain rate $\dot{\varepsilon}_v$ (%/min)	β	Ref
Silica sand No. 8 (air-pluviated)	Quartz-rich sub-angular: $D_{50} = 0.077$ mm, $U_c = 2.43$ $G_s = 2.655$, $e_{max} = 1.335$, $e_{min} = 0.73$	Saturated	$e_c = 1.135$	Drained TC at $\sigma_h' = 400$ kPa	0.00125– 0.25%/min	0.0284	Kiyota (2004)
Jamuna river sand (air-pluviated)	*Containing 5% (by volume) of mica: $D_{50} =$ 0.16, $FC = 7\%$;* $G_s = 2.7$, $e_{max} = 1.173$, $e_{min} = 0.690$	Air-dried	$e_0 = 0.775$ & 0.821	Drained PSC at $\sigma_h' = 100$ & 400 kPa	0.00125– 0.125%/min.	0.0273	Yasin & Tatsuoka (2003)
Crushed concrete	$G_s = 2.65$, $D_{max} = 19$ mm, $D_{50} = 5.84$ mm, $U_c = 18.76$, $FC = 1.32\%$	Moist (w-16.9%, nearly the optimum water content)	$\rho_d = 1.75$ g/cm^3	Drained TC $\sigma_h' = 20$ kPa	0.001– 0.1%/min	0.0536	Umair et al. (2004)
Fujinomori clay	$D_{50} = 0.017$ mm, $U_c \cong 10$, PI = 33, LL = 62%	Saturated	$e_c = 1.53$	Drained TC at $\sigma_h' = 200$ kPa	0.0003– 0.3%/min	0.0525	
		Air-dried ($w_{af} = 4.29\%$; $S_{r.af} = 8.09\%$)	$e_c = 1.093$	Drained TC at $\sigma_h' = 77$ kPa	0.002– 0.2%/min	0.0444	
		Air-dried ($w_{af} = 2.56\%$; $S_{r.af} = 4.68\%$)	$e_c = 1.202$	Drained TC at $\sigma_h' = 80$ kPa	0.003– 0.3%/min	0.0353	
		Saturated	$e_0 = 1.281$	Oedometer tests, $\sigma_v' =$ up to 1,034 kPa	0.0097– 0.97% /min	0.0571	
		Saturated	$e_0 = 1.228$	Oedometer tests, $\sigma_v' =$ up to 1,153 kPa	0.0066– 0.66% /min	0.0545	Li et al. (2004)
		Wet ($w_{af} = 8.54\%$; $S_{r.af} = 16.81\%$, measured after the test)	$e_0 = 1.126$	Oedometer tests, $\sigma_v' =$ up to 1,334 kPa	0.0049– 0.735% /min	0.0437	
		Air-dried ($w_{af} = 2.87\%$; $S_{r.af} = 4.98\%$)	$e_0 = 1.364$	Oedometer tests, $\sigma_v' =$ up to 1,158 kPa	0.0078– 0.78% /min	0.0457	
		Air-dried ($w_{af} = 2.99\%$; $S_{r.af} = 4.92\%$)	$e_0 = 1.548$	Oedometer tests, $\sigma_v' =$ up to 841 kPa	0.016 –0.16% /min	0.0374	
		Air-dried ($w_{af} = 3.00\%$; $S_{r.af} = 6.17\%$)	$e_0 = 1.29$	Oedometer tests, $\sigma_v' =$	0.000026– 0.0038% /min	0.0466	

Material	Grading properties	Wet condition	Density [1]	Test method	Range of vertical strain rate $\dot{\varepsilon}_v$ (%/min)	β	Ref.
				upto 1,000 kPa			
		Air-dried (w_{af} = 3.2%; $S_{r.af}$ = 4.92%)	e_0 =	Oedometer tests, σ_v' = up to 1,000 kPa	0.000025 −0.0038% /min	0.0437	
		Oven-dried (w_{af} = 0.10%; S_r = 0.21%)	e_0 = 0.993	Oedometer tests, σ_v'' = up to 1,445 kPa	0.0082 −0.82%/min	0.0274	Li et al. (2004)
		Oven-dried (w_{af} = 0.61%; $S_{r.af}$ = 1.15%)	e_0 = 1.115	Oedometer tests, σ_v' = up to 1,523 kPa	0.0082 −0.82% /min	0.0286	
Kaolin	D_{50} = 0.0013 mm, PI = 41.6, LL = 79.6%	Air-dried (w_{af} = 0.32%; $S_{r.af}$ = 0.58%)	e_0 = 1.53	Oedometer tests, σ_v' = up to 1,000 kPa	0.0000455 −0.00446% /min	0.0355	
		Air-dried (w_{af} = 0.27%; $S_{r.af}$ = 0.45%)	e_0 = 1.68	Oedometer tests, σ_v' = up to 1,000 kPa	0.0000455 −0.00287 %/min	0.0340	Deng et al. (2004)
Kitan clay (undisturbed)	D_{50} = 0.032 mm, $U_c \cong$ Undefined, PI = 21.6, LL = 31.9%	Saturated	e_c = 1.03	Drained TC at σ_h' = 340 kPa	0.00380 −0.38% /min	.0336	
	D_{50} = 0.0034 mm, $U_c \cong$ Undefined, PI = 41.1, LL = 54.7%		e_c = 0.749	Drained TC at σ_h' = 340 kPa	0.0038 −0.38%/min	0.0338	
	D_{50} = 0.0040 mm, $U_c \cong$ Undefined, PI = 62.1, LL = 86.1%		e_c = 0.736	Drained TC at σ_h' = 470 kPa	0.00038 −0.038% /min	0.0295	Komoto et al. (2003)
	D_{50} = 0.0064 mm, $U_c \cong$ Undefined, PI = 45.7, LL = 65.3%		e_c = 0.607	Drained TC at σ_h' = 470 kPa	0.0004 −0.04% /min	0.0239	
	D_{50} = about 0.008 mm,		e_0 = 0.593	Oedometer tests, σ_h' = upto 5,846 kPa	0.014 −0.28% /min	0.0879	Acosta-Martínez et al., (2003)
Kitan clay (remould)		Saturated	e_c = 0.821	Drained TC at σ_h' = 340 kPa	0.0021 −0.21% /min	0.0553	
			e_c = 0.653	Drained TC at σ_h' = 340 kPa	0.00088 −0.044% /min	0.0448	Komoto et al. (2003)

Material	Grading properties	Wet condition	Density [1]	Test method	Range of vertical strain rate $\dot{\varepsilon}_v$ (%/min)	β	Ref.
			$e_c = 0.682$	Drained TC at $\sigma_h' =$ 470 kPa	0.0021 −0.21% /min	0.0744	
	$D_{50} =$ 0.0064 mm, $U_c \cong$ Undefined, PI = 45.7, LL = 65.3%		$e_c = 0.792$	Drained TC at $\sigma_h' =$ 470 kPa	0.00210 −0.21% /min	0.0806	
Ohimachi clay (undisturbed)	$D_{50} = 0.024$ mm, $U_c = 16.9$, PI = 19.9, LL = 42.9%	Saturated	$e_c = 1.050$	Drained TC at $\sigma_h' = 250$ kPa	0.0004 −0.04%/min	0.0354	Komoto et al. (2003)
			$e_0 = 0.957$	Oedometer tests, $\sigma_h' =$ up to 1,588 kPa	0.0018 −0.18% /min	0.0573	Li et al., (2004)
Ohimachi clay (remoulded)	$D_{50} = 0.024$ mm, $U_c = 16.9$, PI = 19.9, LL = 42.9%	Saturated	$e_c = 0.920$	Drained TC at $\sigma_h' =$ 250 kPa	0.00043 −0.043% /min	0.0457	Komoto et al. (2003)
			$e_0 = 1.046$	Oedometer tests, $\sigma'_v = 995$ kPa	0.12 −0.3%/min	0.0685	Acosta-Martínez et al., (2003)
Pisa clay (undisturbed)	$D_{50} = 0.0011$ mm, PI = 42.91,	Saturated	$e_0 = 0.672$	Oedometer tests; $\sigma'_h =$ up to 1,600 kPa	0.00135– 0.081% /min	0.0789	Li et al. (2004)
Pisa clay (remouded)	Liquid limit = 77.16%	Wet (w_{af} = 10.29%; $S_{r.af}$ = 22.0%)	$e_0 = 0.880$		0.00948 −0.853% /min	0.0606	Tatsuoka (2004)
		Air-dried (w_{af} = 4.03%; $S_{r.af}$ = 9.76%)	$e_0 = 0.700$		0.00884 −0.751% /min	0.0442	
		Oven-dried (w_{af} = 1.07%; $S_{r.af}$ = 2.53%)	$e_0 = 0.735$		0.00884 −0.751% /min	0.0402	

1) e_0 = initial void ratio at the start of oedometer test; and e_c = consolidated void ratio.
2) w_{af} and $S_{r.af}$; measured after each test.

REFERENCES

Abrantes, A.E. and Yamamuro, J.A. (2003): "Effect of strain rate in cohesionless soil, Constitutive modeling of geomaterials Selected contribution from the F.L". DiMaggio Symposium, Eds. Ling et al., CRC Press.

Acosta-Martinez, H.E., Tatsuoka, F. and Li, J.-Z. (2003): "Viscosity in one-dimensional deformation of clay for its modelling", *Proc. 38th Japan National Conf. on Geotechnical Engineering*, JGS, Akita, June.

Adachi, T. and Oka, F. (1982): "Constitutive equations for normally consolidated clays based on elasto-viscoplasticity". *Soils and Foundations*, Vol.22, No.1, pp.57–70.

Adachi, T. and Oka, F. (1995): "An elastoplastic constitutive model for soft rock wit strain softening". *Int. J. Nume. and Anal. Mech. on Geomechanics*, Vol.19, pp.233–247.

Adachi, T., Oka, F. and Mimura, M. (1996): "Modeling aspects associated with time dependent behavior of soils", S-O-A Report, Measuring and Modeling Time

Dependent Soil Behavior, *ASCE Geotech. Special Publication*, Vol.61, pp.61–95.

Akai, K., Adachi, T. and Ando, N. (1975): "Existence of a unique stress-strain-time relation of clays". *Soils and Foundations*, Vol.15, No.1, pp. 1–16.

Anh Dan, L.Q., Koseki, J. and Tatsuoka, F. (2001): "Viscous deformation in triaxial compression of dense well-graded gravels and its model simulation", *Advanced Laboratory Stress-Strain Testing of Geomaterials* (Tatsuoka et al. eds.), Balkema, pp.187–194.

Anh Dan, L.Q., Tatsuoka, F. and Koseki, J. (2004): "Viscous shear stress-strain characteristics of dense gravel in triaxial compression," *Geotechnical Testing Journal* (to be published).

Airey, G.D., Rahimzadeh, B. and Collop, A.C. (2003): "Viscoelastic linearity limits for bituminous materials". *6th international RILEM Symposium on Performance Testing and Evaluation of Bituminous Materials*, Zurich, April 2003.

Balakrishnaier, K., Koseki, J. (2001): "Modeling of stress and strain relationship of dense gravel under large cyclic loadings", *Advanced Laboratory Stress-Strain Testing of Geomaterials* (Tatsuoka et al. eds.), Balkema, pp.195–207.

Barbosa-Cruz, E.R. and Tatsuoka, F. (1999): "Effects of stress state during curing on stress-strain behaviour of cement-mixed sand", *Proc. Second Int. Conf. on Pre-Failure Deformation Characteristics of Geomaterials, IS Torino '99* (Jamiolkowski et al., eds.), Balkema, Vol.1, pp.509–516.

Cazacliu, B. (1996): "Comportement des sables en petites et moyennes déformations; prototype d'essai de torsion compression confinement sur cylindre creux", Thèse de Doctorat, ECP-ENTPE (in French)

Cazacliu, B. and Di Benedetto, H. (1998): "Nouvel essai sur cylyndre creux de sable". *Revue Française de Génie Civil*, Hermès, Vol.2, n°7, pp. 827–55. (in French)

Cristescu, N.D. and Hunsche, U. (1998): *Time effects in rock mechanics*. John Wiley and Sons.

Darve, F. (1978): "Une formulation incrémentale des lois rhéologiques. Application aux sols", Thèse d'Etat, IM Grenoble. (in French)

Deng, J.-L., Tatsuoka, F., Komoto, N. and Koseki, J. (2004): "Ageing and viscous effects on the deformation of clay in 1D compression", *Proc. 39th Japan National Conference on Geotechnical Engineering*, Japanese Geotechnical Society, Niigata.

Di Benedetto, H. and Darve, F. (1983): "Comparaison des lois rhéologiques en cinématique rotationnelle", *Journal de Mécanique théorique et appliquée*, vol. 2, n° 5, pp. 769–798, (in French).

Di Benedetto, H. (1987): "Modélisation du comportement des géomatériaux: application aux enrobés bitumineux et aux bitumes", Thèse de Docteur d'Etat ès Sciences, INPG-USTMG-ENTPE, p.310 (in French).

Di Benedetto, H. and Hameury, O. (1991): "Constitutive law for granular skeleton materials: description of the anisotropic and viscous effects". *Comp. Met. and Ad. In Geomec* (Beer et al. eds.), Rotterdam, Balkema, pp.599–603.

Di Benedetto, H. (1997): "Effets visqueux et anisotropie des sables" *Session de Discu. 1.1 of XIVth Int. Conf. ISSMFE (1997)*, Vol. 4 (1998), Ed. Balkema, Hamburg, pp 2177–79, (in French).

Di Benedetto, H. and Tatsuoka, F. (1997): "Small strain behaviour of geomaterials: Modelling of strain rate effects", *Soils and Foundations*, Vol.37, No.2, pp.127–138.

Di Benedetto, H., De la Roche, C., (1998): "State of the art on stiffness modulus and fatigue of bituminous mixtures", *Bituminous binders and mixtures: state of the art and interlaboratory tests on mechanical behavior and mix design*, E&FN Spon, Ed. L. Francken, pp 137–180.

Di Benedetto, H., Ibraim, E. and Cazacliu, C. (1999a): "Time dependent behavior of sand" ", *Proc. 2nd Int. Symp. on Pre-failure Deformation Characteristics of Geomaterials, IS Torino* (Jamiolkowski et al. eds.), Vol.1, Balkema, pp. 459–466.

Di Benedetto, H., Sauzeat, C. and Geoffroy, H. (1999b): "Modelling viscous effects for sand and behaviour in the small strain domain", *Proc. 2nd Int. Symp. on Pre-failure Deformation Characteristics of Geomaterials, IS Torino* (Jamiolkowski et al. eds.), panel presentation, Vol.2 (published 2001), Balkema, pp. 1357–1367.

Di Benedetto, H. and Neifar, M. (2000): "Loi thermo-viscoplastique pour les enrobés bitumineux". *2nd Eurasphalt & Eurobitume congress*, Barçelone 2000, 9 p. [In French].

Di Benedetto, H., Geoffroy, H. and Sauzeat, C. (2001a): "Hollow cylinder test and modelling of pre-failure behavior of sands", *Int. Conf. Albert Caquot*, Paris, p. 8.

Di Benedetto, H., Geoffroy, H. and Sauzeat, C. (2001b): "Viscous and non viscous behaviour of sand obtained from hollow cylinder tests", *Advanced Laboratory Stress-Strain Testing of Geomaterials* (Tatsuoka et al. eds.), Balkema, pp.217–226.

Di Benedetto, H., Partl, M.N., Francken, L. and De La Roche, C., (2001c): *Stiffness testing for bituminous Mixtures, J. of Materials and Structures*, Vol 34, pp. 66–70.

Di Benedetto, H., Tatsuoka, F. and Ishihara, M. (2002): Time-dependent shear deformation characteristics of sand and their constitutive modelling, Soils and Foundations, 42-2, pp.1–22.

Di Prisco, C. and Imposimato, S. (1996): "Time dependent mechanical behaviour of loose sands", *Mechanics of cohesive-frictional materials*, Vol.1, pp.45–73.

Doubbaneh, E.(1995): "Comportement mécanique des enrobés bitumineux des "petites" aux "grandes" déformations", Thèse de Doctorat, ENTPE-INSA Lyon, (in French)

Feron, C. (2002): "Etude des mécanismes de génération de contraintes et de fissuration par retrait gêné dans les structures à base de matériaux cimentaires", Thèse de Doctorat, ENTPE-INSA Lyon, (in French).

Ferry, J.D. (1980): *Viscoelastic properties of polymers*, New York: John Wiley and Sons, 641 p.

Fodil, A., Aloulou, W. and Hicher, P.Y. (1997): "Viscoelastic behaviour of soft clay", *Géotechnique*, vol.47, n°3, pp.581–591.

Geoffroy, H., Di Benedetto, H., Duttine, A. and Sauzeat, C. (2003): "Dynamic and cyclic loadings on sands: results and modeling for general stress strain conditions", *Proc. 3rd Int. Symp. on Deformation Characteristics of Geomaterials, IS Lyon 03* (Di Benedetto et al. eds.), Balkema, Sept. 2003, pp. 353–363.

Goto. S, Tatsuoka, F., Shibuya, S., Kim, Y.-S. and Sato, T. (1991): "A simple gauge for local small strain measurements in the laboratory", *Soils and Foundations*, Vol.31, No.1, pp.169–180.

Hachi-Benmebareck, F. (2000): "Modèle de référence pour matériaux remaniés en vue d'une caractérisation des erreurs", these de Doctorat, ECP (in French).

Hameuy O. (1995): "Quelques aspects du comportement des sables avec ou sans rotation des axes principaux", thèse de Doctorat, ENTPE-ECP (in French).

Hashiguchi, K. and Okayasu, T. (2000): "Time-dependent elastoplastic constitutive equation based on the subloading surface model and its application to soils", Soils and Foundations, Vol.40, No.4, pp.19–36.

Hayano, K., Matsumoto, M., Tatsuoka, F. and Koseki, J. (2001): "Evaluation of time-dependent deformation property of sedimentary soft rock and its constitutive modelling", Soils and Foundations, Vol.41, No.2, pp. 21–38.

Hirakawa, D., Tatsuoka, F. and Siddiquee, M.S.A. (2003a): "Viscous effects on bearing capacity characteristic of shallow foundation on sand", Proc. 38th Japan National Conf. on Geotechnical Engineering, JGS, Akita, June (in Japanese).

Hirakawa, D., Shibata, Y., Uchimura, T. and Tatsuoka, F. (2003b): "Residual deformations by creep and cyclic loading of reinforced-gravel backfill and their relation", Proc. 3rd Int. Symp. on Deformation Characteristics of Geomaterials, IS Lyon 03 (Di Benedetto et al. eds.), Balkema, Sept. 2003, pp.589–596.

Hirakawa, D., Kongkitkul, W., Tatsuoka, F. and Uchimura, T. (2003c): "Time-dependent stress-strain behaviour due to viscous property of geosynthetic reinforcement", Geosynthetics International (accepted for publication).

Hirakawa, D. (2003): "Study on residual deformation characteristics of geosynthetic-reinforced soil structures", Doctor of Engineering thesis, University of Tokyo (in Japanese).

Hoque, E. (1996): "Elastic deformation of sands in triaxial tests", Dr. Engineering thesis, Univ. of Tokyo.

Hoque, E. and Tatsuoka, F. (1998): "Anisotropy in the elastic deformation of materials", Soils and Foundations, Vol.38, No.1, pp.163–179.

Ibraim, E., (1998): "Différents aspects du comportement à partir d'essais triaxiaux : des petites déformations à la liquéfaction statique", Thèse de Doctorat ENTPE-INSA Lyon, (in French).

Jardine, R.J. (1992): "Some observations on the kinematic nature of soil stiffness", Soils and Foundations, 1992, Vol. 32, No 2, pp. 111–124.

Kaliakin, V. (1988): "An elastoplastic-viscoplastic bounding surface model for isotropic cohesive soils", Proc. Int. Conf. on Rheology and Soil Mechanics (Keedwell eds.), Elsevier Applied Science, pp.147–163.

Kim, D.S. and Stokoe II, K., H. (1994): "Torsional motion monitoring system for small-strain (10-5 to 10-3%) soil testing", Geotechnical Testing Journal, Vol. 17, No 1, pp. 17–26.

Kiyota, T. (2004): "Viscous properties and yield characteristics of loose sand in triaxial tests", Master Thesis, University of Tokyo, Feb. (in Japanese)

Ko. D.H., Itou. H., Tatsuoka. F. and Nishi. T. (2003): "Significance of viscous effects in the development of residual strain incyclic triaxial tests on sand", Proc. 3rd Int. Sym. on Deformation Characteristics of Geomaterials, IS Lyon 03 (Di Benedetto et al. eds.), Balkema, Sept. 2003, pp.559–568.

Kohata, Y., Tatsuoka, F., Wang, L., Jiang, G.L., Hoque, E. and Kodaka, T. (1997): "Modelling the non-linear deformation properties of stiff geomaterials", Géotechnique, Vol.47, No.3, Symposium In Print, pp.563–580.

Komoto, N., Tatsuoka, F. and Nishi, T. (2003): "Viscous stress-strain properties of undisturbed Pleistocene clay and its constitutive modelling", Proc. 3rd Int. Sym. on Deformation Characteristics of Geomaterials, IS Lyon 03, (Di Benedetto et al. eds.), Balkema, Sept. 2003, pp.579–587.

Kondner, R.B., (1963): "Hyperbolic stress-strain response : cohesive soils", Jl of SMF Div., Proc ASCE, Vol. 89, N SMI, Feb., pp. 115–143.

Kongkitkul, W. and Tatsuoka, F. (2004): "Load-strain-time behaviour of geogrid arranged in sand subjected to sustained loading", Proc. 39th Japan National Conference on Geotechnical Engineering, Japanese Geotechnical Society, Niigata.

Kongsukprasert, L., Kuwano, R. and Tatsuoka, F. (2001): "Effects of ageing with shear stress on the stress-strain behavior of cement-mixed sand", Advanced Laboratory Stress-Strain Testing of Geomaterials (Tatsuoka et al. eds.), Balkema, pp.251–258.

Kongsukprasert, L. and Tatsuoka, F. (2003a): "Ageing effects developed during creep loading and during extremely slow straining of cement-mixed gravel", Proc. 38th Japan National Conf. on Geotechnical Engineering, JGS, Akita, June

Kongsukprasert, L. and Tatsuoka, F. (2003b): "Viscous effects coupled with ageing effects on the stress-strain behaviour of cement-mixed granular materials and a model simulation", Proc. 3rd Int. Symp. on Deformation Characteristics of Geomaterials, IS Lyon 03 (Di Benedetto et al. eds.), Balkema, Sept. 2003, pp. 569–577.

Kuwano, R. and Jardine, R.J. (2002): "On measuring creep behaviour in granular materials through triaxial testing", Can. Geotech. J., 39, pp. 1061–1074.

Lade, P.V., Yamamuro, J.A. (1996): "undrained sand behaviour in axisymetric tests at high pressures", Jl of Geotechnical Eng., ASCE, Vol. 122, N°2, pp.120–129.

Leong, W.,K. and Chu, J., (2002) "Effect of undrained creep on instability behaviour of loose sand". Can Geotech J., 39, 1399–1405.

Leroueil, S. and Hight, D.W. (2003) "Behaviour and properties of natural soils and soft rocks". Characterisation and engineering properties of natural soil (Tan et al. eds), Balkema, Vol. 1, pp.29–254.

Lo Presti, D., Shibuya, S. and Rix, J., G. (2001): "Innovation in soil testing", Key Note Lecture, Proc. of the Second International Conference on Pre-failure Deformation Characteristics of Geomaterials, Torino, 1999, Balkema (Jamiolkowski et al., eds.), Vol. 2, pp1027–1076.

Li,J. Z., Tatsuoka, F., Nishi, T. and Komoto, N. (2003): "Viscous stress-strain behaviour of clay under unloaded conditions", Proc. 3rd Int. Sym. on Deformation Characteristics of Geomaterials, IS Lyon 03, (Di Benedetto et al. eds.), Balkema, Sept. 2003, pp.617–625.

Li, J.-Z., Acosta-Martínez, H., Tatsuoka, F. And Deng, J.-L. (2004): "Viscous property of soft clay and its modeling, Engineering Practice and Performance of Soft Deposits", Proc. of IS Osaka 2004.

Mandel, J. (1966): Cours de mécanique des milieux continus, Tomes 1 and 2, Ed. Gauthier et Villars, Paris. (in French)

Marasteanu, M.O. (1999): "Inter-conversions of the linear viscoelastic functions used for the rheological characterization of asphalt binders". Thesis in Civil Engineering, Pennsylvania State University.

Matsushita, M., Tatsuoka, F., Koseki, J., Cazacliu, B., Di Benedetto, H. and Yasin, S.J.M. (1999): "Time effects on the pre-peak deformation properties of sands", *Proc. Second Int. Conf. on Pre-Failure Deformation Characteristics of Geomaterials, IS Torino '99* (Jamiolkowski et al., eds.), Balkema, Vol.1, pp.681–689.

Modaressi, H. and Laloui, L. (1997): "A thermo-viscoplastic constitutive model for clays", *International Journal for Numerical and Analytical Methods in Geomechanics*, Vol.21, pp.313–335.

Mohkam, M. (1983): "Contribution à l'étude expérimentale et théorique du comportement des sables sous chargements cycliques", these de Docteur Ingénieur, ENTPE-USMG, P.231, (in French).

Nakajima, M., Hosono, T., and Shibuya, S. (1994): "Effects of rate of shear stress increment on cyclic behaviour of clay", *Proc. Symposium on Deformaation Characteristics of Geomaterials associated with Dynamic Problems of Ground and Soil-structures, JSSMFE*, 1994, pp. 133–136 (in Japanese).

Namikawa, T. (2001): "Delayed plastic model for time-dependent behaviour of materials", *International Journal for Numerical and Analytical Methods in Geomechanics*, Vol.25, pp.605–627.

Nawir, H., Tatsuoka, F. and Kuwano, R. (2003a): "Experimental evaluation of the viscous properties of sand in shear", *Soils and Foundations*, Vol43, N6, pp.13–31.

Nawir, H., Tatsuoka, F. and Kuwano, R. (2003b): "Viscous effects on the shear yielding characteristics of sand", *Soils and Foundations*, Vol. 43, No 6, pp.33–50.

Nawir, H., Tatsuoka, F. and Kuwano, R. (2003c): "Viscous effects on the shear yielding characteristics of sand and its modelling", *Proc. 3rd Int. Symp. on Deformation Characteristics of Geomaterials, IS Lyon 03* (Di Benedetto et al. eds.), Balkema, Sept. 2003, pp.645–643.

Nawrocki, P.A. and Mroz, Z. (1998): "A viscoplastic degradation model for rocks", *Int. J. Rock Mech. Min. Sci.*, Vol.35, No.7, pp.991–1000.

Neifar, M., Di Benedetto, H. (2001): "Thermo-viscoplastic law for bituminous mixes", *Int. Jl of Road Materials and Pavement Design*, Hermes, Vol 2-1, pp. 71–95.

Neifar, M., Di Benedetto, H. and Dongmo, B. (2003): "Permanent deformation and complex modulus: Simultaneous determination from a unique test", *6th RILEM int. Conference PTEBM*, ed. M. Partl, pp. 8, Zurich.

Oie, M., Sato, N., Okuyama, Y., Yoshida, T., Yoshida, T., Yamada, S. and Tatsuoka, F. (2003): "Shear banding characteristics in plane strain compression of granular materials", *Proc. 3rd Int. Symp. on Deformation Characteristics of Geomaterials, IS Lyon 03* (Di Benedetto et al. eds.), Balkema, pp. 597–606.

Olard, F., Di Benedetto, H., Eckmann, B., Triquigneaux and J-P. (2003): "Linear Viscoelastic Properties of Bituminous Binders and Mixtures at Low and intermediate Temperatures", *Int. jl Road Materials and Pavement Design*, Di Benedetto et al. eds, Hermes Lavoisier, Volume 4, Issue 1.

Omae, S., Sato, N. and Oomoto, I. (2003): "Dynamic Properties of CSG", *Proc. 4th Int. Symp. on Roller*

Compacted Concreter Dams (RCC), Madrid, Spain, 17–19 Nov.

Park, C.-S. and Tatsuoka, F. (1994): "Anisotropic strength and deformations of sands in plane strain compression", *Proc. of the 13th Int. Conf. on S.M.F.F.*, New Delhi, Vol.13, No.1, pp.1–4.

Perzyna, P. (1963): "The constitutive equations for work-hardening and rate-sensitive plastic materials", *Proc. of Vibrational Problems, Warsaw*, 4-3, pp281–290.

Pham Van Bang, D. and Di Benedetto, H. (2003): "Effect of strain rate on the behaviour of dry sand", *Proc. 3rd Int. Symp. on Deformation Characteristics of Geomaterials, IS Lyon 03* (Di Benedetto et al. eds.), Balkema, Sept. 2003, pp. 363–373.

Pham Van Bang, D. (2004): "Comportement instantané et différé des sables des petites aux moyennes déformations: expérimentation et modélisation", Thèse de Doctorat, ENTPE – INSA de Lyon.

Salençon, J. (1983): *Viscoélasticité. Cours de calcul des structures anélastiques*, Presse de l'Ecole Nationale des Ponts et Chaussées, Paris, 88 p., (in French)

Santucci de Magistris, F., Koseki, J., Amaya, M., Hamaya, S., Sato, T. and Tatsuoka, F. (1999), "A triaxial testing system to evaluate stress-strain behaviour of soils for wide range of strain and strain rate", *Geotechnical Testing Journal, ASTM*, Vol.22, No.1, pp.44–60.

Santucci de Magistris, F. and Tatsuoka, F. (2004): "Effects of moulding water content on the stress-strain behaviour of a compacted silty sand", *Soils and Foundations*, Vol.44, No.2 pp. 85–101.

Sato, M., Ueda. M., Hasebe, N. and Kondo, H. (1997a): "Comparison among wave velocities from seismic observation waves and field tests for hard rock mass", *J. of JSCE*, 38, pp. 75–87 (in Japanese).

Sato, M., Ueda. M., Hasebe, N. and Umehara, H. (1997b): "Dynamic elastic modulus of dam concrete earthquake motion", *J. of JSCE*, 35, pp. 43–55.

Sauzéat, C.(2003): "Etude du comportement des sables des petites aux moyennes deformations", Thèse de Doctorat, ENTPE-INSA Lyon, (in French).

Sauzéat, C., Di Benedetto, H., Chau, B. and Pham Van Bang. D. (2003): "A rheological model for the viscous behaviour of sand" *Proc. 3rd Int. Symp. on Deformation Characteristics of Geomaterials, IS Lyon 03* (Di Benedetto et al. eds.), Balkema, pp. 1201–1209.

Schanz, T., Vermeer, P.A. and Bonnier, P.G. (1999): "The hardening soil model: Formulation and verification", Beyond 2000 in *Computational Geotechnics* (Brinkgreve eds.), Balkema, pp.281–296.

Sekiguchi, H. and Ohta, H. (1977): "Induced anisotropy and its time dependency in clays, Constitutive Equations of Soils", *Proc. of Spec. Session 9, the 9th ICSMFE*, Tokyo, pp. 229–238.

Shibuya, S. (2001): "Quasi-elastic stiffness in the behaviour of soft clay", Doctoral of engineering Thesis, University of Tokyo, 152 p.

Siddiquee, M.S.A., Tanaka, T., Tatsuoka, F., Tani, K. and Morimoto,T. (1999): "FEM simulation of scale effect in bearing capacity of strip footing on sand", *Soils and Foundations*, 39 (4), 91–109.

Siddiquee, M.S.A., Tatsuoka, F., Tanaka, T., Tani, K., Yoshida, K. and Morimoto, T. (2001a): "Model tests and FEM simulation of some factors affecting the bearing capacity of footing on sand", *Soils and Foundations*, 41 (2), 53–76.

Siddiquee, M.S.A. and Tatsuoka, F. (2001b): "Modeling time-dependent stress-strain behaviour of stiff geomaterials and its applications", *Proc. Of 10th International Conference on Computer Methods and Advances in Geomechanics*.

Siddiquee, M.S.A., Tatsuoka, F. and Tanaka, T. (2003a): "FEM simulation of viscous effects on the stress-strain behaviour of sand in drained PSC", *Proc. 38th Japan National Conf. on Geotechnical Engineering*, JGS, Akita, June.

Siddiquee, M.S.A., Tatsuoka, F. and Tanaka, T. (2003b): "Implementation of a time dependent constitutive model "TESRA" into the nonlinear FEM scheme", *Proc. 3rd Int. Sym. on Deformation Characteristics of Geomaterials, IS Lyon 03* (Di Benedetto et al. eds.), Balkema, September, 2003, 873–881.

Sugai, M., Tatsuoka, F. and Uchimura, T. (2003): "Effects of ageing and viscosity on the stress-strain behaviour of a cement-mixed soft clay", *Proc. 3rd Int. Sym. on Deformation Characteristics of Geomaterials, IS Lyon 03*, (Di Benedetto et al. eds.), Balkema, Sept. 2003, pp.637–643.

Suklje, L. (1969): *Rheological aspects of soil mechanics*, Wiley-Interscience, London.

Suzuki, M. (1994): "Deformation properties of sedimentary soft rock from cyclic and monotonic loading triaxial tests", Master of eng. Thesis, Univ. of Tokyo. [in japanese]

Tatsuoka, F., and Molenkamp, F. (1983): "Discussion on yield loci for sands", *Mechanics of Granular Materials: New Models and Constitutive Relations*, Elsevier Science Publisher B.V., pp. 75–87.

Tatsuoka, F. and Shibuya, S. (1991a), "Deformation characteristics of soils and rocks from field and laboratory tests", *Keynote Lecture for Session No.1, Proc. of the 9th Asian Regional Conf. on SMFE*, Bangkok, Vol.II, pp.101–170.

Tatsuoka, F., Okahara, M., Tanaka, T., Tani, K., Morimoto, T. and Siddiquee, M.S.A. (1991b): "Progressive failure and particle size effect in bearing capacity of a footing on sand", *Proc. ASCE Geotech. Engineering Congress, 1991, Boulder, ASCE Geotechnical Special Publication*, Vol.27, pp.788–802.

Tatsuoka, F., Sato, T., Park, C.S., Kim, Y.S., Mukabi, J.N. and Kohata, Y. (1994): "Measurements of elastic properties of geomaterials in laboratory compression tests", *GTJ*, 17–1, 80–94.

Tatsuoka, F., Jardine, R.J., Lo Presti, D., Di Benedetto, H. and Kodaka, T. (1999a): "Characterising the Pre-Failure Deformation Properties of Geomaterials", *Theme Lecture for the Plenary Session No.1, Proc. of XIV IC on SMFE*, Hamburg, September 1997, Volume 4, pp. 2129–2164.

Tatsuoka, F., Santucci de Magistris, F., Momoya, F., Maruyama, N. (1999b): "Isotach behaviour and its modelling", *Proc. of the Second International Conference on Pre-failure Deformation Characteristics of Geomaterials, Torino, 1999*, Balkema (Jamiolkowski et al., eds.), Vol. 1, pp. 491–499.

Tatsuoka, F., Santucci de Magistris, F., Hayano, K., Momoya, Y. and Koseki, J. (2000): "Some new aspects of time effects on the stress-strain behaviour of stiff geomaterials", *Keynote Lecture, The Geotechnics of Hard Soils – Soft Rocks, Proc. of Second Int. Conf. on Hard Soils and Soft Rocks, Napoli, 1998* (Evamgelista and Picarelli eds.), Balkema, Vol.2, pp1285–1371.

Tatsuoka, F., Uchimura, T., Hayano, K., Di Benedetto, H., Koseki, J. and Siddiquee, M.S.A. (2001a): "Time-dependent deformation characteristics of stiff geomaterials in engineering practice, the Theme Lecture", *Proc. of the Second International Conference on Pre-failure Deformation Characteristics of Geomaterials, Torino, 1999*, Balkema (Jamiolkowski et al., eds.), Vol. 2, pp. 1161–1250.

Tatsuoka, F., Nishi, T. and Kuwano, R. (2001b): "On the three-component models to describe the viscous effects on the stress-strain behaviour of geomaterials," *Proc. Geotechnical Engineering, the JGS, Tokushima* (in Japanese).

Tatsuoka, F., Ishihara, M., Di Benedetto, H. and Kuwano, R. (2002): "Time-dependent shear deformation characteristics of geomaterials and their simulation", *Soils and Foundations*, 42-2, pp. 103–129.

Tatsuoka, F., Di Benedetto, H., and Nishi, T. (2003): "A framework for modeling of the time effects on the stress-strain behaviour of geomaterials", *Proc. 3rd Int. Symp. on Deformation Characteristics of Geomaterials, IS Lyon 03* (Di Benedetto et al. eds.), Balkema, Sept. 2003, pp. 1135–1143.

Tatsuoka, F. (2004): "Effects of viscous properties and ageing on the stress-strain behaviour of geomaterials." *Proceedings of the GI-JGS workshop (Yamamuro & Koseki eds.), ASCE Geotechnical SPT* (to appear).

Uchimura, T., Tatsuoka, F. and Tanaka, I. (2003): "Long-term residual deformation of prototype geotextile-reinforced gravel structures", *Proc. 3rd Int. Symp. on Deformation Characteristics of Geomaterials, IS Lyon 03* (Di Benedetto et al. eds.), Balkema, Sept. 2003, pp.1353–1361.

Uchimura, T., Tatsuoka, F, Hirakawa, D., Shibata, Y. (2004): "Residual deformations of synthetic-reinforced soil structure subjected to sustained and cyclic loading", *Proceedings of the third European Geosyntheticc Conference: Euro Geo3*, Ed: Flon et al, pp. 403–408.

Umair, A., Uchimura, T., Tatsuoka, F. Msushima, K., and Mohri, H. (2004): "Viscous properties of Recycled Concrete Aggregate in Triaxial Compression", *Proc. 39th Japan National Conference on Geotechnical Engineering*, Japanese Geotechnical Society, Niigata.

Vermeer, P.A. and Neher, H.P. (1999): "A soft soil model that accounts for creep", Beyond 2000 in *Computational Geotechnics* (Brinkgreve eds.), Balkema, pp. 249–261.

Yasin, S.J.M. and Tatsuoka, F. (2003): Viscous property of an air-dried micacious sand in plane strain compression, *Proc. 38th Japan National Conf. on Geotechnical Engineering*, JGS, Akita, June.

Yin, J.-H. and Graham, J. (1994): "Equivalent times and one-dimensional elastic viscoplastic modeling of time-dependent stress-strain behaviour of clays", *Canadian Geotechnical Journal*, Vol.31, pp. 42–52.

Deformation Characteristics of Geomaterials – Di Benedetto et al (eds)
© 2005 Taylor & Francis Group, London, ISBN 04 1536 701 8

From very small strains to failure

J. Biarez
Professor, Ecole Centrale Paris – LMSS-Mat, UMR CNRS, France

A. Gomes Correia
Professor, University of Minho, Guimarães – Portugal

F. Lopez-Caballero
PhD, Ecole Centrale Paris – LMSS-Mat, UMR CNRS, France

ABSTRACT: Numerical methods of solution of many boundary problems in rock and soil mechanics require the knowledge of the entire stress–strain relationship through the entire range of strains from very small to strains at peak and beyond peak strengths – to failure. This paper is a contribution to clarify many of the peculiarities of soil and rock complex behaviour in this large range of strains. For this purpose a method is outlined using two types of artificial materials: grains without and with "glue" and materials without and with fissures. This has been the base to explain many of the features of several natural grounds, from rocks to soils, investigated experimentally along a wide spectrum of stress paths in laboratory tests.

1 INTRODUCTION

The purpose of this paper is to suggest a method to approach the complexity of ground behaviour, by using two types of artificial materials, reproducible and calculable: grains without and with "glue" (e.g. bitumen for bituminous concrete) and materials without and with fissures.

In these materials we have a non linear elasticity influenced by the mean normal stress. Plastic deformations come from the displacement of grains and of the creation or increasing of voids and fissures. These two last aspects were object of well known studies on damage for metals, concretes and rocks and will be not reminded here. Nevertheless, the word failure will be applied here only in that case. Therefore, failure will not be used here for grains without "glue" except if there is failure of grains.

Figure 1 shows a biaxial test on a granular material formed of hexagonal rods (Biarez, 1962). The $q(\varepsilon)$ curve shows a peak due to dilatancy, then a slight softening followed by a even stretch of perfect plasticity with no volume change, beyond an important decrease of $q(\varepsilon)$ corresponding to a localisation associated with large deformations (kinematic discontinuity) to achieve a final even stretch, designated residual, where the planes of hexagons are aligned.

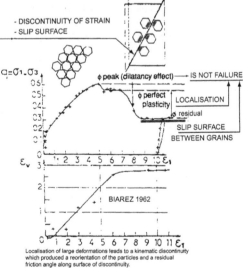

Figure 1.   Biaxial test with hexagonal rods.

"Glues" were also object of numerous theoretical and practical studies, for example in aeronautics. In this paper we shall deal with only usual "glues" of civil engineering: bitumen, injection of grout, cement. "Glues" allow formulations and calculations, knowing the boundary conditions and the constitutive laws of grains, "glues" (elasticity, plasticity, viscosity, ductility, fragility …) and their interfaces.

In this paper complex material behaviour will be simplified in successive classes and their limits, which are found in conventional compression tests.

The constitutive laws of ground behaviour being described by tensors will be represented in related figures using "scalars", and by following three stages:

- Grains without "glue" (unbound materials);
- Grains with "glue";
- Natural grounds (and remoulded).

2 GRAINS WITHOUT "GLUE" – BACKGROUND

This subject was treated exhaustively in the book of Biarez and Hicher (1994). We remind here some results by making reference to figures in the book, here identified by the number of page preceded by BH (e.g. BH p. xxx).

2.1 Grains

We start to define grains as a result of measures made by sieving and sedimentation. The geometry of clay grains is discernible in the microscope. Inside these grains solid sheets exist; they are bounded with water and with adsorbed liquid (water more ions) that bring to the clay a viscous ductile behaviour. The quantity of water inside grains increases with Atterberg's limits and gives elasticity modulus of grains E_g decreasing from the kaolin to the smectite. An elementary picture consists in comparing these grains as sponges, where the volume decreases with the suction after the shrinkage limit. For pedagogic purposes we assume also that these sponges have on their borders a friction and an adsorbed liquid which fabricate a "glue". However, this "glue" can be considered negligible for remoulded soils.

2.2 Geometry of grains packing

The geometry of grains packing (structure) will be simplified with an isotropic function identified by the void ratio and by an anisotropic function defined by the shape of grains and the statistical orientation of tangent plans to grain contacts (BH p. 123). Plastic deformations come generally from the modification of this geometry of grains packing and sometimes from themselves.

2.3 Elasticity

2.3.1 Hertz type calculation for a regular pack of spheres

The method of continuum mechanics can be used for soils if we assume that only grains are continuum (sand). In these conditions, simplified elastic soil behaviour can be described with the same type of parameters used in Hertz formulation (fig. 2). A homogeneous formula can be obtained using the modulus of grains E_g. Elsewhere, the void ratio e express the different types of grain packing, which is simplified by e^{-m}, replacing Hardin formula. The elastic soil modulus can then be expressed by:

$$E = K \cdot e^{-m} \cdot p^n \tag{1}$$

where k decrease and n increase with w_L increase. For $w_L < 50\%$ the following simplified formula can be used:

$$E = \frac{450}{e} \cdot p^{1/2} \tag{2}$$

For the case of $d_{60}/d_{10} < 2$ (for example in sands), the following expression is recommended:

$$E = \frac{380}{e^2} \cdot p^{1/2} \tag{3}$$

2.3.2 Elastic limit

If a clay is compressed under an isotropic stress path from a moisture content w equal to the liquidity limit w_L, triaxial tests usually give an elastic limit in strains ε, around 10^{-5} for a great variety of soils. This corresponds in the $(\sigma_1, \sigma_2, \sigma_3)$ space of stresses to a cone starting from the origin, having for axis the trisectrix

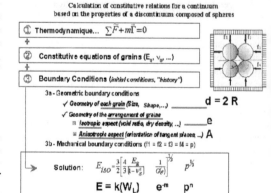

Figure 2. Elastic modulus of soils derived from Hertz approach.

and in the plane q(p) a straight line close to the p axis; p being the mean normal stress and q the shear stress.

If compression follows another stress path starting from the origin characterized by a q/p value, a nearby cone to this straight line will be obtained, keeping the anisotropic geometrical grains distribution. This is obtained on oedometer stress path (BH p. 76). However, for an important unloading, elastic limit and anisotropy will be more complexes.

2.3.3 Secant modulus

We can keep the formula (1) for the secant modulus obtained by triaxial tests with power n increasing with strain (ε_1) until n = 1 at perfect plasticity and m = 0 for the case where the void ratio does not get involved (BH pp. 33, 34 and 42). Figure 3 shows that triaxial test results can be grouped for different e values and confining pressures σ_3, by using n = 1/2 for $\varepsilon = 10^{-5}$ and n = 3/4 for $\varepsilon = 10^{-2}$.

The representation in terms of $\log(E \cdot e^2/\sigma^{3/4})(\log \varepsilon_1)$ in Figure 4 allows a linear interpolation between the precision triaxial tests, with direct strain measurement

in the sample, and the current tests, with measurements outside the triaxial cell (off sample). It is remarkable to observe that all tests with different q(ε) curves are regrouped in these coordinates for $\varepsilon = 10^{-2}$ with n = 3/4. For smaller strains results are scattered since measurements are not enough precise any more.

Figure 5 represents exceptional results of a test which remains homogeneous until the perfect plasticity; in this case tests attend a straight line q = Mp which gives:

$$\frac{E}{p} = \frac{M}{\varepsilon_1} \tag{4}$$

2.4 Normally consolidated behaviour (NC)

Remoulded soils are designated here as normally consolidated soils, where the mechanical properties are independent from the history, so of the loading paths (monotonous decreasing void ratio). The geometry of grains distribution, isotropic and anisotropic, depends only on the tensor of applied stress.

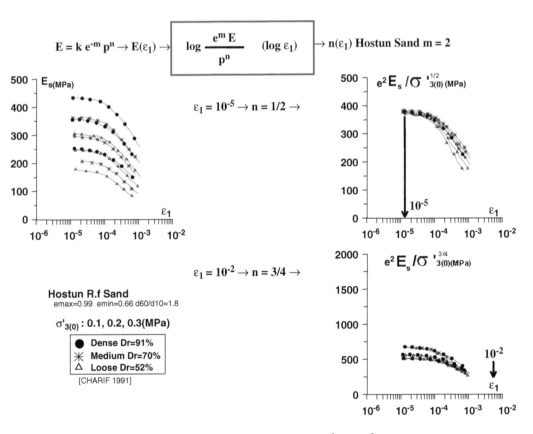

Figure 3. Normalisation of triaxial test results for two strain levels ($\varepsilon = 10^{-5}$ and 10^{-2}).

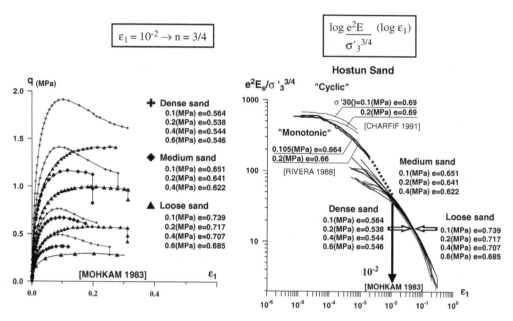

Figure 4. Linear interpolation between "precision" triaxial test and "classical" triaxial test using a **log-log scale**.

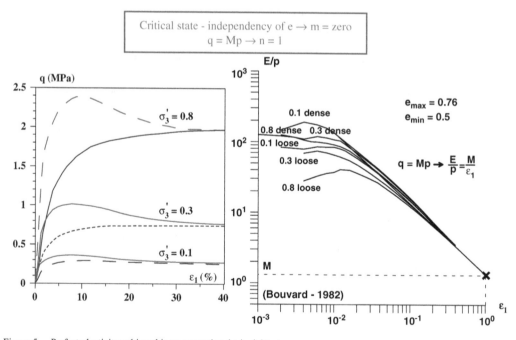

Figure 5. Perfect plasticity achieved in an exceptional triaxial test.

2.4.1 Clays

2.4.1.1 Isotropic compression e(p)

For an isotropic compression starting from a void ratio corresponding to a moisture content of 3 to 4 times w_L, the curve e(p) reaches a void ratio close to zero (BH p. 14). In coordinates $e - \log p$ this curve is locally linear of slope Cc between w_L and w_P (eq. 5).

$$e = e_0 - Cc \cdot \log\left(\frac{p}{p_0}\right) = e_0 - \lambda \cdot \log\left(\frac{p}{p_0}\right) \qquad (5)$$

In the same domain, the compressions lines with q/p constant and oedometric are parallels (BH p. 27), with a maximal distance of the order of $\Delta e \sim 0,1$ between q/p = 0 and q/p = M. This value varies with w_L and w_P, but much less than C_c.

2.4.1.2 Compression e(q) for p constant
For this stress path dilatancy (positive or negative) is defined as the relationship between the isotropic or volumetric strain, characterized by e, and the deviator stress simplified by $q = \sigma_1 - \sigma_3$.

The experimental results are in good agreement with the relations:

$$e = e_0 - C_d \cdot \log\left(1 + \frac{\eta^2}{M^2}\right)$$
$$e = e_0 - \lambda_d \cdot \ln\left(1 + \frac{\eta^2}{M^2}\right) \qquad (6)$$

C_d is in general of the order of 1/3 in agreement with $\Delta e \sim 0,1$ (BH p. 25); $\eta = q/p$; $M = (q/p)_{peak}$.

2.4.1.3 Simplified Roscoe surface e(p, q)
With the previous two relations a surface is obtained which is in a good agreement with the others normally consolidated stress paths: $\sigma_1 = \sigma_3 = $ constant, undrained; e = constant …

This surface of plastic strains does not integrate elastic parameters but if we write $\lambda_d = \lambda - k$ we obtain Cam clay remarkable law which gives NC and OC deviator strains.

2.4.2 Sands
The well known studies on slight dense sands from the early studies of Castro (1969) to the recent ones of Wang (2002) will be not reminded here.

For wet sands (not saturated) it is possible to obtain void ratios greater than the normalised e_{max} (ASTM) and obtain a normally consolidated behaviour similar to the clay (BH p. 22) with $C_c \sim 0.15$ as for the oedometer results presented in Figure 6 (BH p. 78). This value is reasonable having in mind the well known decrease of C_c with w_L given by Skempton for clays:

$$C_c = 0.009(w_L - 13) \qquad (7)$$

2.5 Overconsolidate behaviour (OC)

2.5.1 Isotropic stress path and oedometer
With a clay it is easy to start from w = $1.5 w_L$ under an isotropic compression path C_c until the preconsolidation stress p'_c, then unloading under the overconsolidation path C_s. This is much more difficult on a saturated

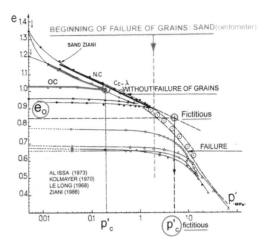

Figure 6. Oedometer results on a wet sand showing a pre-consolidation stress (p'_c) and a fictitious preconsolidation after failure of grains.

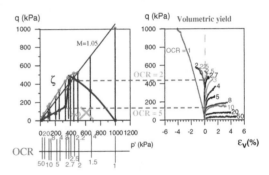

Figure 7. Constant p' triaxial stress path showing no volumetric strain at the beginning of loading – dilatancy line (Hattab, 95).

saturated sand, but easy on a wet sand with $e > e_{max}$ (fig. 6). However, in the test results presented in Figure 6, it must be noticed that failure of grains start after a stress of 2 MPa. For the common void ratios it is necessary to extrapolate the NC line to obtain a fictitious preconsolidation stress p'_c. This stress is very useful because it lets to have the same approach for sands and clays NC and OC. We shall notice here that this stress can be obtained by correlation with the dilatancy.

2.5.2 Triaxial loading path p = constant: clay
Figure 7 shows that for this stress path there is no isotropic or volumetric strain at the begining of loading in the $q(\varepsilon_v)$ curves. The yield points of the beginning

of dilatancy, positive and negative (volumetric yield) give in the q(p) plane the "dilatancy line". Under this line isotropic or volumetric strain depends on p and not on q.

2.5.3 Triaxial path $\sigma_2 = \sigma_3 =$ constant: clay

Let's assume a clay consolidated to 0.8 MPa, then unloaded, for example to 0.2 MPa, on the curve Cs (fig. 8). For a triaxial compression ($\sigma_2 = \sigma_3 =$ constant) the start of loading follows the path Cs until the point E_2, then after the "volumetric yield" the dilatancy leads to the critical void ratio, if strain is homogeneous. In the plane q(Δe) it is observed also a yield point E_2 which gives the same point in the q(p) plane.

If we unload from 0.8 MPa to 0.4 MPa and then a triaxial loading, the path Cs is still followed until E_1, but after this yield point "volumetric yield" a negative dilatancy is observed until reaching the critical void ratio. A "dilatancy line" is also obtained since below this line an isotropic or volumetric strain does not depend on q but only on p. Indeed, Δe(q) results from e(p) by the relation between q and p under this loading path. It must be noticed that this yield point was called "volumetric pseudo elastic limit" because it limits the domain governed by C_s or k, but this line is above the elastic limit.

2.5.4 Triaxial path $\sigma_2 = \sigma_3 =$ constant: sands

In the case of sand (fig. 9), as for clay, the start of curves follows the line Cs until a "volumetric yield",

after that a positive or negative (for high stresses) dilatancy is observed. However, here, as expected, after a certain stress level failure of grains appear. In analogy with the oedometer, a fictitious preconsolidation stress p'_{1c} can be defined.

2.5.5 Anisotropic consolidation and overconsolidation under oedometer stress path $\varepsilon_2 = \varepsilon_3 = 0$ on a remoulded clay, then triaxial compression under different stress paths

The experimental results represented in Figure 10 show that it exists a "volumetric yield" which defines in the q(p) plane a "volumetric yield surface" situated under the "peak surface" of maximum strengths.

2.5.6 Several parameters to characterize overconsolidation

Figure 11 shows the relationships that can be found between the dilatancy β or ψ and the distances, vertical and horizontal, from the initial point: (e_{0c}, p'_{i0}) to the straight lines of critical void ratio and isotropic compression; it's the same with the overconsolidatio OCR = p'_{ic}/p'_{i0}. This can be generalized outer the straight regions C_c of curves in e(log p).

2.5.7 Laws of dilatancy: Rowe–Cam-Clay

Rowe gave a relation between σ'_3/σ'_1 (ε_1) or q/Mp (ε_1) and the variation of volumetric strains ε_v (ε_1) that may be simplified by (BH p. 99).

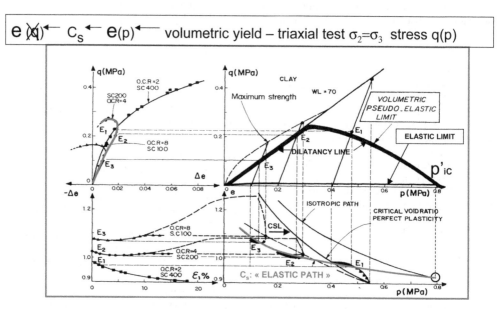

Figure 8. Compression triaxial test results ($\sigma_2 = \sigma_3 =$ constant) on clay showing the volumetric pseudo elastic limit and the dilatancy line.

$$\frac{\sigma_1'}{\sigma_3'} = tg^2\left(\frac{M}{4} + \frac{\phi_{pp}}{2}\right) \cdot \left(1 - \frac{d\varepsilon_v}{d\varepsilon_1}\right) \qquad (8)$$

This gives initially a "volumetric yield" then, for large strains, the perfect plasticity with a critical void ratio. This behaviour is difficult to obtain for overconsolidated soils as a consequence of localisation for large strains or fissures. It is than necessary to clarify the limit of the domain of homogeneous strains.

2.5.8 Benchmark behaviour

The research of homogeneous behaviour until the perfect plasticity in more than one hundred triaxial tests led to the selection of test results presented in Figure 12. This Figure shows pairs of curves parameterised in dilatancy. Generally, as indicated for the "simple clay", the test starts to follow one pair of curves and then shifts at the beginning of the localisation. This research allowed also to obtain benchmarks for straight segments (C_c) of the curves of critical

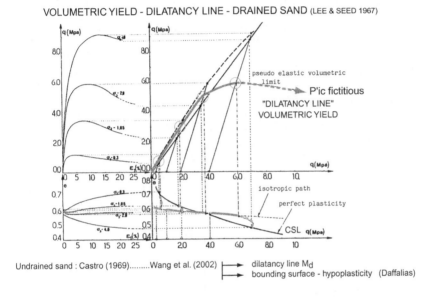

Figure 9.   Compression triaxial test results ($\sigma_2 = \sigma_3 = $ constant) on sand showing the volumetric pseudo elastic limit, the dilatancy line and the fictitious preconsolidation stress.

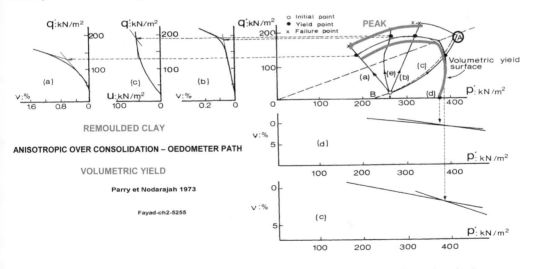

Figure 10.   Oedometer stress path ($\varepsilon_2 = \varepsilon_3 = 0$) showing the volumetric yield surface under the "peak surface".

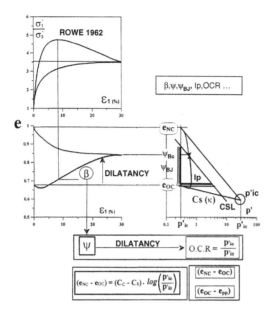

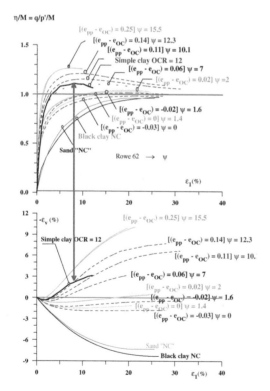

Figure 11. Several parameters associated with dilatancy and overconsolidation.

Figure 12. Benchmark behaviour – Pair of drained compression triaxial curves parameterised in dilatancy.

void ratio and of isotropic behaviour from w_L, w_P, e_{max} and e_{min} (BH p. 100).

3 GRAINS WITH "GLUE"

3.1 Consequences of "glue"

The consequences of "glue" are the following: (1) at the level of the elasticity fig. 13(a): if there is a "glue" with a elasticity modulus E_g equal to that of grains, there is no more significant increase of the contact area of grains, and therefore the role of p disappear, contrary to the case of grains without "glue" where the calculation of Hertz gives a non linear elasticity, since the contact area of grains increases with p; (2) at the level of the plasticity two types of "glues" must be considered fig. 13(b): (a) ductile "glue", where the plasticity is a result of friction during displacement of grains, supplemented by the strength of the "glue"; b) fragile "glue", where the cracks of the "glue" give only place to friction between grains.

3.2 Soil injected by grouts

Figure 14 shows the increases of shear strength of sand created by adding "glue". Fragile (brittle) "glue" formed with cement gives a peak $q(\varepsilon)$ which with

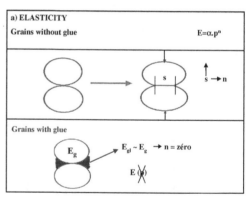

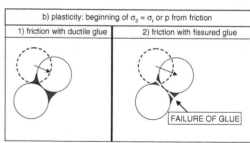

Figure 13. Schematic representation of "glue" and its influence in elastic and plastic behaviour of materials.

damage of "glue" achieves a final even stretch close to the natural sand.

Ductile "glue" formed of silicate and organic hardener gives a maximum shear strength keeping constant until $\varepsilon_1 = 6\%$, then damage of "glue" leads to retrieve practically sand properties. Figure 15 shows agreement with Rowe dilatancy law for $\sigma_3 = 1.2$ MPa: the peak $q(\varepsilon)$ corresponds very well to the maximum of slope β of ε_v (ε_1). However, for lower stresses the peak corresponds to the beginning of dilative behaviour (volume increase $\varepsilon(\varepsilon_1)$); this one is faster than in the benchmark behaviour of grains without "glue" (fig. 12) because of voids developed between blocks created by localisation of cracks.

3.3 Cement concrete

The cement is like a "glue" having the same elasticity modulus as grains; therefore, the modulus of concrete is independent of σ_3 (fig. 16). The beginning of σ_3 effect is evident on $q(\varepsilon)$ curves (fig. 16), probably resulting from creation of fissures or amplification of initial fissures. These yield points identified in the $q(\varepsilon)$ curves give a well defined yield curve in $q(p)$ plane.

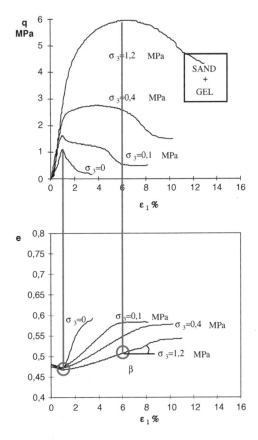

Figure 15.   Influence of stress level in peak strength and volumetric behaviour for a ductile "glue" (sand + gel).

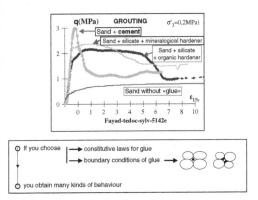

Figure 14.   Influence of the type of "glue" in the stress-strain behaviour (ductile – brittle).

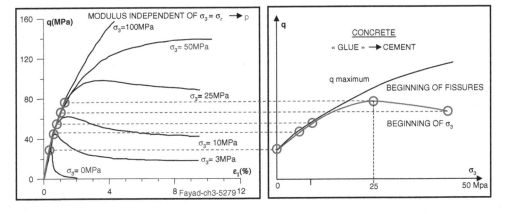

Figure 16.   Illustrative results of the beginning of σ_3 and fissuring influence in the stress–strain behaviour for a cement concrete.

Grains with glue : bitumen ⇒ bituminous concrete

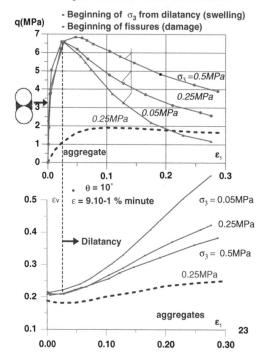

Grains with "glue":
BITUMEN → BITUMINOUS CONCRETE
Constitutive relations of bitumen with viscosity

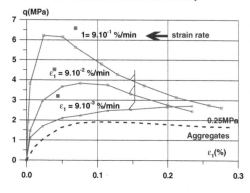

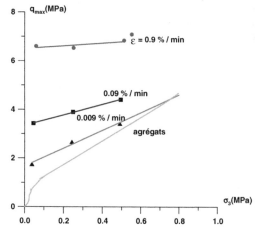

Figure 17. Illustrative results of the beginning of σ_3 influence associated with the initiation of dilatancy and consequently fissuring of bitumen, for a bituminous concrete.

Figure 18. Influence of strain rate on maximum shear strength for a bituminous concrete.

3.4 Bituminous concrete

Figure 17 shows that the beginning of σ_3 influence corresponds to the start of dilatancy, producing tension in bitumen and consequently initiating fissures. Figure 18 shows the influence of the bitumen constitutive law, which includes viscosity, explaining the increase of maximum shear strength with strain rate increase and the decrease of slope of $q(\sigma_3)$ relationship when the strength of the "glue" is big with regard to the friction of grains. Alike, Figure 19 shows the increase of elasticity moduli with strain rate increase and temperature decrease.

3.5 "Capillarity glue" of non saturated soils

Non saturated soils, as grains with "glue", have an unconfined strength, e.g. cohesion in total stresses, and a modulus $E = E_0 + f(p)$. However, contrary to the previous cases, it is not the strength of the "glue" that intervenes but the forces between the grains, which are dependent of the surface tension: $E = g(p + p'_u)$ (Taibi, 1995). Oedometer test results presented in

Figure 20 shows a bend which corresponds to a stress bigger than the maximum loading of the history. If the total stress is increased the void ratio decreases, degree of saturaton increases and $(u_a - u_w)$ decrease until achieving zero in the point where the curve reaches the saturated test if w is constant. However, if $(u_a - u_w)$ is kept constant the non saturated curve remains "parallel" to the saturated curve.

4 NATURAL GROUND

Natural grounds are presented by increasing strength; first the clayey soils with increasing maximum overburden, then materials with increasing $CaCo_3$ and finally rocks.

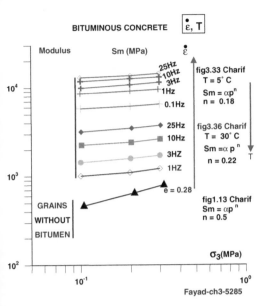

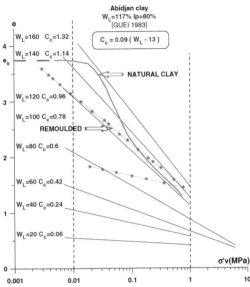

Figure 19. Influence of strain rate and temperature on elastic modulus of a bituminous concrete.

Figure 21. Illustration of a bend in the e (σ'_v) curve for a natural clay, alike an effect of "glue", and comparison with remoulded material.

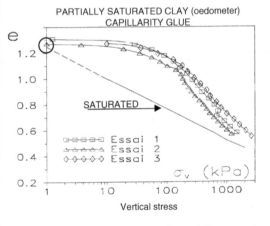

Figure 20. Oedometer test results of a partially saturated clay and its comparison with saturated state under increasing total stresses.

4.1 Clayey soils

Maximum overburdens were chosen from 10 metres to 1 kilometre. The ratio between maximum and minimum history overburden will be defined as over-consolidation. The minimum overburden can be smaller than the actual overburden.

4.1.1 Post last glaciation grounds – PG
Grounds under the water, deposited since the last glaciation, generally, did not undergo overburden

bigger than the current one. Logically they should be compared with the NC remoulded soils. However, they have higher shear strength due to what we shall call arbitrarily "glue". As an illustration, the oedometer results in Figure 21 can be compared with that of Figure 20. The bend in the e (σ'_v) curve could be compared with a kind of supplementary consolidation stress, but this would result in a reduction of void ratio and it could not bring the viscosity of the "glue", which explains the influence of strain rate and temperature on this bend.

The curve of the natural ground (TN) retrieves that of the remoulded ground as in the case of reduction of "glue". The slope in the inflexion point is more important for the "quick clays" as for fragile "glues".

Triaxial results of Figure 22 show greater strength of TN than of remoulded soils, alike grains with and without "glue". It is also noticeable that the beginning of q(ε) curves is independent of σ_3, then with further shearing a clear influence of σ_3.

Numerous tests on the PG, in particular "quick clays" show a peak q(ε) as for the fragile "glues". It is nevertheless possible to see (fig. 23) a even stretch of constant maximum strength until $\varepsilon = 8\%$, alike ductile "glues"; this is confirmed in the q(p') plane by the set of undrained stress paths which attained a straight line giving a cohesion (ordinate at p' = 0) strongly higher than the cohesion of remoulded soils.

135

POST GLACIATION CLAY

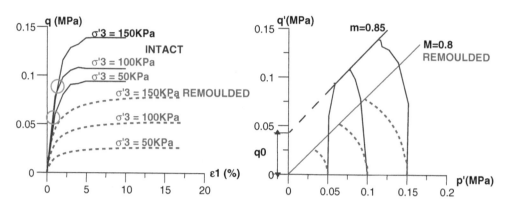

Figure 22. Illustration of the beginning of σ_3 influence on $q(\varepsilon_1)$ curves for natural PG clay, alike behaviour of a ductile "glue" and differences in strengths of natural and remoulded PG clay, alike grains with and without "glue".

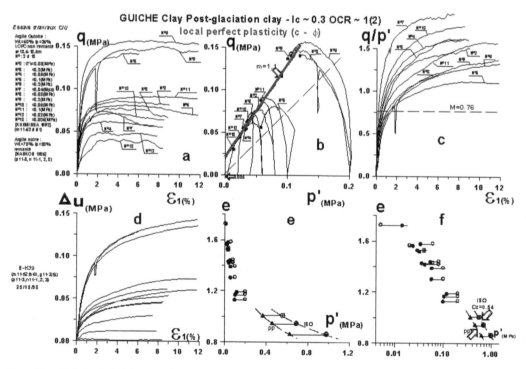

Figure 23. Undrained triaxial test results of a PG clay showing analogies with grains with ductile "glue".

The "state limit line", referred in Leroeuil works, is defined by a sharp bend in the curves e(p') or $\varepsilon_v(\log p')$ (fig. 24), alike the "volumetric yield surface" for grains without "glue". In Figure 24 this line is no more symmetrical to p' axis, as for the remoulded clay consolidated under oedometer path (fig. 10); but, it is bigger as by an effect of "glue" and joins the maximum peak strengths contrary to the

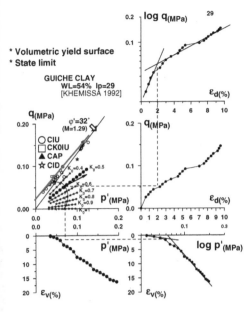

* Volumetric yield surface
* State limit

GUICHE CLAY
WL=54% Ip=29
[KHEMISSA 1992]

Figure 24.   Peculiarities of a PG clay showing the role of "glue".

remoulded soils. This connection to a straight line does not show the usual decrease observed in "quick clays" with "fragile glue".

4.1.2   London clay (Bishop, 1965)

Tests were carried out on samples taken about thirty metres deep in London clay layer that was under a maximum overburden around 300 metres, e.g. with an overconsolidation ratio of OCR = 10. Figure 25 shows that the $q(\varepsilon_1)$ and $\varepsilon_v(\varepsilon_1)$ curves look like those of grains with "glue" of Figure 15 and not that of grains without "glue" of Figure 12, where the peaks $q(\varepsilon)$ correspond to the maximum dilatancy $\varepsilon_v(\varepsilon_1)$. In fact, results in London clay show that peaks for low σ_3 are associated with the beginning of dilatancy with a very important slope.

The final even stretch of $q(\varepsilon_1)$ curves is found in q (p') plane by a straight coming out from origin which is close to the curve of maximum shear strength of a remoulded clay with the same w_L (fig. 26).

The curve of maximum strength joins the straight established with the values of the even stretches for the high confining stress $\sigma_3 = 4.4$ MPa; this one probably destroys progressively the "glue" as shows the $q(\varepsilon)$ curve with a maximum stretch,

BISHOP, 1965 - OCR~10

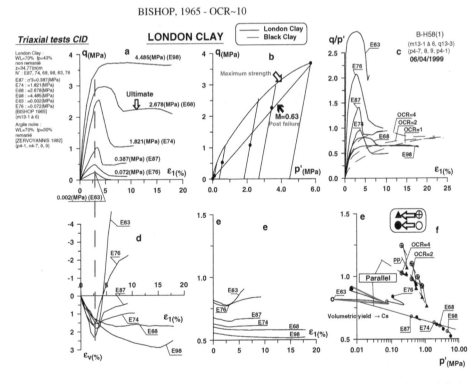

Figure 25.   London clay behaviour showing analogy with grains with "glue" and showing that peaks of strength for low σ_3 are associated with the beginning of dilatancy and the influence of large strains and high stresses to achieve the remoulded behaviour.

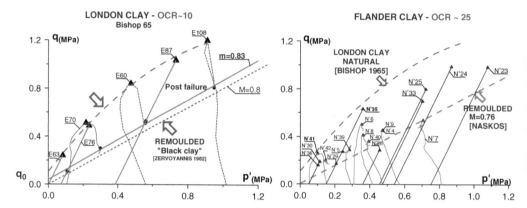

Figure 26. "Flandres" clay (OCR ~25) showing much more scatter on strength than London clay (OCR ~10).

without peak. It must be noticed that this stress is higher than the maximum historical overburden, contrary to the remoulded clays where the intersection point of maximum strength with the value of the even stretch is around half the maximum overburden history.

The maximum strength in $q(p')$ plane presents a sharp bend for low stresses, alike as rocks and "glues" with fissures. It can be assumed that these fissures are homogeneous in the sample, since the drained (fig. 25) and undrained (fig. 26) tests join this maximum curve with small scattering.

4.1.3 Clay of "Flandres" (Josseaume, 1997)

This clay is of the same geological formation as London clay (YPRESIEN), but samples were got 10 metres deep in this layer with erosion above giving an overconsolidation ratio of 25 instead of 10 of the London clay. Figure 26 shows a much more scatter of maximum strengths than London clay. Furthermore, the surfaces of weakness produce sliding through the sample in various directions. This is in agreement with the expected amplification of fissures with the increase of overconsolidation.

4.1.4 Gault clay

This natural ground is a little more ancient than London clay (Ic 1.2 instead of 1.08) but integrates significantly more $CaCO_3$, than "glue", giving a greater maximum strength for $\sigma_3 < 1\,\text{MPa}$ (fig. 27). However, for higher σ_3, the analysis of test results in this figure lets suppose an important decrease of the maximum strength, which becomes lower than that of London clay (Bishop, 1965). This will support the fragile behaviour of $CaCO_3$ which gives a limit curve of fissuring: "damage line".

4.1.5 Chalk marl

In Figure 28 $q(\varepsilon)$ curves seem to show a straight line until $\varepsilon \sim 10^{-2}$, and then a very important decrease of q, which can be attributed to the localisation of fissures, designating failure. Finally a residual even stretch is observed. A loop on log E (log ε_1) shows the important increase of modulus with the decrease of strain level for $\varepsilon < 10^{-3}$.

Tests with other confining pressures σ'_3 show moduli which seem to converge to a value of modulus $E \sim 2000\,\text{MPa}$ for $\varepsilon \sim 10^{-5}$. Two types of behaviour can be assumed before failure as a function of the two types of "glues". The first after a strain of 3×10^{-5} showing at the beginning σ_3 influence and the fissuring that can be supposed to result from a fragile "glue". The second, after a strain of 10^{-3}, where the modulus is relatively constant, probably owing to a relative movement of grains with ductile "glue".

4.2 Sandstone

Figure 29 reveals, alike for concrete (fig. 16), a marked onset of fissures, or its amplification. The yield points in $q(\varepsilon)$ curves give a yielding curve in the q(p) plane, which joins the curve of maximum strength, having a straight segment between $\sigma_3 = 50$ and $\sigma_3 = 350\,\text{MPa}$. This straight segment remains parallel to the line of strengths at large strains passing through the origin.

4.3 Marble

The results on a marble (fig. 30) show, unlike the previous curves, that the line of maximum strengths joins the line of the strengths at large strains passing the origin. It must be noticed that the common point of the lines occurres for $\sigma_3 = 49\,\text{MPa}$ where no more peak appears in the $q(\varepsilon)$ curve.

138

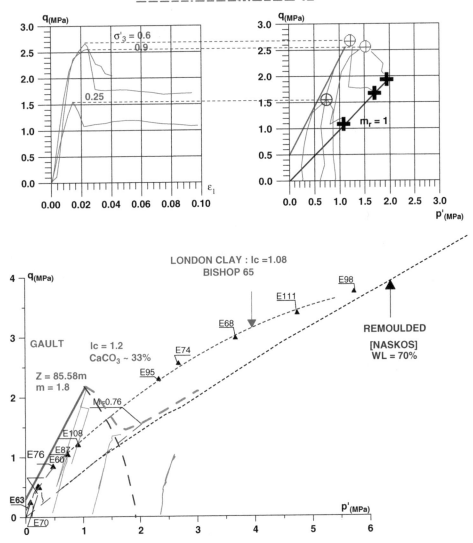

Figure 27. Triaxial tests results on Gault clay showing its analogy with the behaviour of grains with a fragile "glue" and the different behaviour in comparison of London clay by the effect of CaCO₃ (33%).

The initial parts of $q(\varepsilon)$ curves show a modulus independent of σ_3, alike the sandstone and the grains with "glue" when this one has the same strength as the grains.

For σ_3 stresses in the range of zero to 21 MPa no remarkable influence is observed until the beginning of fissuring. These $q(\varepsilon)$ curves continue to grow until a maximum strength, and then drop gradually until a more important decrease within a slight strain, probably due to the localisation of fissures. However, before this sharp bend it is possible to draw a

limit curve where strains and fissures, if present, are homogeneous.

5 FINAL REMARKS

Some simplified aspects of behaviour and their limits for grains without "glue" (sand and remoulded clay) were presented. Then, modifications produced by various types of "glues", in particular in presence of fissuring, were analysed. This approach applied on

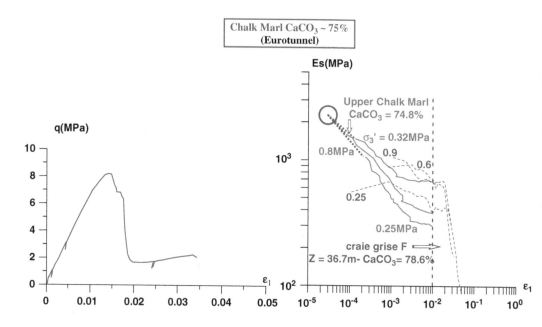

Figure 28. Typical failure behaviour (localisation of fissures) and the influence of strain and stress in the secant modulus for Chalk Marl.

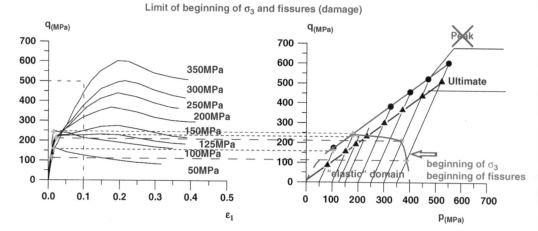

Figure 29. Analogy of sandstone behaviour alike a concrete cement showing clearly the beginning of σ_3 influence and fissures (damage).

reproducible samples allows searching analogies in the complexity of the behaviour of natural grounds, going from soils to rocks.

This study is based on the comparison of the numerical results of more than one hundred tests;

interpretation of results was deliberately simplified to illustrate a method.

This method allows a more generalised analysis theoretically applicable to any material by choosing other grains, other "glues", and other desired fissuring.

140

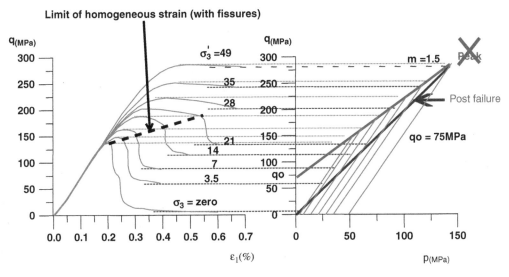

Triaxial Tests on Tennessee Marble
(Wawersik and Fairhurst 1970)

Figure 30.   Analogy of marble behaviour alike grains with "glue" of same strength of grains, showing a limit curve until which strains and fissures are homogeneous.

It would be desirable that this will enhance calculations of continuum mechanics where only grains and "glues" are continuum and where we can put in evidence the role of the boundary conditions and constitutive laws of grains and "glues".

We can also expect calculations of "molecular mechanics" and "ab initio" to clarify, for example, the constitutive laws of clay grains. In fact, we can nowadays create the geometry, knowing the shape of the clay grains, as well as the position of the atoms in the sheets, but we ignore the constitutive laws of the adsorbed liquid.

REFERENCES

This very general paper takes into account numerous publications that cannot be cited here. We just mention the experimental research work carried out at "Ecole Centrale Paris" which results are available in digital format to be able to be compared. In consequence the references in the text can be found in these works mentioned herewith.

Grains without "glue"
Biarez, J. and Hicher, P.Y. 1994. Elementary mechanics of soil behaviour – Saturated-remoulded soils. A. A. Balkema.
Liu, H. 1999. "Etude des comportements ($\varepsilon > 10^{-6}$) des sols naturels et des sols reconstitués au laboratoire". (PhD, ECP).
Saim, R. 1997. "Des comportements repères des grains sans colle a un exemple réel". (PhD, 1ECP), ($\varepsilon > 10^{-2}$).

Grains with "glue".
Bard, E. 1993. "Comportement des matériaux granulaires secs et a liant hydrocarboné" ($\varepsilon > 10^{-2}$). (PhD, ECP).
Charif, K. 1991. "Contribution à l'étude du comportement mécanique en petites et grandes déformation" ($\varepsilon > 10^{-6}$), (PhD, ECP).
Taillez, S. 1998. "Etude experimentale du comportement mécanique des sols granulaires injectés". (PhD, ECP).

Natural ground
Fayad, T. 2000. "Mécanique des grains avec et sans colle pour une comparaison avec des sols saturés et remaniés". (PhD, ECP).

Integrated ground behaviour
Comportement des ouvrages

Deformation Characteristics of Geomaterials – Di Benedetto et al (eds)
© *2005 Taylor & Francis Group, London, ISBN 04 1536 701 8*

Particle-level phenomena and macroscale soil behavior

J.C. Santamarina
Georgia Institute of Technology, Atlanta, Georgia, USA

ABSTRACT: Soils are particulate materials; fluids and microorganisms fill the pore space. Salient particle level phenomena include fluid-mineral interaction, mixed fluid phase interaction, bio-mediated geochemical processes, size-dependent particle level forces, particle-shape effects, columnar force transfer along the granular skeleton, and multiple internal temporal and spatial scales. These phenomena manifest into complex macro-scale soil response including inherent non-linear and non-elastic behavior; stress-dependent stiffness, strength and volume change; porous and pervious material, inherent and stress-induced anisotropy, time-dependent response, and energy coupling among others. Selected particle-level processes and macroscale behaviors are reviewed.

1 INTRODUCTION

The empirical understanding of soil behavior evolved long before its fundamental understanding. The concept of effective stress has its roots in Archimedes (II BC), and it is elaborated in soils and porous media by Terzaghi and Biot (1920's–1930's). Egyptians (XXVII BC) knew how to control friction, which is later analyzed by daVinci (XVI) and Coulomb (XVIII). Reynolds demonstrated dilatancy (XIX); the interplay between shear stress, volume and confining stress is developed by Taylor, Roscoe, and Schofield (1940's–1960's). Suction induced shrinkage was known to the Babylonians (XXIII BC), yet the comprehensive understanding of partial saturation takes place in the 1960's with contributions by Aitchison, Bishop, Morgenstern and Fredlund. The theoretical framework for seepage was advanced by Laplace, the Bernoulli's and Darcy (XVIII-XIX). Long recognized plasticity is investigated by Goldschmidt in the 1920's and Lambe and Mitchell in the 1950's, after colloidal theory is advanced by Guy, Chapman, Stern, Debye and Huckel in the period from 1910 to 1930. The inherent non-linear and non-elastic behavior of soils is explained at the particle level by Hertz (1870's) and Mindlin (1950's) contact theories.

Today, new tools and technological developments open unprecedented possibilities to understand soil behavior and to engineer its response. Such technological advances include discrete element modeling, molecular dynamics, atomic force microscopy, and a wide range of high resolution non destructive techniques and tomographers (e.g., micro-tomographers, magnetic resonance imaging).

The purpose of this manuscript is to present a concise analysis of salient particle-level phenomena and processes in order to comprehend the macroscale soil response. Soil components are analyzed first leading to soil packing and fabric formation. Then, selected macroscale responses are discussed.

2 SOIL COMPONENTS

Soils are particulate materials. Mineral particles form the granular skeleton and the interparticle porosity is filled with one or more fluids. The pore space may also include microorganisms (Figure 1).

2.1 Pore fluid

The fluid that fills the pores left by the mineral particles is either a single-phase fluid, typically water in

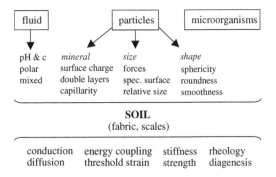

Figure 1. Soils. Particle-level properties and macro-response.

saturated soils, or a mixed-phase fluid, such as water and air (typical in partially saturated soils), or water and an immiscible organic fluid (e.g., contaminated soils or petroleum reservoirs).

When water is the prevailing fluid, the ionic concentration and the pH determine -and reflect- the interaction between the mineral and the water (Figure 2-a). Typically, the pore fluid in soils is an aqueous electrolyte consisting of free water molecules and hydrated ions. The water molecule polarity aided by thermal vibration makes water an effective solvent. Water can dissolve salt until it reaches the "saturation concentration", which for NaCl is 350 g/lt near freezing and 390 g/lt near boiling. For comparison, the ionic concentration is 0.001 g/lt for fresh water, and about 35 g/lt for seawater.

The value of pH is a measure of the relative balance between hydrogens and hydroxyls in the water-base solution. The pH = 7 when the number of hydrogens and hydroxyls is balanced. However, if a substance such as chloric acid is added, the number of hydrogens exceeds that of hydroxyls and pH <7 (Figure 2-b). Alternatively, hydroxyls may be incorporated, for example by adding sodium hydroxide, and the water pH increases, pH >7. The relevance of ionic concentration and pH will become apparent when minerals are taken into consideration.

Molecules at the interface between two immiscible fluids are not randomly oriented because of the different interaction forces they experience to each fluid. This molecular-level effect results in a contractile, membrane-like interface at the macroscale. The "surface tension" in this membrane depends on the properties of the two interactive fluids. Figure 3-a shows the formation of a capillary meniscus at the interface between two spherical particles.

The angle formed between the fluid interface and the mineral surface is called the contact angle. The shallower the contact angle, the more "wetting" the fluid is. Soils surrounded by oil may become "oil-wet", even though most minerals tend to prefer water-wet conditions due to the mineral structure itself. In general, the contact angle is assumed constant for a given fluid-substrate system. However, menisci that form between different particles interact through the fluid pressure, and the resulting contact angles are not the same (Figure 3-b). Given these complexities, the analysis and scaling of microscale effects in partially saturated particulate media is complex.

2.2 Particles

2.2.1 Mineralogy
Each soil grain is typically made of a single mineral, except in some coarse sands and gravels. The mineralogy in coarse grains determines the grain stiffness, strength, hardness, abrasiveness and crushability. For example, the crushing yield stress is >10 MPa for quartz sand and <0.2 MPa for carbonate sands.

The role of mineralogy in fine-grained soils is more subtle, yet more important. Clay minerals are made of layers of silicates and aluminates, and are known as phyllosilicates. Figure 4 shows the chemical structure

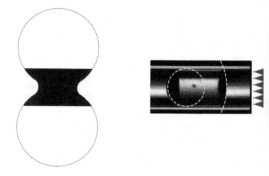

Figure 3. Mixed fluids. Microphotographs: (a) Meniscus between two 1 mm diameter glass beads. (b) Water drop in 1 mm capillary tube subjected to differential pressure (Alvarellos 2003) – Note that the contact angle is not constant when there are interacting menisci.

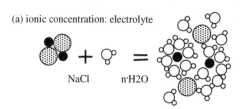

(a) ionic concentration: electrolyte

NaCl   n·H2O

(b) pH

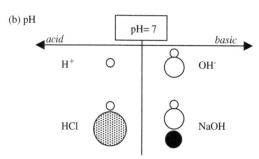

Figure 2. Water – (a) Ionic concentration and (b) pH.

Figure 4. Clay minerals – single sheet. Notice differences among the surfaces. (a) Kaolinite – a particle is made of many similar sheets. (b) Montmorillonite (see Velde 1992).

of kaolinite and montmorillonite. The analysis of these sketches reveals the presence of either oxygen or hydroxyls on the surface of clay minerals, thus the affinity of clay surfaces to either the hydrogens or hydroxyls in water. It also hints to the importance of pH: indeed, the dissolution or precipitation of clay minerals and the surface charge will be controlled by the pH of the pore fluid.

If the clay surface exhibits charge, then there will be counterions in the vicinity of the clay mineral to ensure electro neutrality. Counterions do not bond onto the clay surface in the presence of water because of the thermal agitation of water molecules that continue hydrating and colliding with these ions; instead, counterions remain in the vicinity of the clay surface forming a counterion cloud. The higher the ionic concentration in the pore fluid, the thinner the cloud becomes.

In summary, the pore fluid pH and ionic concentration affect surface charge and the thickness of the counterion cloud near mineral surfaces.

2.2.2 Size (particle-level forces)

Particle-level forces are linear, quadratic or cubic functions of the particle size d (Table 1). Simple force-equilibrium analyses between these forces show that: (1) particles larger than 10-to-100 microns are controlled by the interparticle skeletal forces that result from the applied effective stress; while (2) both capillary and electrical forces prevail for smaller particles.

The drastic change in governing physical forces and related processes among fine and coarse grained soils is recognized by the Unified Soil Classification System USCS, and its emphasis on sieve #200 (76 μm). In the case of coarse soils with no fines, the USCS also requires the determination of the coefficient of uniformity $C_u = D_{60}/D_{10}$: when particles of different size are present, the smaller particles fit between the voids left by the larger particles and soils exhibit lower terminal void ratios e_{max}, e_{min} and those on the critical state line.

Smaller particles that are not part of the load-carrying granular skeleton can be dragged by seeping fluids, and migrate. A small particle d_{small} will be retained at the pore throat formed by larger particles

d_{large} when it is larger than the size of the pore throat $d_{small} \geqslant d_{throat}$, which is about $d_{throat} \sim 0.2$-to-$0.4 d_{large}$ (Figure 5-a). When many small particles migrate, they may reach the pore throat and form a bridge (Figure 5-b). Particle-level experimental results by Valdes (2002) show that migrating particles can be retained by bridge formation when they are about $d_{small} > 0.2 d_{throat}$, therefore $d_{small} \sim d_{large}/(10$-to-$25)$. These observations are the particle-level justification of empirically developed filter criteria (Terzaghi et al., 1996).

2.2.3 Specific surface

Surface related phenomena, such as chemical reactions and viscous drag in seepage are not determined by particle size per se, but by surface area. The specific surface S_s is defined as the ratio between the surface area and the mass of the particle. When a 1 m^3 cube is split into smaller pieces, the total mass of the cube remains constant, however the surface area increases dramatically. In other words, the specific surface increases as the particle size decreases.

Specific surface reflects both particle size and shape. Spherical particles exhibit the least amount of surface for a given volume; however, when a spherical particle is flattened, it develops a surface that is inversely proportional to the thickness of the flattened particle. In general, the specific surface is determined by the smallest dimension of the particle L_{min}, e.g., the thickness of platy particles or the diameter of cylindrical particles. The theoretical maximum value a particle can reach before it looses its mineralogical identity is estimated to be $\sim 2000 \, m^2/g$. The specific surface can range from more than a $1000 \, m^2/g$ for amorphous silica to less than $10^{-4} \, m^2/g$ for coarse sands as shown in Figure 6.

Sieve analysis permits identifying the various fractions in a soil. The resulting distribution is based on mass. Grain size distribution can also be explored in terms of the contribution each fraction makes to the total specific surface of the soil. Figure 7 shows (a) grain size distribution in terms of "cumulative mass", (b) the corresponding density function, i.e., the percentage of each fraction, (c) the density function in terms of surface contributed by each fraction, and (d) the cumulative grain size distribution in terms of surface area. The surface-based distribution is always shifted towards the finer sizes due to their higher

Table 1. Particle-level forces – Note particle-size dependency.

Skeletal	$N = \sigma' d^2$
Weight	$W = (\pi G_s \gamma_w/6)d^3$
Buoyant	$U = (\pi \gamma_w/6)d^3$
Hydrodyn.	$F_{drag} = 3\pi\mu vd$
Capillary	$F_{cap} = \pi T_s d$
El. attraction	$Att = A_h d/24t^2$
El. repulsion	$Rep = 0.0024 \sqrt{c_o} \, e^{-10^{o}t\sqrt{c_o}}d$
Cementation	$T = \pi\sigma_{ten}td$

(a) *no bridge*: $d_{small} \geq d_{throat}$ (b) *bridge*: $d_{small} > \sim 0.2 d_{throat}$

Figure 5. Fines migration and filter criteria at the particle level. (a) Single migrating particle passing condition. (b) Bridge formation in multiple migrating particles.

147

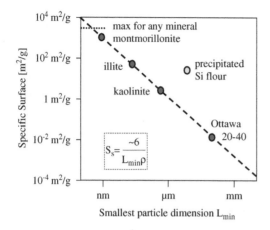

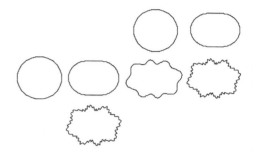

Figure 6. Specific surface and particle size (ρ is the mineral mass density). Precipitated silica flour deviates from the general S_s-vs-L_{min} trend observed for uniform soils; electron microphotographs show that the uniform silt-size particles are made of submicron particles.

The equation shown in Figure 6:

$$S_s = \frac{\sim 6}{L_{min}\rho}$$

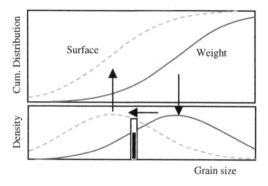

Figure 7. Mass-based and surface-based grain size distributions.

contribution to total surface. The surface-based grain size distribution is a more meaningful representation of the soil for surface-dependent processes, such as seepage, diffusion retardation, chemo-mechanical coupling, and geochemistry in general.

2.2.4 Shape

The emphasis on particle size and relative size in that Unified Soil Classification System conceals the importance of particle shape on soil behavior. Particle shape has three different scales: global shape (sphericity, platiness); bumps and corners (roundness, angularity); and small surface perturbations (smoothness, roughness), as shown in Figure 8.

The shape of particles larger than $\sim 50\,\mu m$ is determined by mechanical action such as fractures, dislocations and collisions. However, the shape of particles finer than $\sim 50\,\mu m$ is controlled by chemical reactions.

Figure 8. Scales in particle shape. Examples of deviations from sphericity: precipitated carbonate; from roundness: crushed sand; from smoothness: forminiferan.

2.3 Microorganisms

The role of microorganisms in soils has been recognized and extensively studied in agriculture and in geoenvironmental remediation, but overlooked in geomechanics.

The size of bacteria is in the order of 1 μm (spores can be as small as 0.2 μm), therefore, there could be $\sim 10^8$ bacteria/mm^3 of pore space. Given the high reproduction rate (duplication every 20-to-60 minutes), microbial activity rapidly adapts to the most diverse environments, from very cold to very hot, from acidic to basic. Hence, there is exceptional bio diversity and ubiquitous presence of microorganisms.

Life requires (1) nutrients, i.e., energy source and carbon for cell mass, (2) water, and (3) proper environmental conditions such as pH, salinity, temperature, and sufficient space. Any of these components can become the limiting factor and restrict life in soils. From this point of view, bacterial activity is space-limited in clayey soils.

While the biological effects on geomechanical behavior are still underexplored, promising areas for further investigation include bio-mediated geochemical reactions and short-term cementation/digenesis, clogging and changes in hydraulic conductivity, and gas generation.

3 PACKING AND FABRIC

Governing forces lead to different fabric formation mechanisms in coarse particles (d $>$ 50-to-100 μm) and in fine particles (d $<$ 10-to-50 μm).

The packing of coarse grain soils is determined by particle shape and the relative size of the particles. Shape affects particle mobility and their ability to attain the minimum energy configuration; hence, non-spherical, angular or rough particles pack in looser states. As discussed earlier, relative size determines the ability of small particles to fit in the voids left by larger particles. Expressions for e_{max} and e_{min}

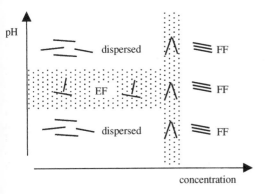

Figure 9. Schematic fabric map applicable to fine-grained soils (controlled by electrical forces). Ionic concentration and pH values in transition zones are mineral specific. EF flocculation may not take place in very thin montmorillonite particles.

as a function of roundness R for $C_u = 1$ follow (Youd, 1973):

$$e_{max} = 0.554 + 0.154 \, R^{-1} \qquad (1)$$

$$e_{min} = 0.359 + 0.082 \, R^{-1} \qquad (2)$$

Non spherical particles tend to sediment with their longest axis on the horizontal plane (minimum potential energy), causing inherent fabric anisotropy. Platy mica particles bridge over voids left by rounded particles beneath it, and enforce the alignment of rounded particles above the sheet. The combination of ordering and bridging makes the void ratio of the soil a function of the percentage of mica and the relative size of mica with respect to the size of round particles (Guimaraes, 2002).

In fine particles, pH and ionic concentration determine fabric formation. At low ionic concentration, edge-to-face aggregation takes place at intermediate pH values, and dispersed structures at either high or low pH. On the other hand, when the ionic concentration is high, the counterion cloud shrinks, the osmotic repulsion decreases and van der Waals attraction prevails causing face-to-face aggregation. A schematic fabric map is presented in Figure 9 (extensive experimental evidence can be found in Palomino 2003). The transition pH and ionic concentration depend on the type of clay mineral as well as the type of ions present in the pore fluid.

4 MACROSCALE SOIL BEHAVIOR

The macroscale response of a soil mass reflects underlying microscale processes that are associated with mechanisms and variables described in the previous sections. In this section, selected macroscale behaviors are analyzed at the microscale, including the mechanical response at small and large-strain, time dependent phenomena, and multiple internal scales.

4.1 Strain-dependent mechanical response

The mechanical behavior of the soil skeleton is strain-dependent, and very different microscale mechanisms develop when small-strain-constant-fabric and large-strain-fabric-changing excitations are imposed.

4.1.1 Small-strain

The mechanical response at small strains takes place at constant fabric, and the deformation concentrates at contacts, the stiffness is maximal, energy losses and volume changes are minimal, and diagenetic effects exert the highest influence.

Deformation at constant fabric means that particles do not lose or change neighboring particles. Therefore, the small strain stiffness (and Poisson's ratio) of the soil reflects contact deformation. In the case of coarse grains, contact stiffness is determined by the deformability of the particle and asperities near the contact (theoretical analog: Hertzian contact). In the case of fine grains the intuitive concept of solid-to-solid contact losses relevance, and interparticle interaction takes place through interparticle electrical forces.

The small-strain shear stiffness for both fine and coarse soils is a power-function of the effective confining stress. In terms of the shear wave velocity V_s,

$$V_s = \alpha \left(\frac{\sigma'_{propagation} + \sigma'_{particle \, motion}}{2 \, kPa} \right)^\beta \qquad (3)$$

where the parameters α and β are experimentally determined. Extensive experimental evidence shows that α and β are inherently interrelated as:

$$\beta = 0.36 - \alpha / 700 \qquad \text{where } \alpha \, [m/s] \qquad (4)$$

The value of the exponent is $\beta = 0.17$–0.20 for dense, rounded sands and can exceed $\beta = 0.3$ for soft clays. The value of β increases for loose sands, soft clays, angular and/or rough grains, and soils with mica. On the other hand, $\beta \rightarrow 0$ for overconsolidated clays and cemented sands before yield. As the degree of cementation increases, the localized deformation at contacts decreases, the soil stiffness increases, the stress sensitivity of the soil stiffness decreases (lower β), and the stress that is required to regain the stress-dependence stiffness of the soil mass increases.

Partial saturation adds capillary forces between particles beyond those carried by the applied effective stress. Therefore, a drying soil will exhibit increasing stiffness unless massive shrinkage fractures develop.

Small-strain anisotropy reflects inherent fabric anisotropy. Post depositional changes due to loading or cementation affect the degree of anisotropy.

4.1.2 Large-strain

The large-strain mechanical response involves fabric change, and contact deformation plays a diminishing effect. Stiffness decreases with increasing strain, and important energy losses take place – mostly of frictional nature.

The increase in mean effective stress causes volume contraction. However, either contraction or dilation results from applied deviatoric stresses as a result of two competing microscale effects: dilation to overcome rotational frustration in particles with high coordination, and chain buckling and collapse when particles have low coordination. At very large strains, the balance between local collapse and local dilation reaches statistical equilibrium, the soil shears at a constant volume φ_{cv}, and the soil fabric approaches the "critical state fabric".

The friction angle of the soil when shearing at constant volume is determined by particle shape, including sphericity, roundness, and roughness. In terms of roundness R (Santamarina & Cho 2004),

$$\varphi_{cv} = 42 - 17R \qquad (5)$$

The constant volume friction angle in sands has been correlated with the angle of repose. However, the angle of repose is different when it is measured for a sand cone φ_{ext}, a planar surface, or an inside flow cone φ_{int} (e.g., when a central plug beneath a filled cylinder is removed). Data presented in Figure 10 show that the internal angle of repose φ_{int} is significantly greater than the external angle φ_{ext}. The internal flow experiences a gradual reduction in cross section (~2D flow gradually becomes 1D as in lateral compression LC), while flow on the external slope experiences a gradual increase in cross section (as in lateral extension LE). At the microscale, interparticle coordination in the annular direction increases in the first case (φ_{int}) while it decreases in the second case (φ_{ext}). Similar friction angle anisotropy is observed in triaxial axial extension AE and axial compression cases AC, both sands and clays (Mayne and Holtz 1985).

The undrained shear strength also exhibits anisotropy. However, in this case, anisotropy is controlled by pore pressure generation following skeletal collapse upon loading. In general, the particle chains that carry load tend to be more stable for continuous loading in the same direction than for sudden transverse loading. Therefore, the undrained shear strength in axial extension is lower than in axial compression (Ladd et al. 1977; Mayne & Holtz, 1985; Zdravkovic & Jardine 1997; Yoshimine et al. 1999).

Particle shape determines the very large-strain residual shear strength where the higher the platiness of particles, the lower the residual friction angle. Nonspherical particles tend to develop anisotropic fabrics. The compressibility of anisotropic fabrics in the vertical and the horizontal directions are quite distinct. These will affect the generation of pore pressure during undrained loading. Therefore, the shear strength of anisotropic fabrics can be dramatically different for dissimilar loading paths (Leroueil & Hight 2003).

4.1.3 Elastic threshold strain

The transition from small to large-strain behavior is not abrupt but gradual. The beginning of modulus degradation denotes the "linear threshold strain" γ_{dt}, which depends on the nature of particle-to-particle interactions (Figure 11). For mechanical Hertzian

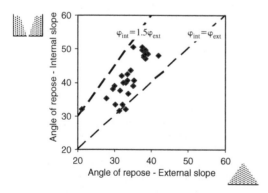

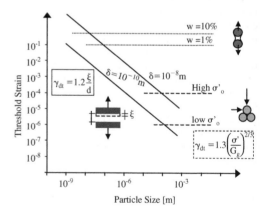

Figure 10. Friction anisotropy (several studies with natural and crushed sands – compiled by G. Narsilio).

Figure 11. Particle-level estimation of elastic threshold strain for different contact mechanisms.

contacts, the amount a particle can deform before detaching depends on the mineral stiffness G_s and the initial flatness of the contact due to the applied initial confining stress σ'. For interparticle interaction of electrical nature, the separation between two particles before ending interaction is related to the thickness of the counterion cloud ξ. Then, the threshold strain is proportional to the ratio between the thickness of the adsorbed layer and the particle size. This ratio also affects soil plasticity, therefore, the linear threshold strain in clayey soils is correlated with the plasticity of the soil (details and discussion in Santamarina et al., 2001; Experimental evidence in Vucetic & Dobry 1991).

4.2 Short Term Effects – Rheology

Time-dependent changes highlight non-equilibrium conditions after the application of load, including gradients of chemical, electrical, mechanical or thermal origin. Macroscale boundary criteria (such as isothermal, constant composition, and constant volume) do not apply at the microscale, and local thermal, chemical, mechanical or electrical transport can take place.

A stable soil under constant boundary conditions is at the bottom of an energy well, and any change will require overcoming some energy threshold (frictional threshold at each contact or energy well for ions in Stern layer). A perturbation may contribute to overcoming the corresponding energy barriers, triggering internal changes. Once the process is initiated, internally liberated energy, for example in the form of acoustic emission, may continue to sustain the process of change until a new global energy well is reached.

More recently, the relevance of strain rate on mechanical soil properties is gaining increased attention. Related macroscale behavior include (Tatsuoka et al. 2003):

– "isotach response" where the strength and stiffness increase when the strain rate increases, such as in low plasticity Pleistocene clays and soft rocks; and
– "viscous evanescence" where the same stress-strain behavior is observed at all strain rates however a transient is measured when the strain rate changes, such as in sands.

Such behavior can have important engineering implications. For example, if the undrained strength increases at a rate of 10% per log cycle of strain rate (Kulhawy & Mayne 1990), then how should in situ and standard laboratory tests ($\dot{\varepsilon} \sim 1-10^{-2}$/min) be analyzed to provide strength data relevant to foundation construction $\dot{\varepsilon} \sim 10^{-7}-10^{-8}$/min (Randolph 2002)?

Several microscale mechanisms are hypothesized next as potential contributors to short term rheological effects. Soil testing to identify the prevailing mechanism requires procedures that vary strain rate,

temperature (activation energy), and pore fluid (dry air, and fluids of different permittivity).

4.2.1 Contact-level mineral creep and friction
Mineral creep and interparticle friction are complex phenomena, which are exacerbated by interparticle interactions throughout the granular skeleton. Salient comments follow.

– Mineral creep at contacts. Numerical micromechanical simulations (Kuhn & Mitchell, 1993; Rothenburg 1992) and photoelastic model studies (Díaz-Rodríguez & Santamarina 1999) show that contact creep cause: force redistribution, changes in load-carrying granular chains, increase the global coordination number, and significant increase in small strain stiffness (Cascante & Santamarina, 1996). In fact, the higher the variability in contact forces the higher the global creep rate.
– The velocity dependency of friction is affected by the presence of water and the development of boundary and hydrodynamic lubrication. Furthermore, a frictional transient is observed upon changes in velocity – similar to macroscale viscous evanescence (Figure 12 – Dieterich 1978)
– Non-linear coupling between friction and noise (noise may be generated by a nearby slippage). Low amplitude vibrations can accelerate thixotropic effects in soils; this is known as rheopexy in clays and cyclic pre-straining in sands.

4.2.2 Fluid migration
Mechanisms related to pore fluid migration apply to both saturated and partially saturated soils:

– Mitchell (1960) measured the rate of thixotropic regain at different temperatures and estimated the activation energy of the process to be 3–4 kcal/mol, which suggests that viscous flow of water as the underlying mechanism.
– Moisture migration in partially saturated soils, is relatively fast -via conduction- in high saturation cases, yet it is slow in low saturation conditions as it takes place through and through the vapor pressure

4.2.3 Surface chemistry and phenomena
The role of adsorbed layers on time-dependent processes has long been recognized, for example,

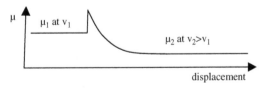

Figure 12. Transient in mineral-to-mineral frictional response when velocity v is increased from v_1 to $v_2 > v_1$.

thixotropic behavior is often related to flocculation prone clay systems (see Van Olphen, 1951, Mitchell, 1960). Mechanisms under this category relate to

- Ion diffusion in adsorbed layers following disturbance by inter-contact relative displacement
- Atomic stick-slip in adsorbed layers (Landman et al. 1996; Persson 1998)
- Pressure solution-precipitation
- Cementation (may be bio-mediated)

4.2.4 *Time*

Time-dependency implies underlying inertial, conduction (includes viscous) or diffusional processes. Simple scale analyses permit estimating the mean time scales:

$$t_{displace} = \frac{\varepsilon}{\dot{\varepsilon}} \qquad \text{imposed} \tag{6}$$

$$t_{inertia} = d\sqrt{\frac{\rho}{\sigma'}} \qquad \text{particle-inertial} \tag{7}$$

$$t_{conduct} = \frac{d}{ki} \qquad \text{local conduction} \tag{8}$$

$$t_{diffuse} = \frac{\varepsilon^2 d^2}{D} \qquad \text{local diffusion} \tag{9}$$

where ε = strain, $\dot{\varepsilon}$ = strain rate, k = conductivity, i = gradient, and D = diffusion coefficient.

Evidence for very high local hydraulic gradients i is available in other poroelastic media such as bones, (Wang et al 2003). The time for local diffusion (Equation 9) is required for counterion reorganization after an imposed disturbance equal to εd. Interparticle repulsion continues changing during this time.

Time dependent effects should be expected when the local, particle level processes cannot keep up with the rate of disturbance. For example, when, $t_{displace} \ll t_{diffuse}$, the expected strain rate is $\dot{\varepsilon} \gg D/\varepsilon d^2$. The diffusion coefficient for ions in the adsorbed layer varies between $D \sim 10^{-10} \, m^2 s^{-1}$ to the bulk $D \sim 10^{-9} \, m^2 s^{-1}$. Therefore, for a strain $\varepsilon = 10^{-2}$, and particle size $d = 10^{-3} \, m$, the predicted strain rate for this phenomenon is $\dot{\varepsilon} > 10^{-2} s^{-1}$.

"Domino" effects throughout the granular skeleton extend the effective duration of contact-level phenomena (Figure 13 – experimental evidence: duration of acoustic emission after a transient excitation). In some respects, the phenomenon resembles queuing: the expected wait in a queue if customers arrive at a certain rate 'c' and are served at rate 's', where both events have some known distribution, e.g., Poisson. If applicable, the mathematics of queuing theory can be invoked to further analyze the effective macroscale time of the process.

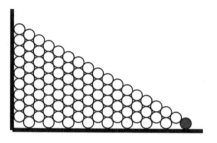

Figure 13. Duration of macroscale processes. If the black marble is suddenly displaced outwards a distance δ, how long will the marble at the top of the pile experience displacement?

4.3 *Variability and scales*

The topic of scales must take into consideration spatial and temporal scales corresponding to the soil mass and the imposed excitations, identification of emergent phenomena (e.g., water gap formation in heterogeneous media subjected to dynamic loading – Fiegel & Kutter 1994; Kokusho 1999), effective properties for design and related engineering implications.

The soil mass and the soil response exhibit very characteristic temporal and spatial scales. Spatial scales L can be identified from the particle level (e.g., roughness and particle size), at intermediate scales (e.g., aggregations and spatial variability in void ratio), and at the level of the engineering system (e.g., layers and macro-heterogeneity). Temporal scales for conduction $t_{cond} \approx L/ki$, diffusion $t_{dif} \approx L^2/D$, and vibration $t_{vib} = 2\pi \sqrt{(m/k^*)} \approx 2\pi L/V_s$ (k^*: stiffness; m: mass; V_s: shear wave velocity) are related to the spatial scale L of the soil mass and/or system, and the corresponding material parameter (similar to Equations 7, 8 and 9). The associated times can range from 0.1 µs for chemical diffusion across the double layers, to years for pressure diffusion consolidation in thick soil layers.

The temporal and spatial scales of the imposed excitation (mechanical, thermal, chemical, electrical or biological) vary from ms to quasi-static, and from cm to regional (Table 2).

The comparison between the internal scales and the scale of the imposed excitation determines the nature of the excitation-soil interaction. In terms of time scales, drained loading applies when the time for load application t_{load} is greater than the time for pressure dissipation ($t_{load} > t_{dif}$), quasi-static analyses apply when $t_{load} \gg t_{vib}$, remediation is diffusion-control when $t_{cond} < t_{dif}$.

In terms of spatial scales, the comparison between the scale of the medium and the excitation affect the choice of analysis, for example, block theory should be used when the scale of the structure (e.g., tunnel diameter) is similar to the joint spacing, otherwise equivalent continuum models apply.

Table 2.	Spatial and temporal scales in the excitation.		
Excitation	Spatial [m]	Time [s]	
In situ testing	10^{-2}	1	
Truck at 100 km/hr	10^{-1}	10^{-2}	
Seismic excitation	10^0-region [a]	10^{-1}–10^2 [b]	
Building construction	10^1 m	10^7	
Landfill	10^2 m	quasi-static	

Notes: (a) spatial scale varies with system under consideration; (b) there are two temporal scales: one related to frequency content and the other to the duration of the event.

In a spatially varying soil mass, the effective properties for design depend on (1) the value of the selected material property for each component k_i present in the region affected by the excitation, (2) the volumetric or gravimetric fraction of each component f_i, and (3) their spatial distribution. Consider proportional properties k between input and output, such as conductivity [gradient→flow rate] and compliance [force → deformation] rather than stiffness. The equivalent "effective" value k_{eq} can be empirically fitted with a power average law:

$$k_{eq} = \left[\sum_i f_i (k_i)^\alpha \right]^{\frac{1}{\alpha}} \quad f_i<1, \ \sum f_i=1.0, \ 1\leq\alpha\leq-1 \quad (10)$$

where k_i is the value of the property for the ith component and f_i its volumetric or gravimetric fraction. The exponent a is related to the spatial arrangement of the components and the physical nature of their interaction. Special cases of the power average expression are listed next for a medium made of two materials:

$$k_{eq} = f_1 k_1 + (1-f_1) k_2 \qquad \alpha=1 \quad parallel \quad (11)$$

$$k_{eq} = \left[f_1 \sqrt{k_1} + (1-f_1)\sqrt{k_2} \right]^2 \quad \alpha=1/2 \quad CRIM \quad (12)$$

$$k_{eq} = k_1 \left(\frac{k_2}{k_1} \right)^{(1-f_1)} \qquad \alpha=0 \quad (13)$$

$$k_{eq} = \left[\frac{f_1}{k_1} + \frac{(1-f_1)}{k_2} \right]^{-1} \qquad \alpha=-1 \quad series \quad (14)$$

Figure 14 shows the predicted trends, and highlights the importance that spatial distribution has on the effective property of the medium. For example, for $f_1 = 0.9$ and $k_2/k_1 = 10$, the effective property of the medium would be $k_{eq} = 1.9 k_1$ when components

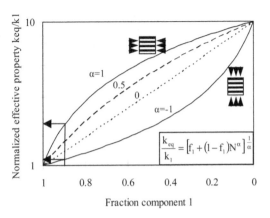

Figure 14. Effective media properties. Power average law.

are in parallel and $k_{eq} = 1.1 k_1$ if components are in series. Upper and lower bounds can be estimated for different phenomena in the absence of spatial distribution information (e.g., Hashin-Shtrikman bounds for stiffness and conduction can be found in Mavko et al. 1998).

5 CLOSING THOUGHTS

Soils are particulate materials, therefore, soils are inherently non-linear and non-elastic; strength, stiffness, and volume change are effective stress dependent; and they are inherently porous and pervious.

The unified soil classification makes a critical distinction between coarse and fine soils. The choice of the Sieve #200 is compatible with the transition in governing interparticle forces and the balance between mechanical and chemical processes within the soil mass. Micromechanical analyses also support the use of C_u for coarse soils and the plasticity chart for fine soils. Additional "index properties" would enhance soil classification, including: (1) specific surface, ph and ionic concentration in clayey soils, and (2) particle shape in sandy and gravely soils.

Macroscale phenomena (such as deformation, strength and strength anisotropy, conduction and clogging), are functions of distinct micro-scale processes. Fundamentally different particle-level mechanisms participate in small and large strain soil deformation. Time scales observed at the macroscale (e.g., strain rate effects) can be significantly different from those observed at the particle-level.

Engineering design is affected by internal temporal and spatial scales. Effective properties for design depend on the properties of the participating materials, their fractions, and their spatial distribution. Enhanced detection and site characterization are needed to properly consider spatial variability.

There is great bio-diversity in the subsurface. Therefore there are unique opportunities to harness this potential. Due to space-limiting conditions, silty and sandy soils are preferred to explore bio-geochemical processes control.

Today's particle level understanding of soil behavior, complemented with new testing and instrumentation technology open unique opportunities to unprecedented developments in engineering soil behavior.

ACKNOWLEDGEMENTS

Support for this research was provided by the National Science Foundation and The Goizueta Foundation.

REFERENCES

Alvarellos, J. 2003. *Fundamental Studies of Capillary Forces in Porous Media*, PhD Thesis, Georgia Institute of Technology, 188 pages.

Cascante, G. and Santamarina, J.C. 1996. Interparticle Contact Behavior and Wave Propagation, *ASCE Geotechnical Journal*, 122: 831–839.

Cho, G.C. and Santamarina, J.C. 2004. Soil Behavior: The Role of Particle Shape. *Proc. Skempton Conference*, March, London.

Díaz-Rodríguez, J.A. and Santamarina, J.C. 1999. "Thixotropy: The Case of Mexico City Soils", *XI Panamerican Conf. on Soil Mech. and Geotech. Eng.*, Iguazu Falls, Brazil, 1: 441–448.

Dieterich, J. H. 1978. Time-dependent friction and the mechanics of stick-slip. *Pure Appl. Geophys.*, 116: 790–806.

Fiegel, G.L., and Kutter, B.L. 1994. "Liquefaction mechanism for layered soils." *J. Geotech. Eng.*, 120: 737–755.

Guimaraes, M. 2002. *Crushed stone fines and ion removal from clays*. PhD Thesis, Georgia Institute of Technology, 238 pages.

Kokusho, T. 1999. "Water film in liquefied sand and its effect on lateral spread." *J. Geotech. Geoenviron. Eng.*, 125: 817–826.

Kuhn, M.R. and Mitchell, J.K. 1993. New Perspectives on Soil Creep, *ASCE J. Geotechnical Engineering*, 119: 507–524.

Kulhawy, F.H. and Mayne, P.W. 1990. *Manual on Estimating Soil Properties for Foundation Design*, Electric power Research Institute, Palo Alto.

Ladd, C.C., Foott, R., Ishihara, K., Schlosser, F. and Poulos, H.G. 1977. Stress-Deformation and Strength, *Proc. 9th ICSMFE*, 2: 421–494.

Landman, U., Luedtke, W.D. and Gao, J.P. 1996. Atomic-Scale Issues in Tribology: Interfacial Junctions and Nano-Elastohydrodynamics. *Langmuir*, 12: 4514–4528.

Leroueil, S. and Hight, D.W. 2003. Behaviour and properties of natural soils and soft rocks. In T.S. Tan, K.K. Phoon, D.W. Hight and S. Leroueil (eds), *Characterization and*

engineering *Peoperties of natural Soils*: 29–254. Rotterdam: Balkema.

Mavko, G.M., Mukerji, T. and Dvorkin, J. 1998. *The Rock Physics Handbook*, New York: Cambridge University Press.

Mayne, P.W. and Holtz, R.D. 1985. Effect of principal stress rotation on clay strength, *Proc. 11th ICSMFE*, San Francisco, 2: 579–582.

Mitchell, J.K. 1960. Fundamentals aspects of thixotropy in soils. *Journal of the Soil Mechanics and Foundations Division,* ASCE, 86: 19–52.

Palomino, A. 2003. *Fabric formation and control in fine grained materials*, PhD Thesis, Georgia Institute of Technology, 193 pages.

Persson, N. J. 1998. *Sliding Friction. Physical Principles and Applications*. NanoScience and Technology, Springer-Verlag.

Randolph, M. 2002, Personal communication

Rothenburg, L. 1992 Effects of Particle Shape and Creep in Contacts on Micromechanical Behavior of Simulated Sintered Granular Media. ASME, *Mechanics of Granular Materials and Powder Systems*, MD 37: 133–142.

Santamarina, J.C., Klein, K. and Fam, M. 2001. *Soils and Waves*, Chichester: John Wiley & Sons.

Tatsuoka, F., Di Benedetto, H. and Nishi, T. 2003. A Framework for Modelling of the time effect on the stress-strain behaviour of geomaterials, In Di Benedetto et al. (eds), *Deformation Characteristics of Geomaterials*: 1135–1143. Lisse: Balkema.

Terzaghi, K., Peck, R.B. and Mesri, G. 1996. *Soil Mechanics in Engineering Practice*, New York: John Wiley & Sons.

Valdes, J.R. 2002. *Fines migration and formation damage – Microscale studies*. PhD Thesis, Georgia Institute of Technology, 241 pages.

Van Olphen, H. 1951. Rheological Phenomena of Clay Sols in Connection with the Large Distribution of Micelles, *Discussions of the Faraday Society*, 11: 82–84. (additional information in his book).

Velde, B. 1992. *Introduction to Clay Minerals: Chemistry, Origins, Uses and Environmental Significance*, New York: Chapman & Hall.

Vucetic, M. and Dobry, R. 1991. Effect of soil plasticity on cyclic response, *Journal of Geotechnical Engineering*, ASCE, 117: 89–107.

Wang, L., Fritton, S.P., Weinbaum, S, and Cowin, S.C. 2003. On bone adaptation due to venous stasis. *Journal of Biomechanics*, 36: 1439–1451.

Yoshimine, M., Robertson, P.K. and Wride, C.E. 1999. Undrained Shear Strength of Clean Sands to Trigger Flow Liquefaction, *Canadian Geotechnical Journal*, 36: 891–906.

Youd, T.L. 1973. "Factors controlling the maximum and minimum densities of sands", *Evaluation of Relative Density and its Role in Geotechnical Projects Involving Cohesionless Soils*, ASTM, Edited by Selig and Ladd, STP 523: 98–112.

Zdravkovic, L. and Jardine, R.J. 1997. Some Anisotropic Stiffness Characteristics of a Silt Under General Stress Conditions, *Géotechnique*, 47: 407–437.

Deformation Characteristics of Geomaterials – Di Benedetto et al (eds)
© *2005 Taylor & Francis Group, London, ISBN 04 1536 701 8*

Integrated ground behavior: in-situ and lab tests

P.W. Mayne

School of Civil & Environmental Engineering, Georgia Institute of Technology, Atlanta, Georgia, USA

ABSTRACT: In complement to drilling and sampling operations for site exploration, results from in-situ tests are increasingly used to derive soil properties and parameters for geotechnical analysis and design. The interpretations of initial state and stress-strain-strength-flow characteristics are calibrated with laboratory test data obtained from high-quality samples, if possible. Yet, inconsistencies often result because no unified framework is available for evaluating all of the in-situ devices, including cone, dilatometer, pressuremeter, and vane. Current interpretation procedures use a hybrid of empirical, analytical, experimental, and/or numerical methods, whereas a comprehensive-integrated numerical simulation of all field tests is needed. In the interim, a semi-analytical-empirical methodology suffices. As the seismic piezocone test with dissipation phases (SCPTù) offers an optimal collection of five separate readings (q_t, f_s, u_b, t_{50}, and V_s) of soil behavior within a single sounding, a few selected case studies are presented to illustrate its versatility.

1 INTRODUCTION

1.1 *General behavior of natural geomaterials*

Soils are extremely complex four-dimensional (x, y, z, t) materials in their constituent components, initial stress state, wide-ranging stiffness, strength, flow characteristics, and rheological behavior. Yet, they must be characterized adequately and properly before any new foundation, embankment, tunnel, or excavation is constructed on or within the ground.

A fully capable solution must consider the anisotropic and preconsolidated initial geostatic state of stress in the ground, the highly nonlinear stress-strain-strength behavior that is directionally-dependent, drainage and flow considerations under dry/saturated, drained/undrained, as well as partially-saturated conditions, as well as address the extreme ranges from nondestructive small-strains to intermediate-strains, peak, post-peak, fully-softened, and final residual strength regimes. Since *Mother Nature* has bequeathed such a wide diversity of particulates, mineralogies, fabrics, cementitious agents, and novel packing arrangements, such a global and universal model may not actually materialize, at least not in the near present. Nevertheless, for the most part, a good combination of drilling, sampling, and field testing has been developed or geotechnical site exploration practices to allow successful civil engineering construction involving large foundations for offshore structures, land reclamation projects, the growth of massive urban centers, and intercoastal highway structures.

Our current understanding of soil behavior has been derived in large part from carefully-controlled laboratory tests on either high-quality field samples or reconstituted specimens. Clever studies have been undertaken to investigate soils at micro-, meso-, and macro-scales and formulate equations that describe the parametric factors under mechanical, hydraulical, thermodynamical, and rheological loading. These include the framework of critical state soil mechanics (e.g., Schofield & Wroth, 1968; Wroth, 1984; Wroth & Houlsby, 1985) to explain the nuances of clay behavior, including undrained to drained loading at normally- to overconsolidated states, contractive to dilative response, porewater pressures, and cyclic loading. As all soil types (clays, silts, sands, gravels) are frictional materials with an effective stress envelope, these CSSM concepts were extended to sands by direct reference to their critical state line (e.g., Been & Jefferies, 1985; Jefferies, 1993; Coop & Airey, 2003). A unified approach is now available (Pestana & Whittle, 1999).

The interwoven relationships between large-strain yielding, shear strength, and intermediate stiffness are now clearly represented in terms of anisotropic yield surfaces approximately centered on the K_0 (NC) line in $t = \frac{1}{2}(\sigma_1 - \sigma_3)$ versus s' = $\frac{1}{2}(\sigma_1' + \sigma_3')$ space (e.g., Diaz-Rodriguez, et al. 1992; Leroueil & Hight, 2003).

As it became appreciated that CSSM could explain the intermediate- to large-strain workings of soils, the research focus shifted towards the behavior of soils at small-strains to detail the initial stiffness and deformational properties of the ground (e.g., Jardine et al.,

1986; Burland, 1989; Tatsuoka & Shibuya, 1992; Tatsuoka, et al. 1997). Here, improved triaxial equipment with local internal strain gages coupled with torsional shear results on improved high-quality sampling methods provided the clues. Series of test comparisons with resonant column equipment were made (e.g., Georgiannou, et al. 1991). Notably, the well-known initial tangent shear modulus (G_{dyn}) that was required in problems involving soil dynamics (e.g., Hardin & Drnevich, 1972) became recognized as the initial stiffness for all stress–strain curves, notably for static monotonic loading and unloading, as well as dynamic loading.

Today, the small-strain shear modulus from nondestructive measurements can be realized as $G_0 = G_{max} = G_{dyn} = \rho_T V_s^2$, where ρ_T = total soil mass density and V_s = shear wave velocity. This fundamental stiffness is relevant as an initial stress state, whereby: e_0 = initial void ratio, σ_{vo}' = initial vertical overburden stress, u_0 = hydrostatic porewater pressure, $K_0 = \sigma_{ho}'/\sigma_{vo}'$ = lateral geostatic stress coefficient, and G_0 = initial shear modulus.

1.2 Fully-integrated ground behavior

To truly implement an integrated approach to characterizing ground behavior, all aspects of the natural geomaterials must be included, including the geologic setting, stress state, and a complete suite of soil parameters-properties, as well as a geostatistical assessment of their variability across the site. The site exploration program should involve geologic field mapping and utilization of geophysical surveys (ground penetrating radar, electromagnetic conductivity, resistivity), careful drilling & sampling to obtain high-quality specimens, indices, laboratory testing, and complementary sets of in-situ tests. A comprehensive and intricate constitutive model (e.g., Whittle & Kavvadas, 1994; Pestana and Whittle, 1999) should be calibrated against the full laboratory database and then used to provide a framework for explanation of the in-situ test results for parameter determination and quantification (Fig. 1). This same constitutive model can be used in closed-form analytical solutions and/or numerical simulation of the full-scale prototype (i.e., foundation, excavation, dam) built in construction. In this manner, the best consistent and reliable predictions of ground behavior can be achieved.

In reality, the above efforts can take years for completion of the comprehensive details. The expense and time needed to procure the necessary high-quality samples at all depths and careful controls and procedures for laboratory stress path testing mandates such a lengthy duration. Perhaps this level of effort can only be undertaken after the establishment of a national experimentation test site (e.g., Bothkennar clay detailed in *Geotechnique*, June 1992). At these special

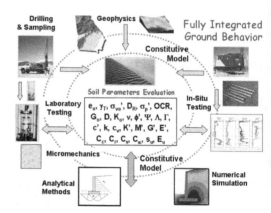

Figure 1. Components of integrated ground behavior.

experimentation sites, full-scale loadings of footings, pilings, and walls are also performed to validate the soil parameter values and permit cross-checking of results and theoretical models (e.g., Lutenegger & DeGroot, 2000). A recent summary of extensive site characterization at well-documented worldwide sites has been made (Tan, et al. 2003), yet it has taken many years to decades to obtain these data.

The adopted constitutive model may require as few as 4 parameters, such as M_c, λ, κ, and G (Wroth & Houlsby, 1985) or as many as 15 separate input parameters (e.g., Whittle, 1993) for each defined soil layer. Clearly, the size of the testing program is related to the framework and constitutive laws chosen.

2 IN-SITU TESTING

2.1 Methods of interpretation

A great number of different in-situ tests are available for site investigation (Robertson, 1986), yet many of these are quite specialized devices and not in general use. Interpretation procedures of the common in-situ tests (Fig. 2) are well-documented, with selected key references given in Table 1 for each test method. Nevertheless, in-situ test interpretation continues to receive attention in research and applied practice because of several reasons: (1) the uncertain nature in the ill-defined soil properties and parameters, (2) sampling disturbance effects on reference laboratory test values, (3) encountering of nontextbook materials, and (4) advent of improved algorithms for evaluating the data.

For most geotechnical projects, the full suite of drilling & sampling, laboratory, and in-situ testing cannot be implemented because of time and costs. Depending upon the nature of geologic setting and level of the proposed construction, perhaps only a

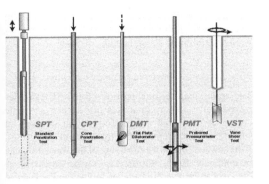

Figure 2. Common physical types of in-situ tests.

Table 1. Selected key references for common in-situ tests.

Test	Method	Reference
SPT	Standard penetration test	Stroud, 1988
CPT	Cone penetration test	Lunne et al., 1997
PMT	Pressuremeter test	Briaud, 1992
VST	Vane shear test	Chandler, 1988
DMT	Flat dilatometer test	Marchetti, 2001
V_s	Geophysics: shear waves	Woods, 1978

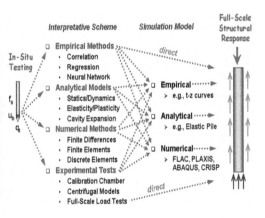

Figure 3. Types of methods for in-situ test interpretation.

select number of lab tests (i.e., index, consolidation, triaxial, permeability) and one or two of the basic in-situ tests (i.e., SPT, CPT, CPTu, DMT, PMT, VST) can be implemented. For these tests, the tasks of soil parameter interpretation can be handled by empirical, closed-form analytical, numerical, or experimental methods (Figure 3). In many cases, an assortment of these different methods are adopted in practical applications.

Empirical interpretative methods are quite common and serve as a reality check on more elegant and desired theoretical relationships. Empirical evaluations are particularly well known for the standard penetration test (SPT) that has no analytical solution. Results of the vane shear test (VST) are traditionally evaluated by limit equilibrium analyses to calculate the undrained shear strength (s_{uv}) of clays, yet now well-known to require an empirical reduction factor to obtain the mobilized strength ($\tau_{max} = \mu \cdot s_{uv}$) used in stability analyses (e.g., Chandler, 1988). Empirical trendlines that were originally drawn by eye have now been replaced by statistical methods using linear regression, multi-regression, autocorrelation, variograms, clustering, and artificial neural networks.

Analytical models can be based in the theorems of plasticity, elasticity, statics, cavity expansion, dislocation theory, or thermodynamics. For interpretation of in-situ strength of clean sands, limit plasticity solutions have seen favor (e.g., Robertson & Campanella, 1983; Lunne, et al. 1994). For interpretations of in-situ results in clays, cavity expansion theory appears to interrelate observed data from piezocone, flat dilatometer, and pressuremeter tests conducted in the same clay deposits (Mayne & Bachus, 1989). Cavity expansion has also proven well in representing penetration in sands (Salgado, et al. 1997).

Numerical methods include those based in finite elements, strain path methods, and discrete (or distinct) elements. Both FEM (Vermeer & van den Berg, 1988) and strain path solutions for penetration in clays have been developed (Whittle & Aubeny, 1993), as well as for porewater pressure dissipation (Houlsby & Teh, 1988). For penetration in sands, numerical simulations have been conducted using finite elements (e.g., Shuttle & Jefferies, 1998) and discrete elements (Huang & Ma, 1994).

Experimental approaches for in-situ test interpretations include the construction of reconstituted deposits of sands or clays with varied setups, stress regimes, and histories. These include small experimental setups with optical measurements or radiography (e.g. Allersma, 1988), large 1-g model calibration chambers (e.g., Schnaid & Houlsby, 1992), or centrifuge deposits subjected to miniature in-situ testing in flight (e.g., Esquivel & Silva, 2000). Field prototypes can also be built, instrumented, installed, and load tested for two levels of interpretation: (1) collection of in-situ test data and information and (2) direct measurement of the full-scale response, e.g., retrievable Imperial College test pile (Jardine, et al. 1998).

2.2 Seismic piezocone testing

The seismic piezocone test (SCPTu) is a hybrid of cone penetration and downhole geophysics to optimize the amount of subsurface data collected during site investigation (Campanella, et al. 1986). During the advance at 20 mm/s, continuous readings of cone tip stress (q_t), sleeve friction (f_s), and penetration porewater pressure at the shoulder (u_2 = u_b) are recorded. With certain equipment, an interchangeable tip component allows alternate measurement of porewater pressures on the cone face (u_1 = u_t), or use of multi-element penetrometers allow simultaneous u_1 and u_2 readings.

At each 1-m interval, a new cone rod is added and the temporary stop allows for two additional measurements: (a) porewater pressure decay with time, Δu(time); and (b) arrival time (Δt_s) of a surface-generated shear wave via a horizontal geophone(s) embedded in the probe. At selected depths, the dissipations are conducted for a duration usually corresponding to when one-half of the excess porewater pressures have decayed, designated t_{50} (Robertson, et al. 1992). If a single geophone is employed, the vertically-propagating and horizontally-polarized waves can be evaluated by a pseudo-interval approach to obtain the variation of shear wave velocity (V_s) with depth (Campanella, 1994). In terms of expediency and economy, the SCPTu provides an optimal means to profile the geostratigraphy and stress–strain-strength-flow soil properties of the ground, as five independent readings are measured with depth at the same location: q_t, f_s, u_b, t_{50}, and V_s. Figure 4 shows a representative 80-m deep SCPTu sounding advanced in northwest Idaho indicating a predominantly soft silty clay profile with interbedded sandy lenses and several sand layers. With downhole testing, the geophone receiver is progressively pushed further and further away from the surface source, thus

the wave arrivals become more difficult to detect. Consequently, for this sounding, the V_s measurements were discontinued beyond depths of 56 m.

2.3 Improved downhole seismic component

The traditional seismic portion of the SCPTu uses paired left- and right-strikes to define a single crossover point that is followed at 1-m intervals at successive test depths (Campanella, et al. 1986). Alternatively, the post-processing of the wave train can be enhanced for pseudo-interval measurements, particularly by signal filtering of high-frequency noise and cross-correlation that matches many thousands of datapoints in the vicinity of the first cycle following shear wave arrival (Campanella & Stewart, 1992). The field procedure is also improved, as only one-sided strikes (either left or right) need be performed, thus expediting the field testing.

True-interval testing with two or more simultaneous geophone records provides a better V_s profile, as small uncertainties and errors in time and depth measurements can cause unreal fluctuations in the pseudo-interval profile (Burghignoli, et al. 1991a; Butcher & Powell, 1996).

In recent developments, a finer resolution in the V_s profile has been obtained by downhole testing at frequent 20-cm intervals using seismic flat dilatometer testing (SDMT) with two levels of geophones, or by a special push-probe dedicated to true-interval

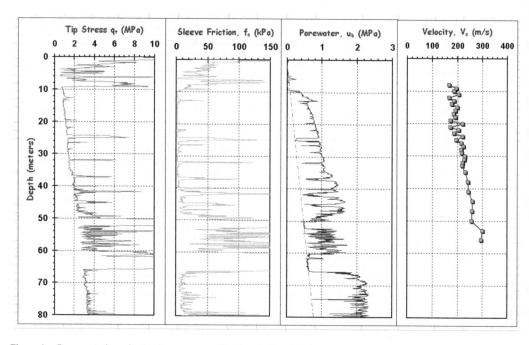

Figure 4.   Representative seismic piezocone sounding in soft silty clay for route 95, sandpoint, northwest Idaho.

downhole testing. Results from a true-interval SDMT in layered silts, sands, and clays in Treporti in the Venetian lagoon region are shown in Figure 5. These are shown in comparison with the conventional pseudo-interval approach from a nearby SCPTu using standard 1-m stops. The enhanced detail in the V_s profile is clearly seen when placed in contrast with conventional downhole readings.

Close-interval V_s data in soft glaciolacustrine silty clays in Evanston, Illinois, north of Chicago are presented in Figure 6. The site is located on the Northwestern University campus, not far from the national geotechnical test site next to Lake Michigan, one of the largest freshwater bodies in the world, thus offering some unusual sediment compared with the vast majority of marine deposits (Finno, et al. 2000). Here, the true-interval data were obtained using a 44 mm diameter push-in geophone probe that was halted each 200-mm for DHT. Also shown are the coarser shear wave profiles developed from standard one-meter pseudo-intervals from an adjacent SCPTu at the site. Again, the true nature and resolution of the V_s variations with depth are better revealed when more frequent readings are made. The profile appears less erratic and more "well-behaved".

2.4 Geostratigraphy evaluated from CPTu

Soil layering and geostratigraphy are evident from CPTu readings plotted with depth. Soil types are assessed either directly based on correlation with adjacent boring records, or indirectly from the CPT data using soil behavioral charts (e.g., Robertson, et al. 1986; Senneset, et al. 1989; Lunne, et al. 1997).

For interpretation of soil parameters, methods are usually divided into "sands" or "clays". As a general rule, sands generally exhibit $q_t > 5$ MPa and $\Delta u = 0$, while intact clays generally show $q_t < 2$ MPa and $\Delta u > 0$. For insensitive soils, the friction ratio (FR = f_s/q_t as %) is low for sands (FR < 1%) and higher for clays (FR > 4%), but in structured clays, the FR decreases with increasing sensitivity (S_t). As depth affects the readings, normalized parameters are preferred (Lunne, et al. 1997), including normalized cone tip stress, $Q = (q_t - \sigma_{vo})/\sigma'_{vo}$, normalized sleeve friction, $F = f_s/(q_t - \sigma_{vo})$, and normalized porewater pressure, $B_q = \Delta u/(q_t - \sigma_{vo})$.

In the case of fissured clays, excess porewater pressures at the shoulder may be negative ($\Delta u_2 < 0$), whereas at the face, they are positive ($\Delta u_1 >> 0$). Thus, dual-element piezocones may be useful in delineating weathered-fractured clay facies from intact clay regions in the soil profile (Mayne, et al. 1990).

Table 2 provides an approximate guide to soil behavioral type using an alternate normalized porewater

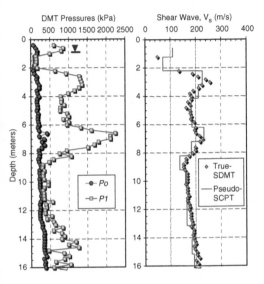

Figure 5. Seismic flat dilatometer sounding with frequent true-interval V_s in venetian lagoon.

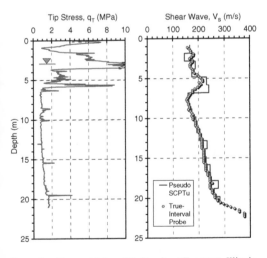

Figure 6. Frequent-interval V_s Results at Evanston, Illinois.

Table 2. Guidelines for evaluating soil type from CPTu data.

Soil behavioral type	Normalized tip stress, Q = $(q_t - \sigma_{vo})/\sigma'_{vo}$	Normalized porewater pressure, U^*	Normalized friction, F = $f_s/(q_t - \sigma_{vo})$
Clean Sand	$Q > 40$	$\Delta u/\sigma'_{vo} \approx 0$	<1%
Intact Clay	$Q < 20$	$\Delta u_1/\sigma'_{vo} > 3$ $\Delta u_2/\sigma'_{vo} > 3$	Often F > 4% but F < 2% for $S_t > 4$
Fissured Clay	$20 < Q < 40$	$\Delta u_1/\sigma'_{vo} > 10$ $\Delta u_2/\sigma'_{vo} < 0$	Generally F > 4%

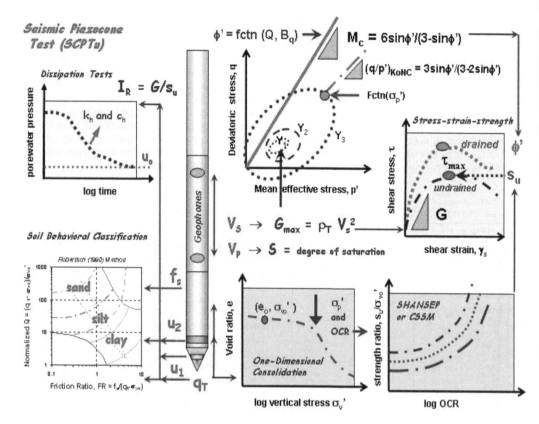

Figure 7. Conceptual framework for interpretation of geostratigraphy and soil parameters from SCPTù results.

pressure parameter, $U^* = \Delta u / \sigma'_{vo}$. Within a spreadsheet worksheet, the above guides are useful for separating which rows of data are interpreted as "sand-like" from which are "clay-like". As many "nontextbook" geomaterials are encountered in practice (Lunne, et al. 1997), it is of particular interest to have methodologies that are somewhat global and applicable to all types of soils, even though exceptions will undoubtably occur.

2.5 Soil parameters evaluated from SCPTù

In this paper, interpretations of select soil parameters are presented, with emphasis on the utilization of data collected by the seismic piezocone. The parameters are grouped according to initial state, stiffness, strength, and flow characteristics, as discussed in the following sections. Where possible, results from field piezocone testing have been related to their laboratory counterparts to allow direct interpretation of SCPTù data. The concept is depicted in Figure 7 in hopes that in-situ testing might be eventually be able to provide a relatively reasonable mapping of different soil parameters in an expeditious manner.

3 INITIAL STATE

The initial conditions in the ground include the effective overburden stress (σ'_{v0}), initial void ratio (e_0), geostatic stress state (K_0), initial stiffness (G_0), and degree of overconsolidation (OCR = σ'_p / σ'_{v0}), where σ'_p = effective preconsolidation stress.

3.1 Initial void ratio

In order to evaluate the magnitude of overburden stresses, a first-order estimate of void ratio (e_0) is necessary for unit weight determinations. From a compiled shear wave database on a variety of worldwide geomaterials (Burns & Mayne, 1996), multiple regression analyses indicated a best fit relationship (n = 756; $r^2 = 0.841$; S.E. = 0.136):

$$e_0 = 120.9 \, V_s^{-1.00} \, z^{0.220} \qquad (1a)$$

where V_s (m/s) and depth z (meters), as presented in Figure 8. An alternate trend using effective overburden

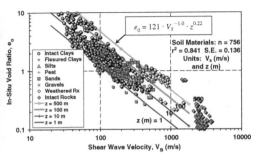

Figure 8.   First-order void ratio estimate from shear wave.

stress (σ'_{vo} in kPa) gave ($n = 727$; $r^2 = 0.790$; S.E. = 0.142):

$$e_0 = 139.4 \ V_s^{-1.08} \ \sigma_{vo}'^{0.181} \qquad (1b)$$

Using stress-normalized shear wave data, Robertson & Fear (1995) showed e_0 of freshly deposited uncemented sands could be estimated from:

$$V_{s1} = (A - B \cdot e_0) \ K_0^{0.125} \qquad (2)$$

where $V_{s1} = V_s \ (\sigma_{atm}/\sigma'_{vo})^{0.25}$, A and B are empirical constants, $K_0 = \sigma'_{ho}/\sigma'_{vo}$ = lateral stress coefficient, and $\sigma_{atm} = 1$ bar $= 100\,kPa$. For practical use, they suggested the K_0 effect could be neglected. Note the regression more or less confirms that void ratio is inversely related to V_{s1}.

3.2   Initial overburden stresses

The initial effective vertical overburden stress ($\sigma'_{vo} = \sigma_{vo} - u_o$) depends the unit weight of the soil layers, groundwater table, degree of saturation, and capillarity. Hydrostatic pressures (u_o) can be determined from borehole data, piezometers, and/or equilibrium readings following complete porewater dissipations.

The total overburden stress is calculated as the accumulation of total unit weights with depth ($\sigma_{vo} = \Sigma \gamma_T \Delta z$). For dry soil above the water table and assuming a narrow range for specific gravity of solids ($G_s = 2.7 \pm 0.1$), the total unit weight is:

$$\gamma_{dry} = G_s \ \gamma_w/(1+e_0) \qquad (3)$$

where $\gamma_w = 9.8\,kN/m^3$ for freshwater and 10.0 for saltwater. The inplace void ratio may be estimated from Figure 8. For saturated soils below the groundwater table:

$$\gamma_{sat} = \gamma_w \ (G_s + e_0)/(1+e_0) \qquad (4)$$

Complications arise when capillarity effects result in saturation some height above the phreatic line, and where desaturation of soils occur, as these can significantly affect the measured shear wave velocities (Cho & Santamarina, 2001). In this regard, the recommendation (Jamiolkowski, 2001) that compression waves (V_p) be measured to evaluate the ambient degree of saturation appears to have merit.

3.3   Initial stiffness

The initial stiffness is calculated in terms of the small-strain shear modulus:

$$G_0 = G_{max} = \rho_T \ V_s^2 \qquad (5)$$

where $\rho_T = \gamma_T/g$ = mass density, γ_T = total unit weight, $g = 9.8\,m/s^2$ = acceleration constant.

The shear wave from downhole testing is a V_s(vh)-type with vertical direction and horizontal polarization. In soft ground, the magnitude of V_s(vh) appears comparable to shear waves obtained by standard crosshole with vertical hammer [V_s(hv)], crosshole with rotary hammer [V_s(hh)], and by Rayleigh wave measurements [V_{sr}], whereas the crosshole values are considerably higher in heavily overconsolidated soils (Butcher & Powell, 1995). Herein, values of shear modulus have been determined per DHT and/or SCPT corresponding to V_s(vh), except where specifically noted.

3.4   Stress history from G_{max}

Intuitively, it might be expected that as a particular soil is subjected to higher prestressing, the void ratio will decrease and consequently, the V_s will be higher. Thus, for a first approximation, the shear wave velocity can be used as an index on the magnitude of preconsolidation stress of clays (Mayne, Robertson, & Lunne, 1998). Figure 9 presents the summary results relating effective preconsolidation stress (σ'_p) vs. small-strain shear modulus of clays.

As expected, fissured clays fall above the trendline for intact clays. The first-order expression is:

$$\frac{\sigma_p'}{\sigma_{atm}} = \left(\frac{1}{158} \cdot \frac{G_{max}}{\sigma_{atm}} \right)^{0.8} \qquad (6)$$

Data from two additional clays are available, both with appreciable cementation: (a) Fucino clay, Italy, having 25 to 50% calcium carbonate content (Burghignoli, et al. 1991) and (b) Cooper Marl of Charleston, South Carolina, having 60 to 80% $CaCO_3$ content (Camp, et al. 2002). The Cooper Marl is a sandy calcareous clay of Oligocene Age with mean values: LL = 78, PI = 38, $w_n = 48 \pm 8\%$. A representative SCPTu for the new Cooper River Bridge is presented in Figure 10 with high porewater pressures below 20 m depths indicative of the marl. Drops in the porewater pressure are at dissipation test depths. Considerable fluctuations are seen in the derived V_s profile that reflect the coarseness associated with standard pseudo-interval DHT method.

Houlsby & Wroth (1991) formulated a more rigorous association, which can be rearranged as:

$$\sigma_p' = (p_o')^{(1-m/n)} \cdot (G^*)^{-1/n} \cdot G^{1/n} \qquad (7)$$

where $G^* = [G/(p_o')^m]_{nc}$ = normalized shear modulus for NC soil, p_o' = effective mean stress, and m and n are exponents. Using this assumed form for σ_p' as a function of both G_{max} and σ_{vo}' in a multiple regression analyses significantly improved the relationship (Figure 11). Most surprisingly, the pausity of data on natural sands with known stress histories also appear to fit. Figure 11 includes results for clean sands and sandy silts that have been lightly overconsolidated by processes of groundwater fluctuations

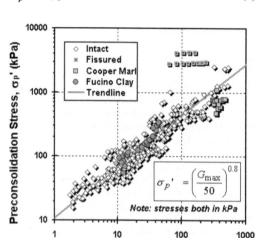

Figure 9. First-order estimate of σ_p' in clays from G_0.

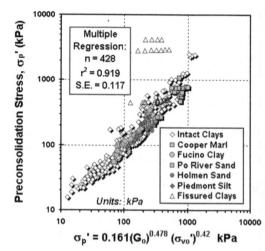

Figure 11. Preliminary σ_p' estimate from G_0 in sands.

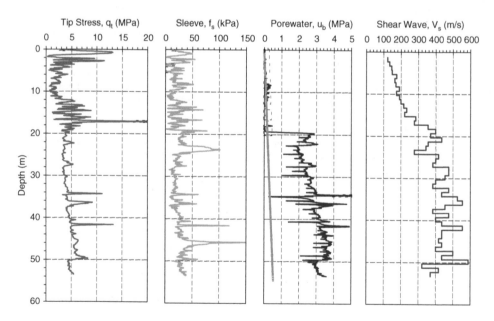

Figure 10. Representative SCPTu for Cooper River Bridge in Charleston, South Carolina. Note: Cooper Marl z > 20 m.

and/or aging, including Holmen, Norway (Lunne, et al. 2003), Po River Sand (Bruzzi, D, et al. 1986), and Piedmont residuum at Opelika, Alabama (Mayne & Brown, 2003). The generalized expression is:

$$\sigma_p' = 0.101 \cdot \sigma_{atm}^{0.102} \cdot G_{max}^{0.478} \cdot \sigma_{vo}'^{0.420} \quad (8)$$

3.5 Stress history from CPT

A simple analytical method built on spherical cavity expansion theory and critical state soil mechanics concepts has shown useful in relating the overconsolidation ratio (OCR = σ_p'/σ_{vo}') in clays to piezocone parameters, including the normalized cone tip resistance $Q = (q_t - \sigma_{vo})/\sigma_{vo}'$, normalized porewater pressure $U^* = (u_2 - u_0)/\sigma_{vo}'$, and normalized effective cone resistance $Q_u = (q_t - u_2)/\sigma_{vo}'$ (Mayne, 1991). A brief summary of the relationships for type 2 piezocones with shoulder filter elements includes:

$$OCR = 2 \cdot \left[\frac{Q}{M\{\tfrac{2}{3}(1 + \ln I_R) + \tfrac{\pi}{4} + \tfrac{1}{2}\}} \right]^{1/\Lambda} \quad (9a)$$

$$OCR = 2 \cdot \left[\frac{U^* - 1}{\tfrac{2}{3} \cdot M \cdot \ln I_R - 1} \right]^{1/\Lambda} \quad (9b)$$

where M = 6sin ϕ'/(3 − sin ϕ'), I_R = G/s$_u$ = rigidity index, Λ = 1 − κ/λ = plastic potential, κ = swelling index, and λ = compression index in e-lnp' space (Wroth, 1984). These can be combined to remove the reliance on rigidity index, such that:

$$OCR = 2 \cdot \left[\frac{Q_u}{(1.95 \cdot M + 1)} \right]^{1/\Lambda} \quad (9c)$$

In the case where Λ = 1, eqn (9a) reduces to the approximate form:

$$\sigma_p' \approx \frac{(q_t - \sigma_{vo})}{M \cdot (1 + \tfrac{1}{3} \cdot \ln I_R)} \quad (10a)$$

Adopting representative values of ϕ' = 30° (M = 1.2) and I_R = 100, eqn(9) simplifies to become:

$$\sigma_p' = 0.33 (q_t - \sigma_{vo}) \quad (10b)$$

which was independently substantiated by a statistical analysis of piezocone-oedometer data involving a variety of different clays (Kulhawy & Mayne, 1990), as presented in Figure 12. More recently, (10) has received corroboration for Pisa clay by Jamiolkowski & Pepe

(2001) and 22 Canadian clays by Demers & Leroueil (2002).

Similarly, porewater pressure data can also be implemented to evaluate σ_p' using simple statistical relationships (Chen & Mayne, 1996):

$$\sigma_p' = 0.47 (u_1 - u_0) \quad (11a)$$

$$\sigma_p' = 0.53 (u_2 - u_0) \quad (11b)$$

$$\sigma_p' = 0.60 (q_t - u_2) \quad (11c)$$

The ability to have redundant methods of assessing σ_p' from CPTu data is actually quite useful, as agreements (or discrepancies) can help the geotechnical engineer investigate whether a particular reading is questionable or in error (e.g., correction from q_c to q_t; loss of saturation of the porous element), or that the investigation has encountered an unusual geomaterial (i.e., highly sensitive, cemented, or structured deposit).

For sands, the evaluation of stress history by CPT is much more elusive and less reliable. A methodology based on statistical analyses of chamber data is of the form (Mayne, 2001):

$$OCR = \left[\frac{0.192 \cdot \left(\dfrac{q_t}{\sigma_{atm}} \right)^{0.22}}{K_{oNC} \cdot \left(\dfrac{\sigma_{vo}'}{\sigma_{atm}} \right)^{0.31}} \right]^{\left(\frac{1}{\alpha - 0.27} \right)} \quad (12)$$

where K_{oNC} = (1 − sin ϕ') = (1 − α) can be adopted.

3.6 Horizontal geostatic stress state

Based on a summary of laboratory studies, the lateral stress coefficient (K_0) for soils that are not highly

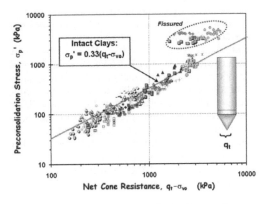

Figure 12.   Preconsolidation stress in clay from net tip stress.

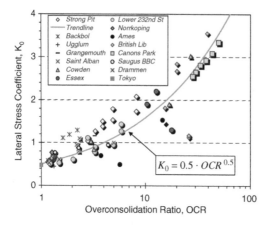

Figure 13. Total stress cell trend of K_0 with OCR for clays.

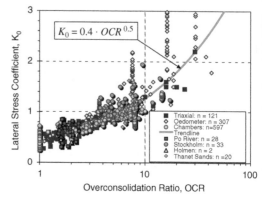

Figure 14. Laboratory and field K_0-OCR trends for sands.

structured or cemented can be calculated from their overconsolidation ratio (OCR):

"Normal" Soils: $K_0 = (1-\sin\phi')\,OCR^{\sin\phi'}$ (13)

In the absence of reliable friction angles of the material, a first-approximation for clays is:

TSC in clays: $K_0 = 0.5\,OCR^{0.5}$ (14)

where Figure 13 shows a summary of field K_0 measurements obtained by total stress cells (TSC) and companion lab oedometer results reconfirms the above. Similar trends from self-boring pressuremeter tests gave (Kulhawy & Mayne, 1990):

SBP in clays: $K_0 = 0.47\,OCR^{0.53}$ (15)

whereas higher exponents up to 1.0 may be found by the hydraulic fracturing technique in the cemented Leda clays and other highly structured clays (Leroueil & Hight, 2003). As a first step, the K_0 in clays can be evaluated from (13) or (14) where OCR is obtained from one or more determinations using (7) through (11).

For sands, instrumented laboratory tests give conflicting results. Oedometer tests and split-ring cells on small reconstituted sand specimens give rather high values of K_0 at high OCRs during rebound, when compared to K_0 triaxial-type test data. Large chamber tests provide values that appear low because zero lateral strains cannot truly be achieved during BC3 boundary conditions in flexible-walled devices (Salgado, Mitchell & Jamiolkowski, 1997). Figure 14 shows the collection of oedometer, triaxial, and chamber data, as

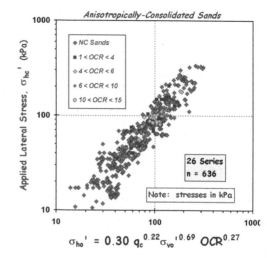

Figure 15. Lateral stress states from CPT chamber tests on uncemented unaged sands.

well as the few field PMT measurements in quartzitic-type sands with an overall trend depicted by:

Uncemented sands: $K_0 = 0.4\,OCR^{0.5}$ (16)

The effective lateral stress applied in chamber tests affects the cone tip stress. Thus, the derived OCR relationship in (12) was obtained from a sorting of the 26 series of CPT chamber test programs on primarily reconstituted quartz to feldspathic sands. This relates the applied $K_c = \sigma_{hc}'/\sigma_{vc}'$ state and OCR to the measured q_t values (Fig. 15) via:

$$K_0 = 0.192\left(\frac{q_t}{\sigma_{atm}}\right)^{0.22}\left(\frac{\sigma_{vo}'}{\sigma_{atm}}\right)^{-0.31} OCR^{0.27}$$ (17)

Of recent, geophysical data have been utilized to deduce the geostatic stress state. The shear wave magnitude depends upon its direction and plane of polarization, thus relating to only two of the three principal stresses. Thus, a set of two of more shear waves in different directions and/or polarizations can be analyzed to deduce the in-situ stress state in the ground. Controlled lab studies using directional-polarized V_s measurements have been performed to develop K_0 expressions (e.g., Yan & Byrne, 1990; Stokoe, et al. 1991; Kokusho, et al. 1995; Hatanaka, et al. 1999), as well as field measurements using combinations and arrangements of downhole, crosshole, surface waves, and other geophysical methods (e.g., Jamiolkowski & LoPresti, 1994; Butcher & Powell, 1995, 1996; Sully & Campanella, 1995). In one formulation (Fioravante, et al. 1998):

$$K_0 = \left[\frac{V_{sHH}}{V_{sHV}} \cdot \frac{C_{sHV}}{C_{sHH}} \right]^{1/n} \quad (18)$$

where V_{sHV} is the shear wave velocity from conventional crosshole tests, V_{sHH} is from special rotary impulse crosshole tests, the experimental ratio $0.90 < (C_{sHV}/C_{sHH}) < 0.98$, and $0.10 < n < 0.15$ (Fioravante, et al. 1998). The shear wave from DHT and SCPTu can also be used, as $V_{sVH} = V_{sHV}$.

4 STRENGTH

4.1 Effective Stress Strength

All soils are frictional materials and therefore characterized by an effective stress frictional envelope that is a fundamental property of the geomaterial. Most commonly, it is reported as the effective friction angle (ϕ'), or alternatively in MIT q-p' space as $\alpha' = \arctan(\sin\phi')$, or in Cambridge University q-p' space as $M = 6\sin\phi'/(3 - \sin\phi')$ for triaxial compression type loading. The effective cohesion intercept (c') is not fundamental, but depends upon the yield surface, stress regime, strain rate, and other factors. The total stress strength (represented by the undrained shear strength, s_u) is really only a subset parameter contained within the regime of critical-state soil mechanics (Wroth & Houlsby, 1985) and reliant on OCR, ϕ', yield surface, strain rate, direction of loading, and more.

With CPT data in clean quartz sands, it is fairly common practice to interpret a effective friction angle using the relationship suggested by Robertson & Campanella (1983), which may be approximated by:

$$\phi' = \arctan \{0.1 + 0.38 \log(q_t/\sigma_{vo}')\} \quad (19)$$

Notably, the CPT data utilized in the original relationship were uncorrected for boundary effects of the

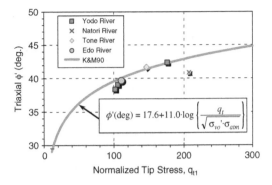

Figure 16. Comparison of measured ϕ' from frozen sand samples (Mimura, 2003) with CPT normalized tip stress.

various size calibration chambers. Recent series of undisturbed samples of carefully-frozen sands have been procured at four river sites and tested under drained triaxial compression conditions (Mimura, 2003). As shown by Figure 16, the derived ϕ' values show excellent agreement with a prior empirical CPT relationship based on statistical analyses of corrected calibration chamber data (Kulhawy & Mayne, 1990):

$$\phi' = 17.6^o + 11.0 \cdot \log\left(q_t / \sqrt{\sigma_{vo}'} \right) \quad (20)$$

where q_t and σ_{vo}' are in atmospheres. An alternate approach is to estimate the relative density of the sand (D_R) and use the critical-state framework proposed by Bolton (1986) to obtain ϕ'. The author believes, however, there are great uncertainties in evaluating D_R from SPT, CPT, and DMT, as the parameter represents very small changes in void ratio between its void limits, e_{max} and e_{min}. Improved correlations for SPT estimates of D_R have been shown by Cubrinovski & Ishihara (2002) using the void ratio difference ($e_{max} - e_{min}$), or its surrogate (D_{50}), as a normalizing index, yet no similar application to the CPT has been made at this time.

Of great interest is the utilization of piezocone results to assess the effective frictional characteristics of all types of soils, since both total stresses and penetration porewater pressures are monitored. An effective stress-limit plasticity solution has been proposed and calibrated by the Norwegian University of Science & Technology (Sandven, et al. 1988; Senneset, et al. 1989), as depicted in Figure 17.

Graphically, the cone resistance number (N_m) is determined as the slope of net cone stress ($q_t - \sigma_{vo}$) vs. effective overburden stress (σ_{vo}') and the negative intercept is attraction, $a' = c' \cot\phi'$. Similarly, the porewater pressure parameter (B_q) can be graphically defined as the slope of excess porewater pressure ($\Delta u = u_2 - u_0$) vs. net cone stress net ($q_t - \sigma_{vo}$).

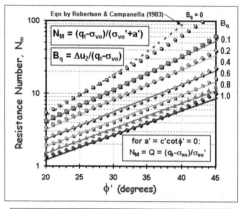

For $0.1 < B_q < 1.0$ and Range : $20° < \varphi' < 45°$

$\varphi' \approx 29.5 \cdot B_q^{0.121} \cdot \left[0.256 + 0.336\, B_q + \log(N_m) \right]$

Figure 17. Effective ϕ' from normalized CPT data using approximation to NTNU method (Senneset et al. 1989).

For the classical bearing capacity case ($\beta = 0$), the theoretical interrelationships are presented as dots in Figure 17 as developed from the following:

$$N_q = \tan^2(45° + \phi'/2)\, \exp[\pi \tan\phi'] \qquad (21a)$$

$$N_u = 6 \tan\phi'\,(1+\tan\phi') \qquad (21b)$$

$$N_m = \frac{N_q - 1}{1 + N_u \cdot B_q} \qquad (21c)$$

where $N_m = (q_t - \sigma_{vo})/(\sigma_{vo} + a')$ and $B_q = \Delta u_2/(q_t - \sigma_{vo})$. For the simplified case where $a' = 0$, then N_m becomes synonymous with normalized cone tip stress, $Q = (q_t - \sigma_{vo})/\sigma'_{vo}$ and an approximate deterministic expression for $B_q > 0.1$ can be provided (shown as lines in Figure 17):

$$\phi' = 29.5° B_q^{0.121} [0.256 + 0.336 B_q + \log Q] \qquad (22)$$

With this expression, values of ϕ' can be calculated incrementally with depth. Interestingly, most site characterization programs adopt a constant value for the entire deposit and do not investigate a variation of ϕ' with depth (for practical reasons).

For sands, where $B_q = 0$, the NTNU expression is very similar to that proposed by Robertson & Campanella (1983), as indicated by the dashed line in upper portion of Figure 17.

4.2 Undrained shear strength

For clays and silts subjected to static loading, the undrained shear strength (s_u) may be mobilized during short term loading. Various modes of undrained

Table 3. Undrained shear strength parameters.

Parameter	Mode	Expression
Normalized undrained shear strength, $(s_u/\sigma_{vo}')_{NC}$	CIUC	$= \frac{1}{2} M(\frac{1}{2})^\Lambda$
	PSC	$= \frac{\sin\varphi'}{2d_p} \cdot \left(\frac{d_p^2+1}{2}\right)^\Lambda$
		where $d_p = 1/(2 - \sin\phi')$
	CK_0UC	$= \frac{\sin\varphi'}{2d_t} \cdot \left(\frac{d_t^2+1}{2}\right)^\Lambda$
		where $d_t = (3 - \sin\phi')/(6 - 4\sin\phi')$
	DSS	$= \frac{1}{2} \sin\phi'$
	PSE	$= (s_u/\sigma_{vo})_{PSC} \cdot (PI + 41)/118$ where PI = plasticity index (%)
	CK_0UE	$= (s_u/\sigma_{vo})_{CKoUC} \cdot (PI + 37)/125$ where PI = plasticity index (%)
Strain rate		$a_{Rate} = 1 + 0.1 \log(\delta\varepsilon/\delta t)$ where rate in % per hour
Overconsolidation		$a_{OCR} = OCR^\Lambda$ where $\Lambda = 0.8$ for "normal" clays and $\Lambda = 1.0$ for structured clays
Continuity		$a_{Cont} = 1$ (intact clays) $a_{Cont} = 2/3$ (moderately fissured) $a_{Cont} = 1/3$ (highly fissured)

strength can be ascertained for analyses that consider strength anisotropy, with the more common including plane strain compression and extension, triaxial compression and extension, and simple shear. For normally-consolidated clays and silts, the different modes of normalized undrained shear strength, $S = s_u/\sigma_{vo}'_{NC}$, can be interrelated via constitutive relationships (e.g., Wroth & Houlsby, 1985; Ohta, et al. 1985; Kulhawy & Mayne 1990), or through empirical correlations with plasticity index (e.g., Jamiolkowski, et al. 1985; Ladd, 1991). An approach for the undrained strength is given by (23) and Table 3:

$$(s_u/\sigma_{vo}') = a_{Rate} \cdot a_{OCR} \cdot a_{Cont} \cdot (s_u/\sigma_{vo})_{NC} \qquad (23)$$

If only a representative undrained strength is needed for analysis, the simple shear mode may be sufficient (Wroth, 1984):

$$(s_u/\sigma_{vo}')_{DSS} = \frac{1}{2} \sin\phi' \, OCR^\Lambda \qquad (24)$$

Empirical data sets have suggested the more basic expression (Ladd, 1991):

$$(s_u/\sigma_{vo}')_{DSS} = 0.22 \, OCR^{0.80} \qquad (25)$$

with slight modifications for plastic clays and varved deposits. However, if the full range of frictional

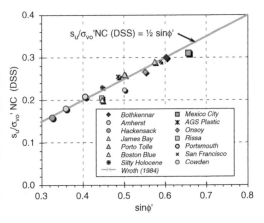

Figure 18. NC DSS normalized undrained strength with ϕ'.

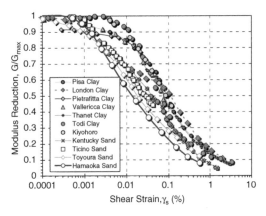

Figure 20. Modulus reduction data vs logarithm shear strain from monotonic torsional shear.

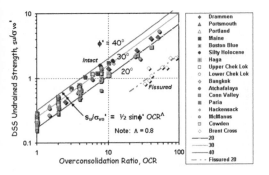

Figure 19. OC DSS undrained strength with OCR and ϕ'.

properties for fine-grained soils is considered as $17° < \phi' < 43°$ (e.g., Diaz-Rodriguez, et al. 1992), then Figures 18 and 19 substantiate the dependency of $(s_u/\sigma'_{vo})_{DSS}$ on ϕ' for this undrained strength mode for normally-consolidated and overconsolidated DSS data, respectively. Notably, the presence of extensive fissures can significantly reduce the undrained shear strength to as low as one-third that of the intact values.

5 STIFFNESS

In most commercial finite element codes involving ground deformation modeling, a simple bi-linear stress–strain curve is adopted for soil behavior that requires only a characteristic stiffness (G) and shear strength (τ_{max}) for the material. In actuality, the stiffness begins at the initial tangent shear modulus (G_{max}) and the full range of stiffness is highly nonlinear from small- to intermediate- to high-strains. Two primary focus areas of the previous conferences and symposia

on *Pre-Failure Deformation Characteristics of Geomaterials* (1995, 1997, 1999) include (1) the collection of high-quality experimental data on a variety of geomaterials, and (2) versatile algorithms for expressing the rate of modulus reduction (G/G_{max}).

For over 3 decades, the resonant column device has provided modulus reduction data for dynamically-loaded soils (e.g., Hardin & Drnevich, 1972). The emphasis of measurement was to define G_{max} and the associated G/G_{max} reduction curves at small- to intermediate strains, as well as damping values (Vucetic & Dobry, 1991). However, rates of loading are quite high for RC tests, thus G/G_{max} curves are not directly appropriate to first-time static loadings. Monotonic torsional shear (TS) tests are able to provide a full range of data from small- to high-strain ranges (e.g., Georgiannou, et al. 1991).

A selection of modulus reduction curves from-monotonic TS performed on different materials is presented in Figure 20, where G = secant shear modulus. Clays are shown by solid dots and sands indicated by open symbols. In general, the clays were tested under undrained loading (except Pisa, tested drained), and the sands were tested under drained shear (except Kentucky clayey sand, tested undrained). Similar curve trends are noted for both drainage conditions.

The G/G_{max} curves can be presented in terms of logarithm of shear strain (γ_s), or alternatively in terms of mobilized shear stress (τ/τ_{max}), as shown in Figure 21. The mobilized shear stress is analogous to the reciprocal of the factor of safety (FS = τ_{max}/τ). With the local strain measurements on triaxial specimens, similar curves of modulus reduction with mobilized deviator stress are developed (e.g., Tatsuoka & Shibuya, 1992).

A simple hyperbola requires only two parameter constants for a nonlinear stress–strain respresentation, however, it can only fit one region of the strain range

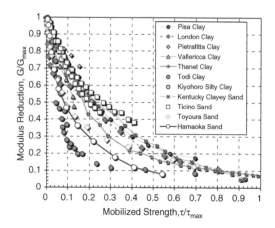

Figure 21. Modulus reduction data vs. mobilized stress from monotonic torsional shear tests.

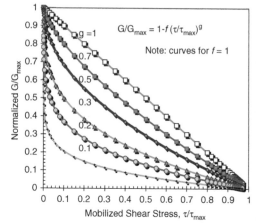

Figure 22. Modulus reduction curves using modified hyperbola of Fahey & Carter (1993).

Table 4. Empirical stress–strain formulations.

Expression type	Type loads	Number inputs	Reference
Power Function	Cyclic	3	Ramberg-Osgood 1946*
Hyperbola	Static	2	Kondner 1963
Parabola	Static	2	Hansen 1963
Mod. Hyperb.	Static	3	Duncan-Chang 1970
Spline Fctns	Static	3rd-OP	Desai 1971
Modified Hyperbola	Cyclic	4	Hardin-Drnevich 1972
Parabola	Static	2nd-OP	Breth et al. 1973
Exponential	Static	4	Daniel, et al. 1975
Mod. Hyperb.	Cyclic	5	Pyke 1979
Logarithmic	Static	5	Jardine, et al. 1986
Hyperbola-Power Fctn	Static	3	Prevost-Keane 1990
Various Type Functions	Static	2 to 6	Tatsuoka-Shibuya 1992
Tanh	Dyn.	4	Ishibashi-Zhang 1993
Mod. Hyperb.	Static	2	Fahey-Carter 1993
Logarithmic	Static	1, 3, or 4	Puzrin-Burland 1996
DEFM	Static	4	Shibuya, et al. 1997
ε -Power Law	Static	3	Atkinson, 2000
Tanh	Dyn.	3	Santos-Correia 2001
Mod. Hyperb.	S/C	6	Tatsuoka, et al. 2003

Notes: S = static; C = cyclic/dynamic.
R-O Model* (Burghignoli, et al., 1991b)
OP = nth-order polynomial.
DEFM = double exponent fitting model.

(small or intermediate or large). Thus other expressions have been sought, as detailed in Table 4. Examples of the derived modulus reduction curves for two selected formulations are presented in Figures 22 and 23: (a) modified hyperbola (Fahey & Carter, 1993), and

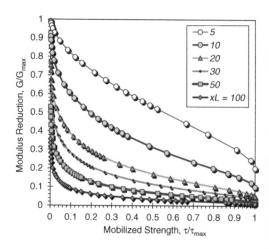

Figure 23. Modulus reduction curves using logarithmic function of Puzrin & Burland (1996).

(b) logarithmic function (Puzrin & Burland, 1996). In terms of the fitting stress–strain data, G/G_{max} vs. mobilized stress level (τ/τ_{max}) plots are visually biased towards the intermediate- to large-strain regions of the soil response. In contrast, G/G_{max} vs. log γ_s curves tend to accentuate the small- to intermediate-strain range.

A unified model for sand and clay (Pestana & Whittle, 1999) provides theoretical relationships for G/G_{max} reduction with log shear strain (Figure 24) which appears in agreement with monotonic data shown in Figure 20. These are also found in good comparison with cyclic data from resonant column tests summarized by Vucetic & Dobry (1991) shown in Figure 25. Careful studies by Shibuya, et al. (1997) and Yamashita, et al. (2001) have shown that differences in

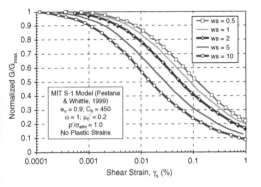

Figure 24. Theoretical G/G_{max} curves from S-1 constitutive model (Pestana & Whittle, 1999).

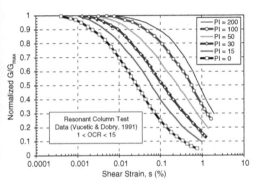

Figure 25. Experimental G/G_{max} curves from cyclic resonant column test data (Vucetic & Dobry, 1991).

monotonic and cyclic reduction curves are predominantly explained by strain rate effects, thus monotonic curves can be used as a basis for all other curves.

Since the SCPTu provides information on both the initial stiffness (G_{max}) at small-strains, as well as shear strength (τ_{max} from s_u and/or ϕ') at large-strains, the difficult question is how to determine stiffnesses at intermediate strain levels for use in calibrating one of the aforementioned stress–strain-strength laws in Table 4. One prospect is the cone pressuremeter (Schnaid & Houlsby, 1992; Ghionna, et al. 1995) that can be hybrid together with seismic piezocone to provide SCPMTu. Another opportunity is the utilization of paired side-by-side in-situ tests, such as companion sets of SCPTu and DMT (Tanaka & Tanaka, 1998), as the latter may provide an equivalent stiffness at intermediate strains.

6 PERMEABILITY AND FLOW PARAMETERS

In fine-grained soils, the permeability (k) and coefficient of consolidation (c_h) can be quantified from

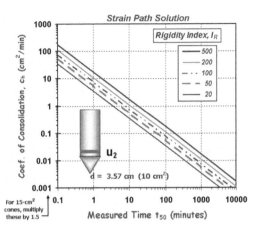

Figure 26. Evaluation of c_h from piezocone dissipation test data (SPM solutions from Houlsby & Teh, 1988).

SCPTù measurements taken during porewater dissipation tests. Most solutions for the interpretation of c_h are based either in cavity expansion theory (e.g., Jamiolkowski, et al. 1985; Burns & Mayne, 1998) or strain path method (e.g., Houlsby & Teh, 1988; Whittle, et al. 2001). The field testing is often carried out until 50% of the excess porewater pressures have decayed (designated by time t_{50}).

For type 2 piezocone data exhibiting monotonic decrease of porewater pressures with time, the solution can be expressed in the form:

$$c_h = \frac{T_{50}{}^{*} \cdot a_c{}^2 \cdot \sqrt{I_R}}{t_{50}} \qquad (26)$$

where a_c = penetrometer radius, $T_{50}^{*} = 0.245$ for SPM (Houlsby & Teh, 1988), and $I_R = G/s_u$ = rigidity index. The solution is graphically depicted in Figure 26.

Calibration of this approach with a large dataset was presented by Robertson, et al. (1992) and indicated that the range of operational I_R was quite large for a variety of clays and silts. In lieu of estimating I_R, the SCE-MCC approach can provide a direct evaluation of I_R by combining (9a), (9b) and (9c) per Burns & Mayne (2002):

$$I_R = \exp\left[\left(\frac{1.5}{M} + 2.925\right)\left(\frac{q_t - \sigma_{vo}}{q_t - u_2}\right) - 2.925\right] \qquad (27)$$

which is quite sensitive to the input values used because of the exponential form. Thus, data from Class I-level penetrometers (Lunne, et al. 1997) of the highest order should be used.

Figure 27 illustrates the full set of five separate types of soil behavioral readings (q_t, f_s, u_2, t_{50}, and V_s)

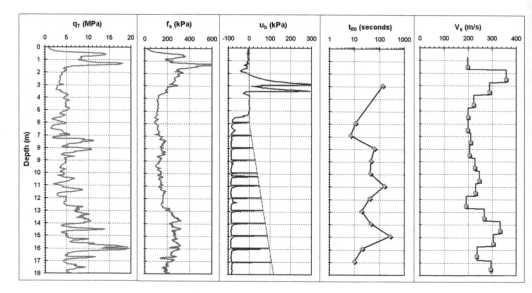

Figure 27.   Seismic piezocone test with dissipation (SCPTù) in piedmont residuum at Atlanta Hartsfield airport.

that are obtained during a SCPTù sounding. Here, the sounding was performed in Piedmont residual fine sandy silts in south Atlanta, Georgia. The negative penetration porewater pressures at the shoulder filter position at depths below the groundwater level of 6 m are common in this geologic setting (Finke, et al. 2001). In these materials, full dissipation occurs to hydrostatic conditions in generally less than 1 to 2 minutes.

The evaluation of hydraulic conductivity (k) or permeability can be made by reference to its relationship with the coefficient of consolidation and constrained modulus (D'):

$$k = \frac{c_h \cdot \gamma_w}{D'}$$

(28)

where D' may be very approximately estimated from net cone resistance (e.g., Kulhawy & Mayne, 1990) or small-strain shear modulus (Mayne, 2001).

Alternatively, permeability can be assessed by a direct relation to the measured t_{50} from the dissipation data using the empirical method of Parez & Faureil (1988), as shown in Figure 28.

7   CASE STUDY APPLICATIONS

In describing the mechanical behavior of soils, an effective stress frictional envelope and limit state curve must be defined for the material, as well as the initial state of the ground. The friction envelope (represented

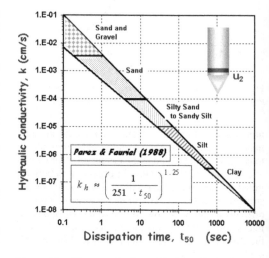

Figure 28.   Direct evaluation of k from dissipation data (method of Parez & Fauriel, 1988).

by ϕ' or M or α') appears little affected by sample disturbance, yet the yield surface can be significantly reduced and altered (e.g., Hight & Leroueil, 2003). The yield surface is best determined by extensive sets of triaxial tests with differing stress paths, yet these are quite time-consuming and expensive. Approximate limit states or yield surfaces can be developed by knowledge of the effective preconsolidation stress and effective ϕ', as detailed by Diaz-Rodriguez et al.

170

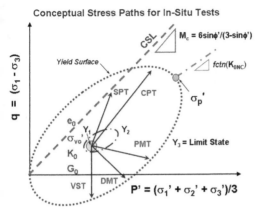

Figure 29. Initial state, yield surfaces, effective frictional envelope, and conceptual in-situ stress paths.

(1991), Leroueil & Hight (2003), and Larsson & Sallfors (1981).

In concept, each of the in-situ tests provides a unique loading situation for the ground. Figure 29 illustrates the significant parameters involved in the yield surface, critical state friction, and K_{oNC} line. Initial state is defined by the parameters: e_0, σ'_{vo}, K_0, and G_0. Herein, the full utilization of seismic piezocone data has been explored to encourage its usefulness in geotechnical site characterization.

7.1 Sandpoint, Idaho

The results from the deep SCPTu sounding from Idaho (Fig. 4) have been used to evaluate the effective preconsolidation stress profile and compare with available oedometer data. The mean index properties for this clay include: $w_n = 44$, LL = 42, and PI = 22. Using the presented relationships in (7), (10), and (11), several profiles of σ'_p are shown in Figure 30 and seen in good agreement with the results from one-dimensional consolidation tests. Generally, the materials are lightly-overconsolidated with $1 < OCR < 2$.

Using the approximate NTNU method from eqn(22), Figure 31 shows reasonable agreement with the CPTu-estimated friction angle at Sandpoint. Each triaxial point represents a set of three CIUC tests with porewater pressure measurements. As a full range of effective confining stresses were applied, the interpreted $\phi' = 33.7°$ at critical-state appears valid (Figure 32).

7.2 Cooper marl, charleston, south carolina

A more difficult geomaterial to characterize is the Cooper Marl that appears to be a highly-structured brittle calcareous clay. In Figure 10, the Cooper Marl

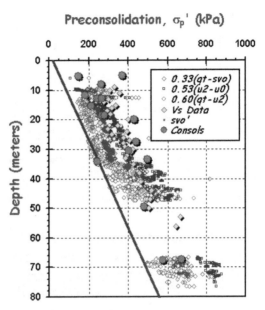

Figure 30. Laboratory and field evaluations of σ'_p profiles at Sandpoint, Idaho.

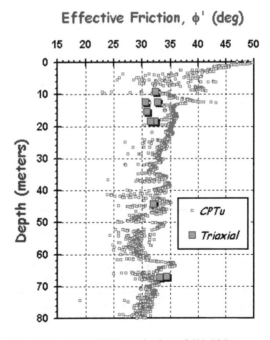

Figure 31. Lab and CPTu evaluations of ϕ' in Idaho.

is encountered below depths of 20 m and extends to depths of about 100 m, serving as the foundation bearing stratum for pilings and deep foundations in the Charleston, SC area.

171

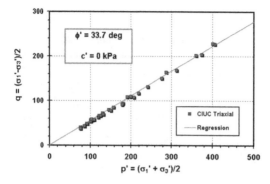

Figure 32. Summary MIT q-p' plot of CIUC triaxial test data from Sandpoint, Idaho.

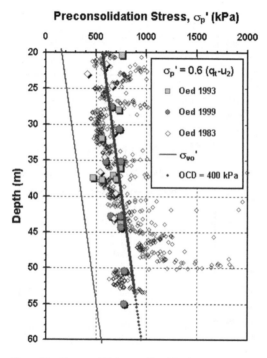

Figure 33. Preconsolidation profile of the Cooper Marl, SC.

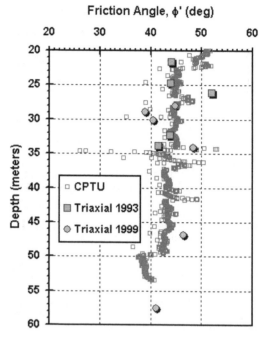

Figure 34. CPTu- and lab-determined ϕ' values in Cooper Marl, South Carolina.

index and friction angle. The rigidity index per (27) suggests $500 < I_R < 2000$ for this material. Thus, the operational coefficient for (10b) would be about 0.17 instead of the normally-assumed 0.33 factor.

The approximate NTNU method from eqn (22) indicates an operational ϕ' of around 40° to 45°, as seen by Figure 34. This is comparable to the frictional characteristics of the Mexico City clay (Diaz-Rodriguez, et al. 1991; Leroueil & Hight, 2003). Also, it is similar to the high ϕ' values of the structured St. Jean Vianney clay of Quebec (Saihi, et al. 2002), at least in the portion of the effective stress regime dominated by the limit state curve. Therefore, the CPTu-interpreted ϕ' may likely represent an operational friction angle, as the CSSM approach for sands (Bolton, 1986), and likely not provide the ϕ'_{cv} corresponding to the large-strain critical state line.

The triaxial test data for Cooper Marl appear to exhibit high friction angles (mean $\phi' = 43°$) even beyond the interpreted preconsolidation stress range (Fig. 35). Perhaps the sampling process has reduced the full peak strengths observed in the overconsolidated limit surface region, as well as lowered the interpreted σ'_p from oedometer tests, as structured materials are particularly susceptible to sample disturbance effects (Hight & Leroueil, 2003). However, a corresponding reduction of the friction angle would not be expected.

The profile of preconsolidation stress has been derived from three sets of consolidation tests and suggest an apparent prestress (OCD = $\sigma'_p - \sigma'_{vo}$ = 400 kPa = overconsolidation difference). Yet, the very high calcite content averaging 70% might imply a quasi-preconsolidation due to cementation effects, as well as aging. Figure 33 shows the σ'_p profile obtained using the effective cone resistance from eqn (11c) with good agreement with lab test results. Similar profiles can be obtained from net tip stress and excess pore-water pressure but eqns (9a) and (9b) must be used as the Cooper Marl exhibits a relatively high rigidity

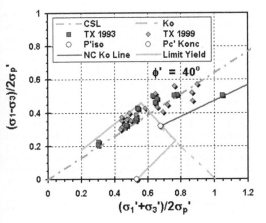

Figure 35. Friction characteristics and approximate limit yield curve for Cooper Marl.

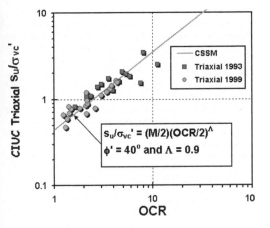

Figure 36. Undrained strength and OCR for Cooper Marl.

An attempted reconstruction of the limit state curve using approximate methods (Larsson & Sallfors, 1981; Leroueil & Hight, 2003) is shown in Figure 35. A fair amount of scatter is noted in the interpreted NC region. Of added support in the interpreted high $\phi' = 40°$, the calculated relationships between normalized undrained shear strengths with OCR appear validated by the triaxial data, as evidenced by Figure 36.

8 NEW DIRECTIONS

Fundamental analytical formulations and comprehensive constitutive models are becoming available to interrelate the interpretations of the in-situ test data with lab results that provide reference benchmark values with which the field methods can be calibrated.

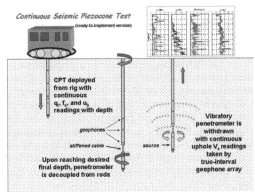

Figure 37. Continuous vs profiling during CPT withdrawal.

Thus, laboratory and in-situ testing are complementary means towards the characterization of geomaterials. For the near future, a recommended set needed enhancements to in-situ testing include:

(a) Dielectric modules for CPT and DMT to allow direct-continuous assessments of the inplace water contents, as this measurement can cross-link the parameters of consolidation, stiffness, strength, and permeability (Santamarina, et al. 2001).

(b) Numerical simulations of all major in-situ test methods (SPT, CPT, CPTu, DMT, PMT, VST) with parametric studies to show influence of variables in a consistent framework.

(c) Validated methods for obtaining the in-situ effective stress friction angles (operational peak secant value, critical-state value, and residual strength value). Perhaps the use of multiple porewater pressure measurements can help facilitate this effort.

(d) Intermediate strain level measurements of stiffness via paired CPT-PMT testing, cone pressuremeter, dilatocone, and other methods.

(e) Collection of continuous V_s profiles with depth, which may be tackled by a variety of approaches: (1) uphole geophysics after completion of the CPT; (2) DHT during CPT using a fixed shaker source at the surface, and (3) both source and receiver(s) traveling down during CPT. A presently-achievable option is depicted in Figure 37.

9 CONCLUSIONS

In-situ and laboratory testing provide complementary data for the site characterization of geomaterials. Reference values are provided by the lab measurements, but require high-quality sampling and are obtained only at discrete points at high cost. The in-situ data are collected quickly and continuously, but need calibration for interpretation.

The seismic piezocone and seismic dilatometer provide information about soil behavior at two opposite ends of the stress–strain curve, thus are quite useful for obtaining subsurface data that can be used both in stability and deformation problems involving geotechnics. Herein, a selection of interpretative methods for the seismic piezocone test with dissipation phases (SCPTù) has been presented as five independent readings on soil behavior are captured by a single sounding (q_t, f_s, u_b, t_{50}, and V_s). Additional work is required in the numerical simulation of all in-situ tests within a common constitutive framework to better represent the installation and imposed loadings, therefore offering consistent and compatible interpretations with lab reference values.

ACKNOWLEDGMENTS

The author thanks the National Science Foundation (Award CMS-0338445) and Mid-America Earthquake Center (Award EEC-9701785) for their support. Any opinions, findings and conclusions or recommendations expressed here are those of the author and do not necessarily reflect those of the NSF or MAE. Appreciation is given to Guy Houlsby for his critique on the draft version and to the following individuals for supplying data: Billy Camp, Alec McGillivray, Paola Monaco, Dean Harris, Abraham Diaz-Rodriguez, and David Hight.

REFERENCES

Allersma, H.G.B. (1988). Optical analysis of stress and strain around the tip of a penetrating probe. *Penetration Testing 1988*, Vol. 2, Balkema, Rotterdam, 615–620.

Atkinson, J.H. (2000). Nonlinear soil stiffness in routine design. *Geotechnique* 50(5), 485–508.

Been, K. and Jefferies, M.G. (1985). A state parameter for sands. *Geotechnique* 35(2), 99–112.

Bolton, M.D. (1986). The strength and dilatancy of sands. *Geotechnique* 36(1), 65–78.

Breth, H., Schuster, E. and Pise, P. (1973). Axial stress–strain characteristics of sand. *Journal of the Soil Mechanics & Foundations Division* (ASCE) 99 (SM8), 617–632.

Briaud, J-L. (1992). *The Pressuremeter*, A.A. Balkema, Rotterdam, 336 p.

Bruzzi, D., Ghionna, V., Jamiolkowski, M., Lancellotta, R. and Manfredini, G. (1986). Self-boring pressuremeter in Po River Sand. *Pressuremeter and Its Marine Applications*, STP 950, ASTM, West Conshohocken/PA, 57–74.

Burghignoli, A., et al. (1991a). Geotechnical characterization of Fucino clay. *Proc. 10th ECSMFE* (I), Firenze, 27–40.

Burghignoli, A., Pane, V., and Cavalera, L. (1991b). Modeling stress–strain-time behavior of natural soils: monotonic loading, *Proc. 10th ECSMFE* (III), Firenze, 959–979.

Burland, J.B. (1989). Small is beautiful: the stiffness of soils at small strains. *Canadian Geotechnical J.* 26(4), 499–516.

Burns, S.E. and Mayne, P.W. (1996). Small- and high-strain soil properties using the seismic piezocone. *Transportation Research Record 1548*, National Acad. Press, D.C., 81–88.

Burns, S.E. and Mayne, P.W. (1998). Monotonic and dilatory pore pressure decay during piezocone tests. *Canadian Geotechnical Journal* 35(6), 1063–1073.

Burns, S.E. and Mayne, P.W. (2002). Analytical cavity expansion-critical state model for piezocone dissipation in fine-grained soils. *Soils & Foundations* 42(2), 131–137.

Butcher, A.P. and Powell, J.J.M. (1995). The effects of geological history on the dynamic stiffness in soils. *Proceedings, XI ECSMFE*, Vol. 1, Copenhagen, 1.27–1.36.

Butcher, A.P. and Powell, J.J.M. (1996). Practical considerations for field geophysical techniques used to assess ground stiffness. *Advances in Site Investigation Practice*, Thomas Telford, London, 701–714.

Camp, W.M., Mayne, P.W. and Brown, D.A. (2002). Drilled shaft axial design values in calcareous clay. *Deep Foundations 2002*, Vol. 2, GSP 116, ASCE, Reston, 1518–1532.

Campanella, R.G., Robertson, P.K., and Gillespie, D. (1986). Seismic cone penetration test. *Use of in situ tests in geotechnical engineering* (GSP 6), ASCE, Reston, 116–130.

Campanella, R.G. and Stewart, W.P. (1992). Seismic cone analysis using digital signal processing for dynamic site characterization. *Canadian Geot. J.* 29(3), 477–486.

Campanella, R.G. (1994). Field methods for dynamic geotechnical testing. *Dynamic Geotechnical Testing II* (STP 1213), ASTM, West Conshohocken/PA, 3–23.

Chandler, R.J. (1988). The in-situ measurement of the undrained shear strength of clays using the field vane. *Vane Shear Strength Testing in Soils*. STP 1014, ASTM, West Conshohocken/PA, 13–44.

Chen, B. S-Y. and Mayne, P.W. (1996). Statistical relationships between piezocone measurements and stress history of clays. *Canadian Geotechnical J.* 33(3), 488–498.

Cho, G.C. and Santamarina, J.C. (2001). Unsaturated particulate materials: particle-level studies. *Journal of Geotechnical & Geoenv. Engrg.* 127(1), 84–96.

Clayton, C.R.I., Hight, D.W., and Hopper, R.J. (1992). Progressive destructuring of Bothkennar clay: implications for sampling and reconsolidation procedures. *Geotechnique* 42(2), 219–239.

Coop, M.R. and Airey, D.W. (2003). Carbonate sands. *Characterization and Engineering Properties of Natural Soils*, Vol. 2, Swets & Zeitlinger, Lisse, 1049–1086.

Cubrinovski, M. and Ishihara, K. (2002). Maximum and minimum void ratio characteristics of sands. *Soils & Foundations* 42(6), 65–78.

Daniel, A.W.T., Harvey, R.C. and Burley, E. (1975). Stress–strain characteristics of sand. *Journal of Geotechnical Engineering* 101 (GT5), 508–512.

Demers, D., and Leroueil, S. (2002). Evaluation of preconsolidation pressure and the overconsolidation ratio from piezocone tests of clay deposits in Quebec. *Canadian Geotechnical Journal* 39(1), 174–192.

Desai, C.S. (1971). Nonlinear analyses using spline functions. *Journal of Soil Mechanics & Foundations Division* (ASCE) 97 (SM10), 1461–1480.

Diaz-Rodriguez, J.A., Leroueil, S. and Aleman, J. (1992). Yielding of Mexico City clay and other natural *clays*. *Journal of Geotechnical Engineering* 118(7), 981–995.

Duncan, J.M. and Chang, C.Y. (1970). Nonlinear analysis of stress and strain in soils. *Journal of the Soil Mechanics & Foundation Divison* (ASCE) 96 (SM5), 1629–1653.

Esquivel, E.R. and Silva, C.H. (2000). Miniature piezocone for use in centrifuge testing. *Innovations & Applications in Geotechnical Site Characterization*, GSP No. 97, ASCE, Reston/Virginia, 118–129.

Fahey, M., and Carter, J.P. (1993). A finite element study of the pressuremeter test in sand using a nonlinear elastic plastic model. *Canadian Geotechnical Journal* 30(2), 348–362.

Fernandez, A.L. and Santamarina, J.C. (2001). Effect of cementation on the small-strain parameters of sands. *Canadian Geotechnical Journal* 38(1), 191–199.

Finke, K.A., Mayne, P.W., and Klopp, R.A. (2001). Piezocone penetration testing in Atlantic Piedmont residuum. *Journal of Geotechnical & Geoenv. Engrg.* 127(1), 48–54.

Finno, R.J., Gassman, S.L. and Cavello, M. (2000). The NGES at Northwestern Univ., *National Geotechnical Experimentation Sites*, GSP 93, ASCE, Reston/Virginia, 130–159.

Fioravante, V., Jamiolkowski, M., LoPresti, D., Mandfredini, G. and Pedroni, S. (1998). Assessment of the coef. of earth pressure at rest from V_s. *Geotechnique* 48(5), 657–666.

Georgiannou, V.N., Rampello, S., and Silvestri, F. (1991). Static and dynamic measurements of undrained stiffness on natural overconsolidated clays. *Proceedings, 10th European Con. Soil Mechanics and Foundation Engrg* (1), Florence, 91–95

Ghionna, V.N., Jamiolkowski, M., Pedroni, S. and Piccoli, S. (1995). Cone pressuremeter tests in Po River sand. *The Pressuremeter and Its New Avenues*, Balkema, 471–480.

Hansen, J.B. (1963). Discussion, *Journal of Soil Mechanics & Foundations Division* 89 (SM4), 241–242.

Hardin, B.O. and Drnevich, V.P. (1972). Shear modulus and damping in soils: Design equations and curves. *J. Soil Mechanics and Foundations Div.* 98 (SM7), 667–692.

Hatanaka, M., Uchida, A. and Taya, Y. (1999). K_0-value of in-situ gravelly soils. *Proc. 11th Asian Regional Conference on Soil Mechanics & Geotechnical Engineering*, Vol. 1, Balkema, 77–80.

Hight, D. and Leroueil, S. (2003). Engineering *Characterization and Engineering Properties of Natural Soils* (1). Swets and Zeitlinger, Lisse, 255–360.

Houlsby, G.T. and Teh, C.I. (1988). Analysis of the piezocone in clay. *Penetration Testing 1988*, Vol. 2, Balkema, Rotterdam, 777–783.

Houlsby, G.T. and Wroth, C.P. (1991). The variation of shear modulus of a clay with pressure and overconsolidation ratio. *Soils & Foundations* 31(3), 138–143.

Huang, A-B. and Ma, M.Y. (1994). An analytical study of cone penetration tests in granular material. *Canadian Geotechnical Journal* 31(1), 91–103.

Ishibashi, I. and Zhang, X. (1993). Unified dynamic shear moduli and damping ratios of sand and clay. *Soils & Foundations*, Vol. 33(1), 182–191.

Jamiolkowski, M., Ladd, C.C., Germaine, J. and Lancellotta, R. (1985). New developments in field and lab testing of soils. *Proc. 11th ICSMFE* (2), San Francisco, 57–154.

Jamiolkowski, M. and LoPresti, D.C.F. (1994). Validity of in-situ tests related to real behavior. *Proceedings 13th ICSMGE*, Vol. 5, New Delhi, 51–55.

Jamiolkowski, M. (2001). Where are we going? *Prefailure Deformation Characteristics of Geomaterials*, Vol. 2, Balkema, Rotterdam, 1251–1262.

Jamiolkowski, M. and Pepe, M.C. (2001). Vertical yield stress of Pisa clay from piezocone tests. *Journal of Geotechnical & Geoenvironmental Engineering* 127(10), 893–897.

Jardine, R.J., Potts, D.M., Fourie, A.B., and Burland, J.B. (1986). Studies of the influence of non-linear stress–strain characteristics in soil-structure interaction. *Geotechnique* 36(3), 377–396.

Jardine, R.J., Chow, F.C., Matsumoto, T. and Lehane, B.M. (1998). A new design procedure for driven piles and its application to two Japanese clays. *Soils & Foundations* 38(1), 207–219.

Jefferies, M.G. (1993). Norsand: a simple critical state model for sand. *Geotechnique* 43(1), 91–104.

Kokusho, T., Yoshida, Y. and Tanaka, Y. (1995). Formulation of shear wave velocity in gravelly soils. *Earthquake Geotechnical Engineering*, Balkema, Rotterdam, 245–250.

Kondner, A.M. (1963). Hyperbolic stress–strain response: cohesive soils. *ASCE Journal of the Soil Mechanics and Foundations Division* 89(1), 115–143.

Kulhawy, F.H. and Mayne, P.W. (1990). Manual on estimating soil properties for foundation design. *Report EPRI EL-6800*, Electric Power Research Institute, Palo Alto, 306 p.

Lacasse, S., Berre, T., and Lefebvre, G. (1985). Block sampling of sensitive clays. *Proceedings, 11th International Conference on Soil Mechanics and Foundations Engineering* (2), San Francisco, 887–892.

Ladd, C.C. (1991). Stability evaluation during staged construction. Terzaghi Lecture, *ASCE Journal of Geotechnical Engineering* 117(4), 540–615.

Larsson, R. and Sallfors, G. (1981). Hypothetical yield envelope at stress rotation. *Proc. 10th ICSMFE*, Vol, 1, Stockholm, 693–696.

Larsson, R. and Mulabdić, M. (1991a). Shear moduli in Scandinavian clays: *Report No. 40*, SGI, Linköping, 127 p.

Larsson, R. and Mulabdić, M. (1991b). Piezocone tests in clays: *Report No. 42*, SGI, Linköping, 240 p.

Leroueil, S. and Hight, D.W. (2003). Behavior and properties of natural soils and soft rocks. *Characterization and Engineering Properties of Natural Soils* (1). Swets and Zeitlinger, Lisse, 29–254.

Lo Presti, D.C.F., Jamiolkowski, M., Pepe, M. (2003). Geotechnical characterization of the subsoil of Pisa Tower. *Characterization and Engineering Properties of Natural Soils* (1). Swets and Zeitlinger, Lisse, 909–946.

Lunne, T., Lacasse, S. and Rad, N.S. (1994). General report: SPT, CPT, PMT, and recent developments in in-situ testing. *Proc. 12th ICSMFE* (4), Rio de Janeiro, 2339–2403.

Lunne, T., Robertson, P.K., Powell, J.J.M. (1997). *Cone Penetration Testing in Geotechnical Practice*. Blackie Academic EF Spon/Routledge Pub., New York, 312 p.

Lunne, T., Long, M. and Forsberg, C.F. (2003). Characterization & engineering properties of Holmen sand. *Characterization & Engineering Properties of Natural Soils*, Vol. 2, Swets & Zeitlinger, 1121–1148.

Lutenegger, A.J. and DeGroot, D., editors (2000). *National Geotechnical Experimentation Sites*, GSP No. 93, ASCE, Reston/Virginia, 396 p.

Marchetti, S., Monaco, P., Totani, G. and Calabrese, M. (2001). The flat dilatometer test in soil investigations.

Proceedings, Intl. Conf. on In-Situ Measurement of Soil Properties & Case Histories, Bali, 95–131.

Mayne, P.W. and Bachus, R.C. (1989). Penetration pore pressures in clay by CPTu, DMT, and SBP. *Proceedings, 12th ICSMFE*, Vol. 1, Rio de Janeiro, 291–294.

Mayne, P.W., Kulhawy, F.H. and Kay, J.N. (1990). Observations on the development of porewater pressures during piezocones. *Canadian Geotech. J.* 27(4), 418–428.

Mayne, P.W. (1991). Determination of OCR in clays by piezocone tests using cavity expansion and critical state concepts. *Soils and Foundations* 31(1), 65–76.

Mayne, P.W. (1993). In-situ determination of clay stress history by piezocone model. *Predictive Soil Mechanics*, Thomas Telford, London, 483–495.

Mayne, P.W., Robertson, P.K. and Lunne, T. (1998). Clay stress history evaluated from SCPTu. *Geotechnical Site Characterization* (2), Balkema, 1113–1118.

Mayne, P.W. (2001). Stress-strain-strength-flow parameters from enhanced in-situ tests. *Proceedings, the International Conference on In-Situ Measurement of Soil Properties and Case Histories*, Bali, Indonesia, 27–48.

Mayne, P.W. and Brown, D.A. (2003). Site characterization of Piedmont residuum of North America. *Characterization & Engineering Properties of Natural Soils*, Vol. 2, Swets & Zeitlinger, Lisse, 1323–1339.

Mayne, P.W., Puzrin, A.M. and Elhakim, A.F. (2003). Field characterization of small- to high-strain behavior of clays. *Soil and Rock America 2003*, Vol. 1, Proceedings, 12th Pan Am Conference, Verlag Glückauf, Essen, 307–313.

Mimura, M. (2003). Characteristics of some Japanese natural sands. *Characterization and Engineering Properties of Natural Soils* (2), Swets and Zeitlinger, Lisse, 1149–1168.

Ohta, H., Nishihara, A., and Morita, Y. (1985). Undrained stability of K_0-consolidated clays. *Proc. 11th ICSMFE*, Vol. 2, San Francisco, 613–616.

Parez, L. and Faureil, R. (1988). Le piézocône. Améliorations apportées à la reconnaissance de sols. *Revue Française de Géotech*, Vol. 44, 13–27.

Pestana, J. and Whittle, A.J. (1999). Formulation of a unified constitutive model for clays and sands. *Intl. J. Numerical & Analytical Methods in Geomechanics* 23(12), 1215–1243.

Prevost, J.H. and Keane, C.M. (1990). Shear stress–strain curve generation from simple material parameters. *Journal of Geotechnical Engineering* 116(8), 1255–1263.

Puzrin, A.M., and Burland, J.B. (1996). A logarithmic stress–strain function for rocks & soils. *Geotechnique* 46(1), 157–164.

Pyke, R. (1979). Nonlinear soil models for irregular cyclic loadings. *Journal of Geotech. Engrg.* 105 (GT6), 715–726.

Rinaldi, V.A. and Santamarina, J.C. (2003). Cemented soils: behavior and conceptual framework. Submitted for review.

Robertson, P.K. and Campanella, R.G. (1983). Interpretation of cone penetration tests: sands. *Canadian Geotechnical Journal* 20(4), 719–733.

Robertson, P.K. (1986). In-situ testing and its application to foundation engineering. *Canadian Geot. J.* 23(4), 573–594.

Robertson, P.K, Campanella, R.G., Gillespie, D. and Greig, J. (1986). Use of piezometer cone data. *Use of In-Situ Tests in Geotechnical Engineering*, GSP 6, ASCE, 1263–1280.

Robertson, P.K, Sully, J.P., Woeller, D.Jl, Lunne, T., Powell, J.J.M. and Gillespie, D.G. (1992). Estimating coefficient of consolidation from piezocone tests. *Canadian Geotechnical Journal* 39(4), 539–550.

Robertson, P.K. and Fear, C.E. (1995). Application of CPT to evaluate liquefaction potential. *Proceedings, Intl. Symposium on CPT*, Vol. 3, Swedish Geot. Society, 57–79.

Saihi, F., Leroueil, S., LaRochelle, P. and French, I. (2002). Behavior of stiff sensitive Saint-Jean Vianney clay. *Canadian Geotechnical Journal* 39(5), 1075–1087.

Salgado, R., Mitchell, J.K., and Jamiolkowski, M. (1997). Cavity expansion and penetration resistance in sand. *Journal of Geotechnical & Geoenvironmental Engineering* 123(4), 344–354.

Sandven, R., Senneset, K. and Janbu, N. (1988). Interpretation of piezocone tests in cohesive soils. *Penetration Testing 1988*, Vol. 2 (ISOPT), Balkema, Rotterdam, 939–953.

Santamarina, J.C., Klein, K.A. and Fam, M.A. (2001). *Soils and Waves*, John Wiley & Sons, New York, 488 p.

Santos, J.A.D. and Correia, A.G. (2001). Reference threshold shear strain of soil. *Proceedings*, 15th ICSMGE, Vol. 1, Istanbul, 267–270.

Schnaid, F. and Houlsby, G.T. (1992). Measurement of properties of sand in a calibration chamber by the cone pressuremeter test. *Geotechnique* 42(4), 587–601.

Schofield, A.N. and Wroth, C.P. (1968). *Critical State Soil Mechanics*, McGraw-Hill, London, 310 p.

Senneset, K. Sandven, R., Janbu, N. (1989). Evaluation of soil parameters from piezocone tests. *Transportation Research Record* 1235, 24–37.

Shibuya, S., Mitachi, T., Fukuda, F. and Hosomi, A. (1997). Modeling of strain-rate dependent deformation of clay at small strain. *Proc. 14 ICSMFE* (1), Hamburg, 409–412.

Shibuya, S., and Tamrakar, S.B. (1999). In-situ and laboratory investigations into engineering properties of Bangkok clay. *Characterization of Soft Marine Clays*, Balkema, Rotterdam, 107–132.

Shuttle, D. and Jefferies, M. (1998). Dimensionless and unbiased CPT interpretation in sand. *Intl. J. for Numerical & Analytical Methods in Geomechanics*, Vol. 22, 351–391.

Stokoe, K.H., Lee, J.N-K., and Lee, S.H-H. (1991). Characterization of soil in calibration chambers with seismic waves. *Calibration Chamber Testing*, Elsevier, 363–376.

Sully, J.P. and Campanella, R.G. (1995). Evaluation of in-situ anisotropy from crosshole and downhole shear wave velocity measurements. *Geotechnique* 45(2), 267–282.

Tan, T.S., Phoon, K.K., Hight, D.W. and Leroueil, S., editors (2003). *Characterization and Engineering Properties of Natural Soils*, Vols. 1 & 2, Swets & Zeitlinger, Lisse.

Tanaka, H. and Tanaka, M. (1998). Characterization of sandy soils using CPT & DMT. *Soils & Foundations* 38(3), 55–65.

Tatsuoka, F. and Shibuya, S. (1992). Deformation characteristics of soils and rocks from field and laboratory tests. *Report of the Institute of Industrial Science*, Vol. 37(1), Univ. of Tokyo, 136 p.

Tatsuoka, F., Jardine, R.J., LoPresti, D.C.F., DiBenedetto, H. and Kodaka, T. (1997). Theme lecture: Characterizing the pre-failure deformation properties of geomaterials. *Proceedings, 14th ICSMGE* (4), Hamburg, 2129–2164.

Tatsuoka, F., Masuda, T. Siddiquee, M.S.A. and Koseki, J. (2003). Modeling the stress–strain relations of sand in cyclic PS. *Journal Geotechnical & Geoenv. Engrg.* 129(6), 450–467.

Vermeer, P.A. and van den Berg, P. (1988). Analysis of CPT for undrained clay. Penetration Testing 1988, Vol. 2, Balkema, Rotterdam, 1035–1041.

Vucetic, M. and Dobry, R. (1991). Effect of soil plasticity on cyclic response. *Jour. of Geotechnical Engng* 117(1), 89–107.

Whittle, A.J. (1993). Evaluation of a constitutive model for overconsolidated clays. *Geotechnique* 43(2), 289–313.

Whittle, A.J. and Aubeny, C.P. (1993). The effects of installation disturbance on interpretation of in-situ tests in clay. *Predictive Soil Mechanics*, Thomas Telford, UK, 742–767.

Whittle, A.J. and Kavvadas, M.J. (1994). Formulation of MIT-E3 constitutive model for overconsolidated clay. *Journal of Geotechnical Engineering* 120(1), 173–224.

Whittle, A.J., Sutabutr, T., Germaine, J.T. and Varney, A. (2001). Prediction and interpretation of pore pressure dissipation for piezoprobe. *Geotechnique* 51(7), 601–617.

Woods, R.D. (1978). Measurement of dynamic soil properties. *Earthquake Engineering & Soil Dynamics*, Vol. I, ASCE Conference, Pasadena, 91–178.

Wroth, C.P. (1984). The interpretation of in-situ soil tests. *Geotechnique* 34(4), 449–489.

Wroth, C.P. and Houlsby, G.T. (1985). Soil mechanics: property characterization & analysis procedures. *Proceedings, 11th ICSMFE*, Vol. 1, San Francisco, 1–56.

Yamashita, S., Kohata, Y., Kawaguchi, T. and Shibuya, S. (2001). International round robin test organized by TC-29. *Advanced Lab Stress–Strain Testing of Geomaterials*, Swets & Zeitlinger, Lisse, 65–110.

Yan, L. and Byrne, P.M. (1990). Simulation of downhole and crosshole seismic tests. *Canadian Geot. J.* 27(4), 441–460.

Modelling
Modélisation

Deformation Characteristics of Geomaterials – Di Benedetto et al (eds)
© 2005 Taylor & Francis Group, London, ISBN 04 1536 701 8

Diffuse modes of failure in geomaterials

G. Servant, F. Darve & J. Desrues
Laboratoire Sols Solides Structures, INPG-UJF-CNRS, Grenoble, France

I.O. Georgopoulos
NTUA, Athens, Greece

ABSTRACT: Various kinds of failure are noticed in practice. Localized failure modes with shear band formation, diffuse modes with chaotic displacement fields and instabilities of geometrical origin are detected. In this paper, it is proposed to focus on diffuse modes. Firstly some undrained triaxial tests, which are axially force controlled, on loose sand are presented and their diffuse failure at the peak value of q is discussed (q characterizes the second stress invariant). Then, the case of axisymmetric q constant loading paths is considered. It is shown that the application of Hill's sufficient condition of stability to these paths can predict diffuse modes of failure, which have been detected experimentally and which remain unexplained until now.

1 INTRODUCTION

In practice, failure appears as induced by instabilities. These instabilities can have a geometric origin as in the well known case of buckling. Slender triaxial samples collapse due to geometrical instabilities. Other instabilities have a material origin and two different modes are possible in this class. The localized failure mode is characterized by shear band formation, due to localization of plastic strains. This failure mode is well described by the localization condition, which corresponds to the vanishing values of the determinant of the acoustic tensor:

$$\det(\mathbf{nLn}) = 0 \tag{1}$$

where $\mathbf{L}$ is the elasto-plastic tensor:

$$d\sigma_{ij} = L_{ijkl}d\varepsilon_{kl} \qquad (i, j, k, l = 1, 2, 3) \tag{2}$$

and $\mathbf{n}$ is the normal to the shear band (Rice 1976; Darve 1984). On the contrary, diffuse modes of failure are characterized by the absence of localization and a chaotic displacement field. This aspect is illustrated for continuous media in section 2 of this paper and for discrete materials (Schneebeli material constituted by piled cylinders) in Darve et al. (2003). The most known examples of diffuse failure are given by liquefaction phenomenon in continuum mechanics and by grain avalanches in discrete mechanics. For all these kinds of failure, the most basic definition of stability as proposed by Lyapunov (1907) is satisfied.

Indeed at a failure state, a small perturbation of the loading (as a small additional force) is changing drastically the subsequent deformation.

From a theoretical point of view, the notion of failure corresponds to the existence of limit states. In classical views a limit stress state is reached asymptotically and it is not possible to go through for the stress state.

If so, a limit stress state will be characterized in elasto-plasticity by:

$$d\sigma = 0 \quad \text{and} \quad \|d\varepsilon\| \text{ undefined} \tag{3}$$

By writing the elasto-plastic relation of Equation 2 in the six-dimensional space:

$$d\sigma_\alpha = M_{\alpha\beta}d\varepsilon_\beta \qquad (\alpha, \beta = 1,\dots,6) \tag{4}$$

conditions 3 imply for incrementally linear constitutive relations:

$$\begin{cases} \det \mathbf{M} &= 0 \\ \mathbf{M}d\varepsilon &= 0 \end{cases} \tag{5}$$

$\det \mathbf{M} = 0$ corresponds to the so-called "plastic limit condition" and $\mathbf{M}d\varepsilon = 0$ to the "flow rule", since the first equation defines a limit surface in the six-dimensional stress space and the second equation gives the direction of $d\varepsilon$ (but not its norm) when the plasticity criterion is reached.

However, the experiments show (see section 2) that some failure states can be described neither by localization condition of Equation 1 nor by plasticity criterion in Equation 5. For that, let us consider a tri-axial undrained loading on loose sands. The deviatoric stress q ($q = \sigma_1 - \sigma_3$) is passing through a maximum (before an eventual liquefaction), which is located strictly inside Mohr-Coulomb plastic limit surface. If the loading is axially force controlled, at this maximum a complete sudden failure of the whole sample occurs. Thus this maximum is a proper failure state. However it does not fulfill Mohr-Coulomb plastic limit condition (given by conditions 5). In the same way, it does not satisfy localization condition 1, because for loose sands the localization condition is very close to the Mohr-Coulomb limit (for loose sands shear bands develop practically on Mohr-Coulomb surface).

Indeed it is now necessary to consider this "new" class of failures without localization and not located on the Mohr-Coulomb limit. In this perspective let us recall Hill's sufficient condition of stability (Hill 1958). A stress–strain state is called "stable" if, for any ($d\sigma$, $d\varepsilon$) linked by the constitutive relation, the second order work is strictly positive:

$$d^2W = d\sigma : d\varepsilon > 0$$

In axisymmetric conditions it comes:

$$d^2W = d\sigma_1 d\varepsilon_1 + 2d\sigma_3 d\varepsilon_3 \qquad (6)$$

and for undrained (isochoric) conditions:

$$d\varepsilon_1 + 2d\varepsilon_3 = 0 \qquad (7)$$

Equations 1 and 6 imply:

$$d^2W = dq d\varepsilon_1 \qquad (8)$$

Relation 8 proves that the peak value of q in undrained loading is an "unstable" state in Hill's sense (Darve & Chau 1987).

Thus a conjecture was to consider that Hill's condition of stability is a proper criterion of diffuse failure.

For an incrementally linear rate-independent constitutive relation given by $d\sigma = \mathbf{M}d\varepsilon$ in the six-dimensional spaces, the expression of second order work becomes:

$$d^2W = {}^t d\varepsilon \mathbf{M} d\varepsilon = {}^t d\varepsilon \mathbf{M}^s d\varepsilon \qquad (9)$$

where superscripts t and s indicate respectively a transposed vector and the symmetric part of the matrix.

With realistic assumptions for the eigenvalues, it comes:

$$d^2W = 0 \Leftrightarrow \det \mathbf{M}^s = 0 \qquad (10)$$

In associated elasto-plasticity, the constitutive matrix is symmetric. In this case, Equations 5 and 10 coincide and Hill's condition does not give any new insight.

But geomaterials are strongly non-associated materials, thus the matrix $\mathbf{M}$ is non-symmetric and, after linear algebra, condition 10 is satisfied before plastic limit condition 5. Thus a whole stress domain of instabilities can be expected from the theory of non-associated elasto-plasticity.

The boundaries of this unstable domain have been determined and plotted for dense and loose sands in axisymmetric conditions, plane strain conditions and more general 3D conditions (Darve & Laouafa 2000, Laouafa & Darve 2002).

Besides, the second order work criterion is essentially a directionally dependent quantity. Indeed for incrementally non-linear relations like

$$d\varepsilon = \mathbf{N}(\mathbf{u})d\sigma \qquad (11)$$

with $\mathbf{u} = d\sigma/\|d\sigma\|$, it comes successively

$$d^2W = {}^t d\sigma \mathbf{N} d\sigma = \|d\sigma\|^2({}^t\mathbf{u}\,\mathbf{N}(\mathbf{u})\,\mathbf{u}) \qquad (12)$$

Thus the second order work takes negative values inside cones of unstable directions. The equations of the boundaries of these cones have been also determined and plotted (Darve & Laouafa 2000, Laoufa & Darve 2002).

It has just been seen that the maximum of q for undrained loading on loose sand is a proper failure state. In a more essential manner this is a bifurcation state.

Because the stress–strain energy is equal to

$$\sigma_1\varepsilon_1 + 2\sigma_3\varepsilon_3 \equiv q\varepsilon_1 + \varepsilon_v\sigma_3 \qquad (13)$$

with $\varepsilon_v = \varepsilon_1 + 2\varepsilon_3$, $q - \varepsilon_1$ and $\varepsilon_v - \sigma_3$ are conjugate variables and the constitutive relation can be written under the following form

$$\begin{bmatrix} dq \\ d\varepsilon_v \end{bmatrix} = \mathbf{P} \begin{bmatrix} d\varepsilon_1 \\ d\sigma_3 \end{bmatrix} \qquad (14)$$

For undrained loading, the isochoric condition implies $d\varepsilon_v = 0$. At the peak value of q, $dq = 0$. Thus the maximum of q is characterized by

$$\begin{cases} \det \mathbf{P} = 0 \\ \mathbf{P} \begin{bmatrix} d\varepsilon_1 \\ d\sigma_3 \end{bmatrix} = \begin{bmatrix} 0 \\ 0 \end{bmatrix} \end{cases} \qquad (15)$$

The first term of Equations 15 is a bifurcation condition, while the second term corresponds to a "failure rule". There are an infinite number of solutions in ($d\varepsilon_1$, $d\sigma_3$), which all are verifying Equations 15.

Equations 15 can be viewed as generalizations of Equations 5. From the peak value of q, the path is no more controllable in Nova's sense (Nova 1994).

In sections 3 and 4 of this paper, the notions, which have been previously introduced, are applied to the specific case of q-constant loading paths. This case has been chosen essentially because some experiments (see subsection 3.2) have detected unexpected diffuse failure modes for loose sands. The theoretical analysis (subsection 3.4) exhibits indeed such failure modes and gives the conditions of their initiation. Then section 4 confirms these results by numerical modelling of these paths for sands at different initial densities and at different initial mean pressures.

Finally let us notice that, in discrete mechanics, a discrete form of second order work can be defined by

$$d^2W = \overrightarrow{dF} \cdot \overrightarrow{dl} + dCd\Omega \qquad (16)$$

where $\overrightarrow{dF}$ is the incremental force on a particle inducing displacement $\overrightarrow{dF}$ and dC the incremental torque inducing rotation $d\Omega$ (in two-dimensional conditions).

Grain avalanches have been analyzed by a discrete element method. Some nice correlations have been exhibited between the spatial zones with local bursts of kinetic energy and the ones with local negative values of the discrete second order work. Global kinetic energy (for the whole body) and global second order work (for all the particles) have been also computed. Plotted versus a loading parameter, the bursts of global kinetic energy appear as very well correlated to the negative minima of global second order work (Darve, Servant, and Laouafa 2003).

2 DIFFUSE MODES OF FAILURE, EXPERIMENTAL RESULTS

2.1 Scope-motivation

In this section we will try to investigate the diffuse modes of failure in geomaterials from an experimental point of view. In particular, we are mostly interested in diffuse modes of failure in sands, which may appear in common triaxial compression tests. For this reason, a series of triaxial compression tests were performed. These tests resemble the common triaxial compression tests but are modified in such a way that after the peak in the stress–strain curve the test is stress-controlled and not strain-controlled, as this is the usual case in most of the common triaxial compression tests.

2.2 Experimental evidence

In the following, a short description of the Triaxial Apparatus, which was used, is given. Basic information and details on the procedure of the testing or the concerned apparatus, can be found in the classical book by Bishop & Henkel (1962) and further information is available in more recent textbooks as for exemple Head (1986), and Bardet (1997).

• *A short description of the triaxial apparatus*
The Triaxial Apparatus is located in the Laboratoire 3-S, in Institut National Polytechnique de Grenoble, in France. It is manufactured by Wykeham Farrance and it is basically a strain-controlled apparatus. The user can select a deformation rate during the devia-toric charge. The axial load is measured by an externally installed load cell/ring, whose capacity is ~30 kN. An LVDT is used for measuring the axial deformation of the specimen and it is placed on the top of the cell. The cell pressure is controlled by a mechanical valve, which is connected to the data acquisition system to enable a continuous record of the cell pressure. As far as the pore pressure is concerned, a pressure gauge is installed at the base plate, next to the base plate tap. Finally, the volume variation of the specimen is measured by a manometric volume gauge.

• *Testing program*
A series of five undrained triaxial compression tests were performed in order to investigate the aforementioned diffuse modes of failure. Loose specimens prepared by moist tamping technique (5% water content) from Hostun Sand S_{28} were subjected to undrained triaxial compression. All specimens were fully saturated (degree of saturation $S_r = 99.4\% - 99.7\%$) and isotropically consolidated up to $p' = 800\,kPa$. The confining pressure was kept constant during the test, equal to $\sigma_c = 850\,kPa$. Table 1 summarizes the five undrained triaxial compression tests.

• *Experimental results and discussion*
The series of undrained triaxial compression tests on loose Hostun Sand S_{28} were performed in order to verify the modes of failure after the peak in the stress–strain curve. All five specimens were subjected to a load control compression after reaching the peak in the stress–strain curve, or $p' - q$ plane.

Table 2 summarizes the results of these triaxial tests. In Table 2 the first column is the No. of the test. The second column is the maximum mobilized friction angle ϕ in degrees, while the third one indicates whether or not the sand specimen lost completely its resistance, meaning, whether it was fully liquefied or

Table 1. Specimen's parameters-Hostun Sand S_{28}.

No. Test	Void Ratio e	Porosity n	Saturation degree S_r	Deformation rate
CUSPR01	1.070	0.517	99.4%	2.0 mm/min
CUSPR02	1.106	0.525	99.5%	2.0 mm/min
CUSPR03	1.140	0.533	99.6%	Various rates
CUSPR04	1.138	0.532	99.6%	1.0 mm/min
CUSPR05	1.158	0.537	99.7%	1.0 mm/min

not. Below, all five curves in the $p'-q$ plane are shown (Figure 1). In this figure, the sequence of symbols interrupts when the specimen enters in the unstable mode (sudden acceleration of the strain rate). Depending on the post-peak control, either total liquefaction or near liquefaction is observed. In the $p'-q$ plane, the difference between these two situations is not evident, since in both cases the stress point tends toward the origin of the axes, either reaching it or stopping very close. It is the same in Figure 2, where one can see the evolution of the pore pressure in both cases: near liquefaction shows pore pressure approaching the cell pressure, but not reaching it; total liquefaction corresponds to the case when pore pressure comes to equal the cell pressure (tests CUSPR02 and 03). But the most significant difference is shown by the photographs in

Table 2. Results-Hostun Sand S_{28}.

No. Test	Maximum mobilized friction angle ϕ	Total liquefaction
CUSPR01	23:5°	NO
CUSPR02	16.2°	YES
CUSPR03	19.3°	YES
CUSPR04	19.7°	NO
CUSPR05	16.3°	NO

Figures 3 and 4. In Figure 3 we observe complete destruction of the specimen when complete liquefaction is reached, while in Figure 4 it can be seen that the specimen has recovered enough strength after its unstable deformation phase to remain solid and observable.

What could generally be said from the above tests, is that shifting from strain-controlled condition to stress-controlled after the peak in the stress–strain curve leads to a sudden loss of resistance of the sand specimen. This can either be expressed by total loss of effective stresses, i.e. total liquefaction (Figure 3), or by loss of most of its resistance (nearly liquefied), Figure 4.

As far as the mode of failure is concerned, the above two photos illustrate the lack of any observable organization in the deformation mode: Figure 3 shows complete liquefaction, while Figure 4 does not show any obvious strain localization band or pattern of bands. Instead, a kind of apparently unstructured heterogeneous mode of deformation is observed. This is also clearly stated from the evolution of the pore pressure of the sand specimens during the test. In Figure 2, the pore pressure of the specimen is plotted against the total axial deformation. In this Figure specimens 2 and 3 were totally liquefied as the pore pressure at the end of the test reaches the pressure of the cell. Specimens No. 1, 4 and 5 were partially liquefied, as it is clear from the above figure that the

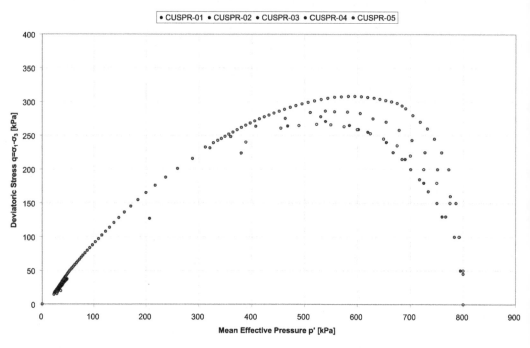

Figure 1. Consolidated undrained triaxial compression tests.

184

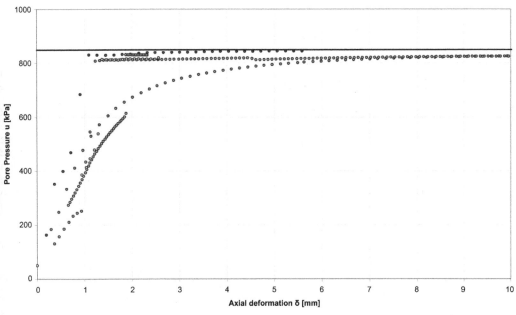

Consolidated Undrained Triaxial Compression Test
Hostun S$_{28}$, σ$_3$=850kPa, u$_0$=50kPa

● CUSPR-01 ● CUSPR-02 ● CUSPR-03 ● CUSPR-04 ○ CUSPR-05 ——Cell Pressure

Figure 2.   Evolution of specimen's pore pressure against specimen's deformation.

Figure 3.   Total liquefaction of sand specimen.

Figure 4.   Partial liquefaction of sand specimen.

185

pore pressure at the end of the test was very close to the cell pressure. In tests No. 2, 3, 4 and 5 we can observe a jump in the evolution of the pore pressure, at the moment where we pass to the region where we are load controlled, and more precisely when the system starts accelerating.

2.3 Conclusion

The above work was performed in order to investigate the modes of failure in Geomaterials in undrained compression tests on very loose specimens of Hostun S_{28} sand. No strain localization could be observed in any of the five specimens tested, so they are considered having undergone diffuse modes of failure.

It should nevertheless be mentioned that in triax-ial tests on loose, contractant, sand specimens, it is usually difficult to decide whether a mode of failure is diffuse or localized. Using X-Ray Computed Tomography on sand specimens tested in drained conditions (Desrues et al. 1996) have shown that complex strain localization patterns can be hidden to the naked eye observation of the outer membrane of the specimens. Thus a more extensive triaxial pro-gram is needed so as to verify and investigate the deformation of the specimen around the q peak.

In a pioneering paper by Han & Vardoulakis (1991), it has been shown that in undrained displacement-controlled tests on loose sand specimens "no localiza-tion was observed", while for load-controlled tests and for large strains "the deformation localizes inside the rapidly deforming zone". Mokni & Desrues (1999), in undrained displacement-controlled tests, do observe localization in loose and very loose sand, occurring when the stress state reaches the line of maximum mobilized friction determined from drained tests on the same sand. Finno et al. (1997) in displacement-controlled undrained test on loose sand observe localiza-tion also; they state that "the stress state when the localization begins is very close to, yet precedes that corresponding to the maximum mobilized friction". In the early stages of the tests, they described non per-sistent shear bands. In summary, for displacement-controlled undrained tests on loose specimens, all these experiments show that, when localization occurs, it is very close to the line of maximum mobilized friction.

The conclusions concerning load-controlled tests may seem in contradiction with the present results. However, they lead to two remarks. First, diffuse and localized modes can coexist and can appear succes-sively inside a given sample, since the theory of elasto-plasticity (Bigoni & Hueckel 1990) shows that, if both exist, the diffuse mode will always precede the localized one. Secondly it is generally admitted that for higher stress–strain rates the localized failure is dominant in experiments. So for load-controlled tests, with much higher strain rates than in displacement

controlled tests, localized strains can appear after a phase of diffuse failure.

3 Q-CONSTANT LOADING PATHS, EXPERIMENTS AND ANALYSES

3.1 Test program

During the past 20 years, to clarify the liquefaction phenomenon, many authors (Casagrande 1975, Lade 1992, Sasitharan et al. 1993) have performed experi-mental studies on loose sands mainly under undrained loading. The monotonic undrained test results on loose sands have shown that the effective stress paths are characterized by a peak in the stress bissector plane (Castro 1969, Méghachou 1993). As it is recalled in the introduction, further analyses (Darve 1996) have shown that this stress peak was unstable according to Lyapunov's definition of stability (Lya-punov 1907) and Hill's condition of stability (Hill 1958).

However, it appears that instability can also occur under drained conditions. Indeed, Eckersley (Eckersley 1990) carried out many instrumented laboratory flowslides and the main conclusion was, excess pore pressures were generated during, rather than before, slope movements, and liquefaction was therefore a result of shear failure rather than the cause. It means that the flowslide took place under drained condi-tions. Specimens collapses during standard drained load controlled triaxial tests were first observed by Begemann et al. (1977).

In order to investigate the behavior of loose sand under drained loading paths, it is worth, as suggested by Brand (1981) & Brenner et al. (1985), to simulate the failure of slopes caused by water infiltration. This phe-nomenon implies a reduction in the effective mean stress. So, the resulting stress path could be idealized as paths with constant shear stress but decreasing effective mean stress. This is the so-called Constant-Shear Drained (C.S.D.) test (Anderson & Riemer 1995).

The decrease in mean effective stress might be obtained by increasing the pore pressure if the sample is saturated (as proposed by Sasitharan et al. 1993), i.e. a load controlled test. Whereas an increase of the volume of the sample would lead to the same conse-quence but by controlling the displacement.

To resume, constant shear axisymmetric stress load-ing paths consist in, from an isotropic state of stresses, loading the sample following a conventional drained tri-axial compression path ; then, maintaining a constant value of the deviatoric stress $q (q = \sigma_1 - \sigma_3$ where σ_1 and σ_3 are respectively the axial and lateral stresses) by an incrementally isotropic unloading de-fined by $d\sigma_1 = d\sigma_2 = d\sigma_3 =$ negative constant (Fig. 1a). This type of experiment can be carried out at different levels of shearing and/or different initial densities. Gajo et al.

(2000) has also performed constant q drained test with preshear in order to examine the influence of initial anisotropy (Fig. 1b).

The next subsection is then devoted to present the main results of the literature.

3.2 Experimental results

First, let us consider the experimental program of Sasitharan et al. (1993) who compared drained and undrained collapses of saturated loose sand subjected to constant shear paths. The authors noticed a catastrophic failure with either a slight modification in sample's void ratio or no volume change. From Figures 6 and 7, it has been concluded that the postpeak portion of a constant void ratio stress path defines the state boundary surface[1]. Indeed the q-constant stress loading paths are reaching asymptotically this surface

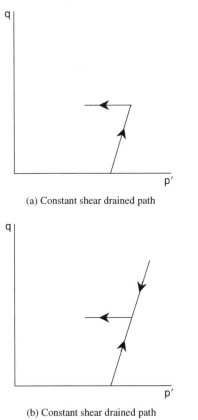

(a) Constant shear drained path

(b) Constant shear drained path
    with preshear

Figure 5.   Constant q stress controlled tests.

[1] Let us recall here that this state boundary surface is located inside the domain given by the Mohr-Coulomb criterion.

(Fig. 7). This approach has also been used to predict the failure of the Nerleck berm (Sladen et al. 1985). The collapses were then interpreted as the consequence of an instantaneous and undrained mechanical response of the material.

But, in the experiments carried out by Skopek et al. (1994), because of the unsaturated (dry to be precise) character of the sample, the authors concluded that the structural collapse of loose dry sand is a result of progressive destabilization of grain assembly structure.

Such behavior has been interpreted by di Prisco & Imposimato (1997) by associating the mechanical response with the microstructural evolution of the granular system. It is interesting to note that their approach can fill the lack presented by "classical" elastoplastic and elastoviscoplastic models to simulate the considered instability.

The unstable behaviour of sand along a C.S.D. path was also examined experimentally by Chu et al. (2003). An instability in the form of a rapid increase of plastic strains was observed. With no pore water pressure change during the whole test, this instability

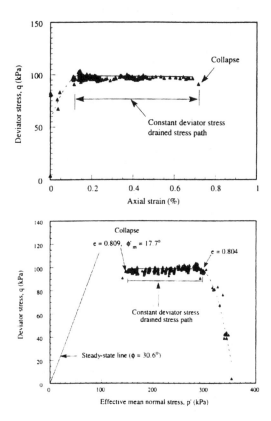

Figure 6.   Deviator stress versus axial strain and resulting stress path with constant deviator stress (Sasitharan et al. 1993).

187

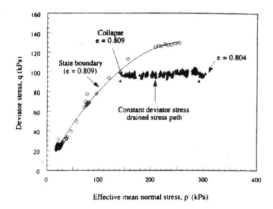

Figure 7. State boundary and constant deviator stress drained path (Sasitharan et al. 1993).

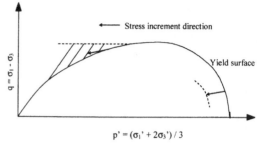

Figure 8. Instability zone according to Chu (1993).

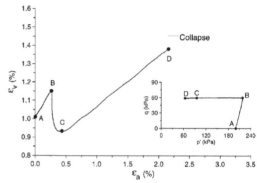

Figure 9. Typical volumetric versus axial strain behavior for q constant stress path on a loose sand, ε_v positive means contraction (Gajo et al. 2000).

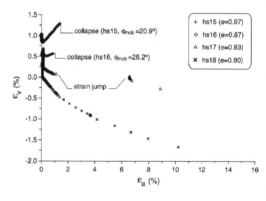

Figure 10. Volumetric versus axial strain behavior of sands, influence of the initial void ratio (Gajo et al. 2000).

is observed under fully drained conditions. The explanation is then far from an undrained liquefaction. They also performed strain controlled drained tests by imposing a strain increment ratio defined by $(d\varepsilon_v/d\varepsilon_1)_i$ (where ε_1 is the axial strain). That makes possible to define an instability mechanism ruled by the inequality: $(d\varepsilon_v/d\varepsilon_1)_i - (d\varepsilon_v/d\varepsilon_1)_s \leq 0$. The strain increment ratio of the soil $(d\varepsilon_v/d\varepsilon_1)_s$ is the maximum $d\varepsilon_v/d\varepsilon_1$ ratio determined from a drained test. According to the authors, this inequality can define a zone in the $q-p'$ stress plane where instability can occur. This unstable domain for a stress state located on the yield surface is represented by a shaded zone on Figure 8 and, the constant-q drained path is included inside this zone. The criterion presented herebefore can also be expressed in terms of dilatancy angle: the liquefaction state can be reached if the material dilatancy angle is lower than the imposed dila-tancy angle (Darve & Pal 1997).

Let us now observe the typical volumetric behavior of sands under constant-q stress paths. Figure 9 represents the volumetric versus axial strain behavior of loose sand under a constant-q stress path. For the loose sand, collapse occurs at point D of Figure 9. According to the observations by Gajo et al. (2000), the sample suffered total, catastrophic failure, with large strains. The experimentalists also noticed, like Sasitharan, a mobilized friction angle at collapse much lower than the expected failure strength given by a Mohr-Coulomb condition. It is also of interest to conclude from Figure 10 that, for sands with an initial void ratio lower than 0.80, the sample suffered no instability. Figure 11 emphasizes the influence of the initial anisotropy on the instability occurence. It is clear that, the higher the anisotropy is, the sooner the instability appears.

In order to perform constitutive modelling some constitutive relations are of course needed. The aim of the next subsection is thus to define the different models we used in the subsequent computations.

3.3 Constitutive models

Because the failures considered in this paper are mainly of plastic type, rate-independent elasto-plastic relations have to be utilized. Let us recall the general

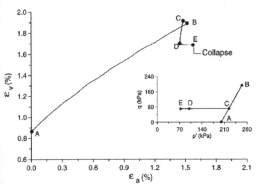

Figure 11. Volumetric versus axial strain behavior of sands, effects of preshearing on a loose sand (Gajo et al. 2000).

expression of rate-independent constitutive relation given in section 1

$$d\varepsilon = \mathbf{N_h}(\mathbf{u})d\sigma$$

with $\mathbf{u} = d\sigma/\|d\sigma\|$ and where $\mathbf{h}$ are state variables taking into account memory effects (Darve 1990, Darve et al. 1995). Polynomial serie expansions for $\mathbf{N(u)}$ give

$$\mathbf{N(u)} = \mathbf{N^1} + \mathbf{N^2}\,\mathbf{u} + \mathbf{N^3}\,\mathbf{u}\,\mathbf{u} + \ldots \tag{17}$$

The incrementally non-linear relations "of second order" are obtained by taking into account only the first both terms,

$$d\varepsilon_{ij} = N^1_{ijkl}d\sigma_{kl} + \frac{1}{\|\mathbf{d\sigma}\|}N^2_{ijklmn}d\sigma_{kl}d\sigma_{mn} \tag{18}$$

In this paper loading paths are considered only in fixed stress–strain principal axes. If so and by considering other constitutive assumptions (Darve 1990), it comes in these principal axes

$$\begin{pmatrix} d\varepsilon_1 \\ d\varepsilon_2 \\ d\varepsilon_3 \end{pmatrix} = \frac{1}{2}[\mathbf{N^+} + \mathbf{N^-}]\cdot\begin{pmatrix} d\sigma_1 \\ d\sigma_2 \\ d\sigma_3 \end{pmatrix} + \\ \frac{1}{2\|d\sigma\|}[\mathbf{N^+} - \mathbf{N^-}]\cdot\begin{pmatrix} (d\sigma_1)^2 \\ (d\sigma_2)^2 \\ (d\sigma_3)^2 \end{pmatrix} \tag{19}$$

with

$$\mathbf{N^+} = \begin{bmatrix} \dfrac{1}{E_1^+} & -\dfrac{\nu_2^{1+}}{E_2^+} & -\dfrac{\nu_3^{1+}}{E_3^+} \\[2mm] -\dfrac{\nu_1^{2+}}{E_1^+} & \dfrac{1}{E_2^+} & -\dfrac{\nu_3^{2+}}{E_3^+} \\[2mm] -\dfrac{\nu_1^{3+}}{E_1^+} & -\dfrac{\nu_2^{3+}}{E_2^+} & \dfrac{1}{E_3^+} \end{bmatrix}$$

and $\mathbf{N^-}$ defined in an identical manner. The superscripts $(+)$ and $(-)$ mean respectively "compression" $(d\sigma_i > 0)$ and "extension" $(d\sigma_i < 0)$ in the axial direction "i" of "generalized" triaxial paths (two lateral constant stresses σ_j and σ_k). E_i "generalized" tangent Young's moduli and ν_i^j "generalized" tangent Poisson's ratios defined by

$$E_i = \left(\frac{\partial\sigma_i}{\partial\varepsilon_i}\right)_{\sigma_j,\sigma_k}$$
$$\nu_i^j = -\left(\frac{\partial\varepsilon_j}{\partial\varepsilon_i}\right)_{\sigma_j,\sigma_k}$$
$$\nu_i^k = -\left(\frac{\partial\varepsilon_k}{\partial\varepsilon_i}\right)_{\sigma_j,\sigma_k}$$

The "octo-linear" relation, based on the same definitions for $\mathbf{N^+}$ and $\mathbf{N^-}$, is given by

$$\begin{pmatrix} d\varepsilon_1 \\ d\varepsilon_2 \\ d\varepsilon_3 \end{pmatrix} = \frac{1}{2}[\mathbf{N^+} + \mathbf{N^-}]\cdot\begin{pmatrix} d\sigma_1 \\ d\sigma_2 \\ d\sigma_3 \end{pmatrix} \\ + \frac{1}{2}[\mathbf{N^+} - \mathbf{N^-}]\cdot\begin{pmatrix} |d\sigma_1| \\ |d\sigma_2| \\ |d\sigma_3| \end{pmatrix} \tag{20}$$

While Equation 19 can be viewed as a non-linear (quadratic) interpolation between the material responses to generalized triaxial loading, Equation 20 corresponds to a linear interpolation. In other words, Equation 19 is a thoroughly incrementally non-linear relation, while relation 20 is incrementally piece-wise linear. Both are homogeneous of degree one, which means that they describe only rate independent strains.

3.4 Theoretical analyzes with the octo-linear model

The incrementally piece-wise linear model (Eq. 20) (Darve & Labanieh 1982) is composed of 8 linear relations between $d\varepsilon$ and $d\sigma$ Each relation corresponds to one "tensorial zone". The application of this model to the axisymmetrical case leads to the expression below

$$\begin{bmatrix} d\varepsilon_1 \\ \sqrt{2}d\varepsilon_3 \end{bmatrix} = \frac{1}{2}(\mathbf{Q^+} + \mathbf{Q^-})\begin{bmatrix} d\sigma_1 \\ \sqrt{2}d\sigma_3 \end{bmatrix} \\ + \frac{1}{2}(\mathbf{Q^+} - \mathbf{Q^-})\begin{bmatrix} |d\sigma_1| \\ \sqrt{2}|d\sigma_3| \end{bmatrix} \tag{21}$$

with:

$$\mathbf{Q^\pm} = \begin{bmatrix} \dfrac{1}{E_1^\pm} & -\sqrt{2}\dfrac{\nu_3^{1\pm}}{E_3^\pm} \\[2mm] -\sqrt{2}\dfrac{\nu_1^{3\pm}}{E_1^\pm} & \dfrac{1-\nu_3^{3\pm}}{E_3^\pm} \end{bmatrix} \tag{22}$$

189

As we have seen yet, the upper index $(+)$ means "compression" (that is to say $d\sigma_i > 0$) whereas $(-)$ is equivalent to "extension" $(d\sigma_i > 0)$. Thus, from now on, there exists only 4 linear relations between $\mathbf{d}\varepsilon$ and $\mathbf{d}\sigma$ depending on the signs of $d\sigma_1$ and $d\sigma_3$.

Let us now follow the reasoning used for the undrained loading and presented in the introduction. Considering stress σ and strain ε, they are work-conjugate in the Hill's sense (Hill 1968), so that in the axisymmetrical case

$$W = \sigma_1\varepsilon_1 + 2\sigma_3\varepsilon_3$$

which can be written by taking into account the volumetric strain $\varepsilon_v = \varepsilon_1 + 2\varepsilon_3$

$$W = \varepsilon_v\sigma_3 + q\varepsilon_1 \tag{23}$$

Thus, conjugate quantities are respectively: ε_v to σ_3 and q to ε_1. It makes possible the expression of the second order work in

$$d^2W = d\varepsilon_v d\sigma_3 + dq d\varepsilon_1$$

which becomes

$$d^2W = d\varepsilon_v d\sigma_3 \tag{24}$$

because of the constant shear axisymmetric stress loading path which imposes $dq = 0$. Equation 24 can now be interpreted in a suitable manner to deduce the development of instabilities. From the definition of the loading path (see section 3.1), it is clear that the quantity $d\sigma_3$ never vanishes since it is a loading parameter equal to a given negative constant.

On the contrary, if ε_v passes through an extremum value, at this moment $d\varepsilon_v$ will be equal to zero. So, the second order work will also vanish. And, from this point this quantity will stay lower than or equal to zero. Indeed, if we suppose a volume variation controlled loading, a small additional volume variation will induce a sudden failure of the sample.

Thus, this extremum of volume variation is unstable according to the Lyapunov's definition of stability (or non-controllable in Nova's sense, 1994).

To develop an analysis in the same framework as in section 1 for undrained loading, let us re-write the constitutive equation under the following form (similar to Eq. 14)

$$\begin{bmatrix} d\varepsilon_v \\ dq \end{bmatrix} = \mathbf{T} \begin{bmatrix} d\sigma_3 \\ d\varepsilon_1 \end{bmatrix} \tag{25}$$

Because of the definition of the loading path which is located in the negative area of the stress plane

$(d\sigma_1 < 0$ and $d\sigma_3 < 0)$, the expression of the octo-linear model (see Eq. 21) leads to

$$\begin{bmatrix} d\varepsilon_v \\ dq \end{bmatrix} = \begin{bmatrix} \dfrac{2\left(1-\nu_3^{3-}-2\nu_1^{3-}\nu_3^{1-}\right)}{E_3^-} & 1-2\nu_1^{3-} \\ 2\dfrac{E_1^-}{E_3^-}\nu_3^{1-}-1 & E_1^- \end{bmatrix} \begin{bmatrix} d\sigma_3 \\ d\varepsilon_1 \end{bmatrix} \tag{26}$$

We can, once again, express a generalization of equations 5 because

$$\det \mathbf{T} = 0 \tag{27}$$

is a bifurcation criterion at the peak value of $d\varepsilon_1 + 2d\varepsilon_3$. From Equation 26, it comes

$$2\frac{E_1^-}{E_3^-}\left(1-\nu_3^{3-}-\nu_3^{1-}\right)+1-2\nu_1^{3-}=0 \tag{28}$$

and, the rupture rule corresponds to

$$\mathbf{T} \begin{bmatrix} d\sigma_3 \\ d\varepsilon_1 \end{bmatrix} = \begin{bmatrix} 0 \\ 0 \end{bmatrix} \tag{29}$$

or for the octo-linear model to,

$$E_1^- d\varepsilon_1 + \left(2\frac{E_1^-}{E_3^-}\nu_3^{1-}-1\right) d\sigma_3 = 0 \tag{30}$$

These equations are exactly equivalent to Equations 15 for the undrained loading.

From Equation 30 we can conclude to the existence of an unlimited number of solutions for $(d\varepsilon_1, d\sigma_3)$, all satisfying relation 30. The unstable direction is given by

$$d\sigma_1 = d\sigma_3$$

which is a stress path parallel to the hydrostatic line.

3.5 Conclusions

We can now conclude to the unstable behavior of loose sand subjected to constant shear drained tests. The fact that instability can develop is neither the consequence of touching the steady state line nor reaching a limit state defined in the stress plane by the drained Mohr-Coulomb limit condition, but rather, the presence or not of an extremum of volumetric deformation during the loading. And, in order to let the instability giving rise to a sudden failure, the loading must be volumetric strain controlled.

We have seen in section 3.2 that this maximum exists for a loose sand (Fig. 9).

In the light of these conclusions, the collapse observed by Gajo et al. (2000) is rather a diffuse collapse due to the imperfections of the sample than a sudden loading induced failure. Indeed, the authors performed the constant shear test by decreasing the mean effective stress instead of increasing the sample volume. Thus, the instability was potentially existing from point C in Figure 9 but not effective because of the loading mode.

The next section is devoted to investigate the influence of the initial void ratio and the effect of different levels of initial shear stress at which we perform the hydrostatic unloading. The aim is to validate numerically our previous remarks.

4 Q-CONSTANT LOADING PATHS, MODELLING

The octo-linear and the non-linear models which have been presented in section 3.3 are both used to simulate constant shear drained tests. We perform these computations with two sets of parameters, one for a loose sand, the second for a dense sand. Previous calculations have been made so that the sand parameters are calibrated on the Hostun sand. The loose Hostun sand possesses an initial void ratio equal to 0.87 whereas it is equal to 0.55 for the dense Hostun sand.

4.1 Loose Hostun sand

The diagrams presented hereafter are composed of two parts. The first one corresponds to the classical drained compression test and, the second one is the constant shear part itself where q is held constant (see Fig. 1a). The initial isotropic pressures are ranging from 100 kPa to 500 kPa through a mean value of 300 kPa. As we mentioned yet, the aim of these simulations is to verify the presence or not of an extremum in the volume variations during the tests.

If this extremum appears, the second order work should be negative from this state (see section 3.4). The computed points are then replaced by different symbols (circles, triangles, squares) depending on the level of shearing.

On the layouts, the representation of the relative volume variation $(\Delta V)/(V_0)$ is preferred to the evolution of ε_v but, $\varepsilon_v = -(\Delta V)/(V_0)$ thus, a minimum for ε_v is a maximum for $(\Delta V)/(V_0)$ and conversely.

Figure 12 shows the results obtained for the loose Hostun sand for the lower initial isotropic pressure and with the octo-linear model. The first diagram of Figure 12 is the loading program and, the four subsequent diagrams correspond to the material response. If we focus on the diagrams presenting $(\Delta V)/(V_0)$ versus ε_1 we note that the expected behavior is observed. Indeed, while the lateral pressure is monotonously decreasing, the relative volume variation is passing

through a maximum value. Of course, the second order work is negative from this maximum.

It is not out of interests to state from Figures 13 and 14 that this result is independent of the initial isotropic pressure. One can notice on these two figures that the snap back of the axial strain in the $(\Delta V)/(V_0)$ versus ε_1 plane is only observed when the deviator stress value is much smaller than the spheric part of the stress. A last remark on the diagrams will be that the contractancy of the sand seems to be linear in the drained triaxial phase of the loading. But this is only an artefact caused by the scale of the plotting which is quite large (up to 18 % of axial deformation).

The diagrams of Figure 15 represent the comparison between the computations with the octo-linear model and with the non-linear one. This time, only one shear level has been tested because these diagram can obviously allow us to extend our previous conclusions to the non-linear model.

4.2 Dense Hostun sand

Figure 16 shows the results obtained for dense Hostun sand. It is clear that the volume variation doesn't possess any maximum. It implies that the second order work is always positive. Similar conclusions have already been made by Gajo et al. (2000) (see Fig. 10) who didn't observe any instabilities for an initial void ratio lower than 0.80.

4.3 Conclusions

As in an undrained triaxial test, a loose sand loaded in a specific way can exhibit an unstable behavior on a drained triaxial path.

Many studies have been carried out in order to explain the unstable behavior of loose or dense sand in axisymmetric conditions, and plane strain conditions. Darve (1996) proved that Hill's condition of stability can be a proper criterion of diffuse failure.

This paper allows us to extend this conclusion to the specific case of q-constant loading paths. Indeed, several experiments have been performed on a loose sand (Chu 2003) and also on soft rocks (Zhu 2002) that exhibit sudden failure without strain localization pattern. The numerical computations which have been done here can explain the conditions of instability.

Furthermore, the first section of this paper recalls that the second order work is essentially a directional quantity. Some papers (Darve & Laouafa 2000) focus on this aspect and it has been concluded that the second order work vanishes inside cones of stress direction.

These cones of unstable stress directions are plotted in the bissector stress plane in Figure 13a for the loose sand and in Figure 13b for the dense sand. The previous results can be interpreted by noticing that the cones of unstable stress directions include the

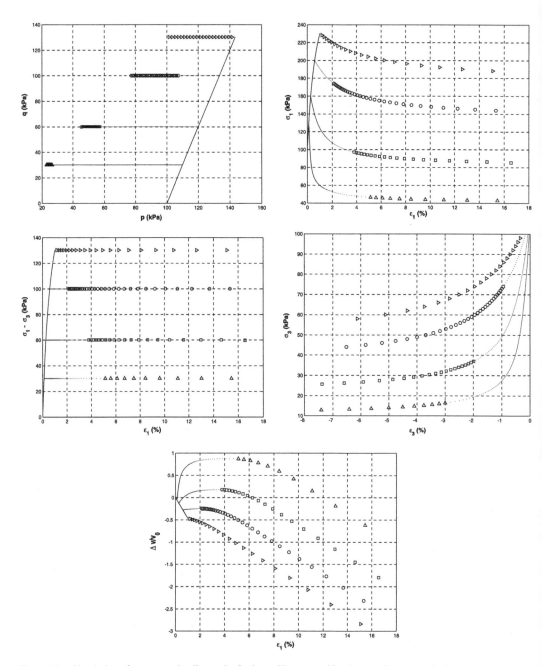

Figure 12. Simulation of q constant loading paths for loose Hostun sand by the octo-linear constitutive model. Points are replaced by other symbols when the second order work takes negative values from the maximum of volume variations. The initial isotropic pressure is equal to 100 kPa.

isotropic direction ($d\sigma_1 = d\sigma_3 < 0$) from certain deviatoric stress levels (points C and D of Fig. 13a) whereas this direction is never included in the cones for the dense sand (Fig. 13b).

In addition, it is necessary to remark here that the cones continuously opens from point A to D on Figure 13a. Thus, one can observe that if the cone of unstable directions plotted at point C includes the constant q

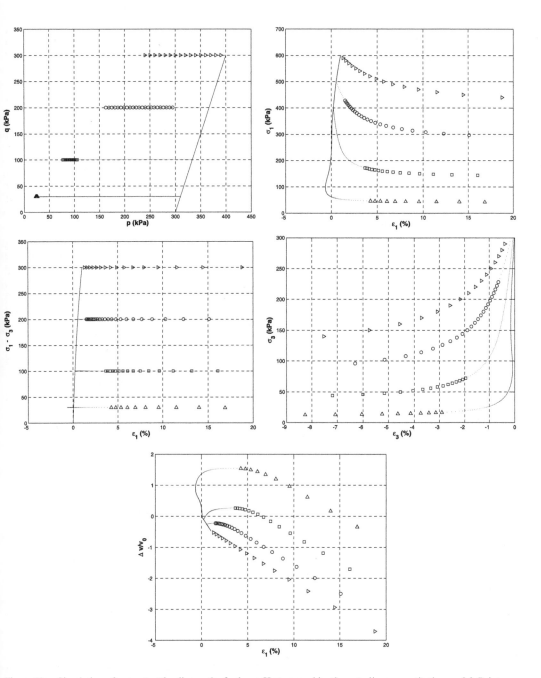

Figure 13. Simulation of q constant loading paths for loose Hostun sand by the octo-linear constitutive model. Points are replaced by other symbols when the second order work takes negative values from the maximum of volume variations. The initial isotropic pressure is equal to 300 kPa.

direction, it means that, as soon as we begin the constant shear part of the loading, the second order work will be negative. Figures 12–14 perfectly illustrate this statement.

5 CONCLUSIONS

Essentially because of the non-associated character of geomaterials plastic strains, the failure is an intricate

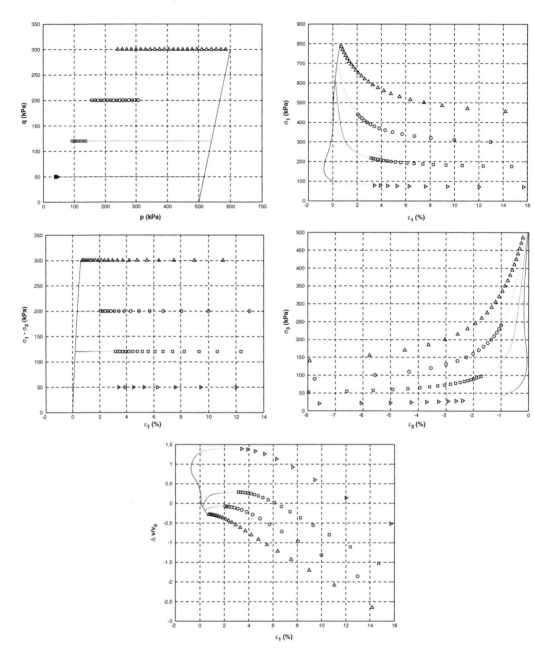

Figure 14. Simulation of q constant loading paths for loose Hostun sand by the octo-linear constitutive model. Points are replaced by other symbols when the second order work takes negative values from the maximum of volume variations. The initial isotropic pressure is equal to 500 kPa.

question. The traditional view of single failure surface in stress plane is not at all verified experimentally.

Indeed, section 2 of this paper has shown an example of failure which does not verify neither empirical Mohr-Coulomb limit condition nor any localization condition (which is very close to Mohr-Coulomb limit surface in the case of loose sands). This example is constituted by axially load controlled undrained

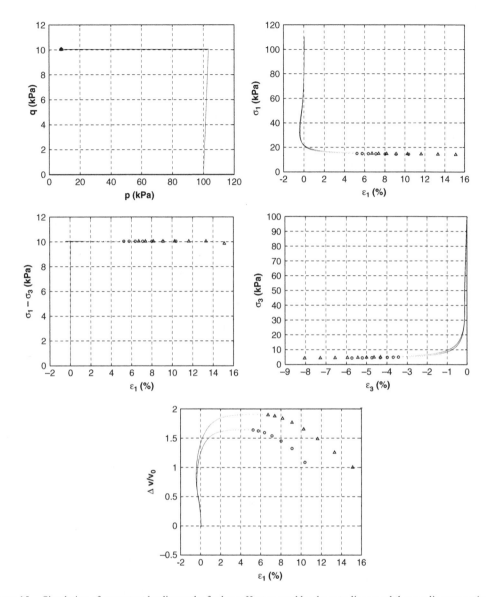

Figure 15. Simulation of q constant loading paths for loose Hostun sand by the octo-linear and the non-linear constitutive models. Points are replaced by other symbols (respectively triangles for the octo-linear model and circles for the non-linear one) when the second order work takes negative values from the maximum of volume variations. The initial isotropic pressure is equal to 100 kPa, the shearing level is equal to 10 kPa.

triaxial tests. From this case, two conclusions can be drawn out:

- because there is apparently not any shear band formation around the peak value of q but rather a chaotic displacement field, this mode of failure can be called "diffuse",
- Hill's condition of stability (given by the sign of second order work) is able to predict this failure mode.

In introduction, a general methodology was briefly recalled and illustrated by the undrained case for loose sands, in order to show that failure states are essentially characterized by bifurcation points associated to a bifurcation criterion and a failure rule. This general framework was applied to the case of constant q loading paths in subsection 3.4 from a theoretical point of view and in section 4 from a numerical view point.

195

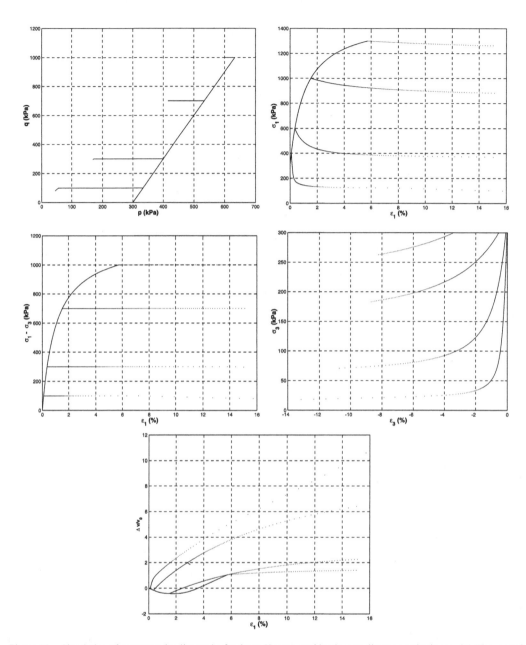

Figure 16. Simulation of q constant loading paths for dense Hostun sand by the octo-linear constitutive model. The second order work is never vanishing.

From section 3 it can be concluded that constant q loading paths on loose sands are unstable from the maximum of volume variations. A volume variations controlled q-constant loading path is defined by:

$$\begin{cases} d\varepsilon_1 + 2d\varepsilon_3 = & \text{constant} < 0 \text{ (loading programme)} \\ d\varepsilon_2 = d\varepsilon_3 & \text{(axisymmetry condition)} \\ d\sigma_1 = d\sigma_3 & (q\text{-constant condition)} \end{cases}$$

This control mode leads to a sudden diffuse failure at the maximum of volume variations. The bifurcation criterion is given by Equation 27 and the failure rule by Equation 29. It appears that the bifurcation criterion is the same as in an undrained loading (given by Eq. 15). From this result, a very interesting conclusion can be deduced by considering two curves in the q versus p' plane. A first stress curve corresponds to the

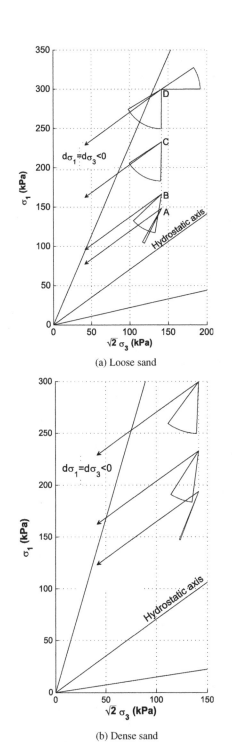

(a) Loose sand

(b) Dense sand

Figure 17.   Cones of unstable stress directions for a drained triaxial test for the loose sand (Fig. a) and for the dense sand (Fig. b) simulated by the octo-linear model and comparison with the directions of the constant shear path.

maxima of volume variations for q-constant loading paths. The second one is a stress curve which relates the peak values of q for undrained loading. Both these curves will coincide approximately by neglecting the influence of both the previous different stress–strain histories. This theoretical conjecture has to be verified experimentally.

Section 4 presents various numerical simulations of constant q paths. The influences of the initial isotropic pressure and of the initial sand density are emphasized. Concerning the initial isotropic pressure, the results show that, in q versus p' plane, the first stress points where the second order work is vanishing are approximately on a single stress curve. That means that, for q-constant paths, there exists an intrinsic instability curve (independent of the initial stress conditions). Concerning the initial density, it appears that the unstable states are related to loose sands and disappear for dense sands. The basic explanation has been given on Figure 17: the hydrostatic stress direction is not included into the cone of unstable stress directions for dense sands, while it is for loose sands.

A comparison has been also carried out between an incrementally piece-wise linear constitutive relation and a thoroughly non-linear relation. The results show that the slight quantitative difference does not change the qualitative conclusions.

This general framework, which has been presented here in the case of sands, is applied more generally to natural media (naturals soils, debris flows, fractured cliffs, snow) for numerical simulations of natural hazards, which are gravity driven (landslides, debis flows, rockfalls, snow avalanches).

ACKNOWLEDGEMENTS

The authors would like to acknowledge the EU project Degradation and Instabilities in Geomaterials with Application to Hazard Mitigation (DIGA) in the framework of the Human Potential Program, Research Training Networks (HPRN-CT-2002-00220). This research has been developed in the framework of a french research federative structure called RNVO ("Natural Hazards and Structure Vulnerability").

REFERENCES

Anderson, S. and M. Riemer (1995). Collapse of saturated soil due to reduction in confinement. *Journal of Geotechnical Engineering 121*(2), 216–220.

Bardet, J. (1997). *Experimental Soil Mechanics*. Prentice Hall.

Begemann, H., H. Koning, and J. Linderberg (1977). In *Critical density of sand*. Tokyo, Japan.

Bigoni, D. and T. Hueckel (1990). On uniqueness and strain localization in plane strain and plane stress elastoplasticity. *Mechanics Research Communications 17*(1), 15–23.

Bishop, A. and D. Henkel (1962). *The Measurment of Soil Properties in the Triaxial test*. Edward Arnold, London.

Brand, E. (1981). Some thoughts on rain-induced slope failures. In *Proceedings of the* 10th *ICSMFE*, Volume 3, pp. 373–376. Stockholm.

Brenner, R., H. Tam, and E. Brand (1985). Field stress path simulation of rain induced slope failure. In *Proc. XI ICSMFE*. San Francisco.

Casagrande, A. (1975). Liquefaction and cyclic deformation of sands – a critical review. In *Proc. 5th Pan American Conference on Soil Mechanics and Foundation Engineering*.

Castro, G. (1969). Liquefaction of sand. *Harvard Soil Mechanics Series 81*. Cambridge, MA.

Chu, J. (2003). Recent progress in experimental studies on instability of granular soil. In Labuz & Drescher (Ed.), *Int. Workshop on Bifurcations and Instabilities in Geomechanics*. Swets & Zeitlinger.

Chu, J., S.-C. Lo, and I. Lee (1993). Instability of granular soils under strain path testing. *J. Geotechnical Eng. 119*(5), 874–892.

Darve, F. (1984). An incrementally non-linear constitutive law of second order and its application to localization. In Desai and Gallagher (Eds.), *Mechanics of Engineering Materials*, pp. 179–196.

Darve, F. (1990). Incrementally non-linear constitutive relationships. *Geomaterials Constitutive Equations and Modelling F. Darve ed., Elsevier Applied Science*, 213–238.

Darve, F. (1996). Liquefaction phenomenon of granular materials and constitutive instability. *Int. Journal of Engineering Computations 7*, 5–28.

Darve, F. and B. Chau (1987). Constitutive instabilities in incrementally non-linear modelling. *Constitutive Laws for Engineering Materials C.S. Desai (ed.)*, 301–310.

Darve, F., E. Flavigny, and M. Méghachou (1995). Yield surfaces and principle of superposition revisited by incrementally non-linear constitutive relations. *Int. J. Plasticity 11*(8), 927–948.

Darve, F. and S. Labanieh (1982). Incremental law for sands and clays. simulations of monotonic and cyclic tests. *Int. J. Num. Anal. Meth. in Geomech. 6*, 243–275.

Darve, F. and F. Laouafa (2000). Instabilities in granular materials and application to landslides. *Mech. Cohes. Frict. Mater. 5*(8), 627–652.

Darve, F., F. Laouafa, and G. Servant (2003). Discrete instabilities in granular materials. *Italian Geotechn. J. 37(3)*, 57–67.

Darve, F. and O. Pal (1997). Liquefaction: a phenomenon specific to granular media. In Berhinger & Jenkins (Ed.), *Proc. 3th Int. Conf on Powders & Grains*, pp. 69–73. Balkema.

Darve, F., G. Servant, and F. Laouafa (2003). Continuous and discrete modelling of failure modes in ge-omechanics. In Labuz and Drescher (Eds.), *Bifurcation and Instabilities in Geomechanics*, pp. 65–78. Swets and Zeitlinger.

Darve, F., G. Servant, F. Laouafa, and H. Khoa (2004). Failure in geomaterials: continuous and discrete analyses. to appear in Computer Meth. Appl. Mech. Eng.

Desrues, J., R. Chambon, M. Mokni, and F. Mazeroll (1996). Void ratio evolution inside shear bands in triaxia sand specimens studied by computed tomography *Geotechnique 46*(3), 529–546.

di Prisco, C. and S. Imposimato (1997). Experimental analy sis and theoretical interpretation of triaxial load con trolled loose sand specimen collapses. *Mech. Cohes. Fric Mater. 2*, 93–120.

Eckersley, J. (1990). Instrumented laboratory flowslides *Géotechnique 40*(3), 489–502.

Finno, R., W. Harris, M. Mooney, and G. Viggiani (1997) Shear bands in plane strain compression of loose sand *Géotechnique 47*(1), 149–165.

Gajo, A., L. Piffer, and F. De Polo (2000). Analysis o certain factors affecting the unstable behaviour o saturated loose sand. *Mech. Cohes.-Frict. Mater. 5* 215–237.

Han, C. and I. Vardoulakis (1991). Plane strain compressio experiments on water-saturated fine-grained sand *Géotechnique 41*(1), 49–78.

Head, K. (1986). *Manual of Soil Laboratory Testing, Volum 3: Effective Stress Tests*. John Wiley & Sons.

Hill, R. (1958). A general theory of uniqueness and stabilit in elastic-plastic solids. *J. of the Mech. and Phys. o Solids 6*, 239–249.

Hill, R. (1968). On constitutive inequalities for simpl materials. *J. Mech. Phys. Solids 16*, 229–242.

Lade, P. (1992). Static instability and liquefaction of loos fine sandy slopes. *Journal of Geotechnical Engineerin 118*(1), 51–69.

Laouafa, F. and F. Darve (2002). Modelling of slope failur by a material instability mechanism. *Comp and Geo techn. 29*(4), 301–325.

Lyapunov, A. (1907). Problème général de la stabilité de mouvements. *Annales de la faculté des sciences d Toulouse 9*, 203–274.

Méghachou, M. (1993). *Stabilité des sables lâches. Essai et modélisations*. Ph. D. thesis.

Mokni, M. and J. Desrues (1999). Strain localization meas urements in undrained plane-strain biaxial tests o Hostun RFsand. *Mech. Cohes. Frict.*

Mater. 4, 419.

Nova, R. (1994). Controllability of the incremental respons of soil specimens subjected to arbitrary loading pro grammes. *J. Mech. behav. Mater. 5, No 2*, 193–201.

Rice, J. (1976). The localization of plastic deformatior *Theoretical and applied mechanics, W.T. Koiter e North-Holland publishing Company*, 207–220.

Sasitharan, S., P. Robertson, D. Sego, and N. Morgenster (1993). Collapse behavior of sand. *Canadia Geotechnical Journal 30*(4), 569–577.

Skopek, P., N. Morgenstern, P. Robertson, and D. Seg (1994). Collapse of dry sand. *Can. Geotech. J. 31* 1008–1014.

Sladen, J., R. D'Hollander, and J. Krahn (1985). Back analy sis of the nerlerk berm liquefaction slides.

Can. Geotech. J. 22(4), 579–588.

Zhu, W. (2002). poster. *Gordon conference on Roc deformation*.

Case history and field or physical model measurements
Expériences tirées de la pratique et mesures sur modèles
réduits ou ouvrages

Deformation Characteristics of Geomaterials – Di Benedetto et al (eds)
© 2005 Taylor & Francis Group, London, ISBN 04 1536 701 8

Lessons learned from full scale observations and the practical application of advanced testing and modelling

R.J. Jardine & J.R. Standing
Imperial College, London, UK

N. Kovacevic
Geotechnical Consulting Group, London, UK

ABSTRACT: Field observations from thirteen recent case histories from the UK and France are described which are used to assess an approach for predicting ground movements and soil-structure interaction. High quality site investigations, including very careful sampling and advanced testing are seen to be crucial to success. Providing there are adequate geotechnical data, the non-linear FE predictive approach advocated by the Authors is shown to be sufficient for most engineering purposes, considering typical project time scales. Factors that merit particular attention include: the geological model; sampling and testing quality; anisotropy and pressure dependency; a fully non-linear modelling approach; addressing behaviour from very small to large strains; the 'structure' of natural and artificial geomaterials; the possibility of brittle failure; three dimensional geometry and construction details; permeability and groundwater variations. It is shown, however, that the present approach does not capture some important long term effects of time, creep and ageing. Comments are also made on the use of such advanced techniques in combination with an 'observational approach', and on pos-sible avenues for further research and development.

1   INTRODUCTION

Full scale field observations have played a central role in the development of soil mechanics and geotechnical engineering. Carefully recorded case histories exposed the severe shortcomings of conventional procedures for estimating ground movements and soil-structure interaction. The field data discussed at meetings such as the Brighton ECSMFE in 1979 stimulated the improvements in measurement and analytical techniques that have been reported and debated in specialist conferences in Florence (1991), Hokkaido (1994), London (1997), Turin (1999)[*] and now Lyon (2003).

Papers have been presented at each of these meetings that report the gradual updating and practical application of a simplified approach developed at Imperial College London (Jardine et al. 1984, 1986, 1991 and Jardine & Potts 1988). This paper discusses a selection of thirteen recent case histories and attempts to use these to underline both successes and shortcomings, focusing on aspects of behaviour that are

becoming of increasing interest to geotechnical experimentalists, modellers and practising civil engineers.

2   A SIMPLIFIED NON-LINEAR APPROACH

The key features of this approach are outlined below under three headings.

2.1   Careful site characterisation

An integrated approach is required such as that recently described by Hight (1998) and Hight & Leroueil (2002). The vital requirements are:

- Engineering geology studies and full descriptions of the ground profiles
- High quality sampling if possible. Wherever feasible, block samples are recommended, or careful rotary coring for stiff clays/soft rocks and sharp-edged piston sampling for soft clays
- In-situ testing to provide profiles of key parameters through piezocone (CPTU) soundings, seismic velocity and other geophysical measurements, pressuremeter tests etc. Particular reliance is placed on the CPTU data in sands where undisturbed sampling is rarely feasible

[*]With the substantial conference volumes edited by AGI (1991); Shibuya et al. (1994); Jardine et al. (1998) and Jamiolkowski et al. (1999).

2.2 Stress path laboratory testing

The latter has been central to improvements in modelling stress-strain behaviour; see Jardine (1995), Tatsuoka et al. (1999), Jardine et al. (2001). Test programmes are specified for the most important layers that involve:

- Samples that are usually reconsolidated to the (typically anisotropic) in-situ stresses
- Local strain measurements, and often local pore water pressure sensors
- Pause stages being specified prior to any change in the imposed stress path (particularly prior to shearing) which allow the samples to age under stress until the ratio of their creep rates to the subsequently applied strain rate becomes small (typically 0.01 to 0.02)
- Shearing to failure at relatively slow rates under various stress conditions

The test programmes usually involve triaxial compression and extension, but other triaxial or torsional shear modes may be deployed to help explore any anisotropy. In some cases supplementary bender element, resonant column or small strain cyclic tests are performed; see for example Hight et al. (1997). When good quality samples cannot be taken, for example with sands, reliance is placed on tests on reconstituted samples, combined with data from in-situ testing.

2.3 Boundary value analysis with simplified soil models

The authors and their colleagues have been involved in more than one hundred cases where the simplified non-linear treatment proposed by Jardine (1985) has been applied in practice. Predictions for ground movements and soil-structure interaction have been made with the code Imperial College Finite Element Program (ICFEP) written by Professor David Potts, and in many cases field performance has also been monitored sufficiently closely for the predictions to be assessed critically.

In this approach stress path test data are used to calibrate non-dimensional, highly non-linear, elastic stiffness expressions that are substituted into classical elastic-plastic models such as the Tresca, Mohr-Coulomb, or Modified Cam Clay formulations. Advances in computing power have allowed increasingly complex geometries and circumstances to be considered, as summarised by Potts & Zdravkovic (1999, 2001) and Potts (2003). Key points regarding the simplified non-linear approach include:

- The stiffness relationships that are most frequently used are effective stress based, with global 'isotropic' secant shear stiffness G and bulk stiffness

K' being linked to shear and volumetric strains as follows:

$$3\frac{G}{p'} = A + B.\cos\left\{\alpha.\left[\log_{10}\left(\frac{\varepsilon_D}{\sqrt{3}.C}\right)\right]^\gamma\right\} \quad (1)$$

$$\frac{K'}{p'} = R + S.\cos\left\{\delta.\left[\log_{10}\left(\frac{\varepsilon_V}{T}\right)\right]^\mu\right\} \quad (2)$$

where:
p' is the mean effective stress: $p' = (\sigma_1' + \sigma_2' + \sigma_3')/3$
ε_D is the deviatoric strain invariant:

$$\varepsilon_D = \sqrt{2/3}\left[(\varepsilon_1 - \varepsilon_2)^2 + (\varepsilon_2 - \varepsilon_3)^2 + (\varepsilon_3 - \varepsilon_1)^2\right]^{1/2}$$

ε_D is related to the shear strain invariant ε_s by $\varepsilon_D = \sqrt{3}\varepsilon_S$, and in undrained triaxial tests by $\varepsilon_a = \varepsilon_S$
ε_v is the volumetric strain: $\varepsilon_v = \varepsilon_1 + \varepsilon_2 + \varepsilon_3$
$A, B, C, R, S, T, \alpha, \gamma, \delta$, and μ are material properties found from laboratory (or in-situ) test data

- Tangent stiffness expressions derived from Equations 1 and 2 are used by the FE code, with the tangent stiffnesses at each particular point being updated continually, depending on the current strains (ε_D, ε_v) and p' values. If the strains fall below the specified minima ($\varepsilon_{D,min}$ or $\varepsilon_{v,min}$) the stress point is assumed to fall within the Y_1 pseudo-elastic region and tangent stiffnesses vary only with the mean effective stress, p'. Minimum fixed ratios of 3G/p' and K'/p' also apply once specified upper strain limits ($\varepsilon_{D,max}$ or $\varepsilon_{v,max}$) are exceeded. Finally, the calculated stiffnesses may be prevented from falling below specified absolute stiffness values (G_{min} or K_{min}); this feature can be useful when considering problems involving low (or tensile) stresses
- Implicit in Equations 1 and 2 is the assumption that stiffness scales linearly with mean effective stress p'. While this is a reasonable approximation at larger strains, it is generally recognised that a lower fractional power applies at very small strains (particularly for sands). Care needs to be taken to ensure that the stiffness parameters are appropriate to the stress levels anticipated in-situ. In some special cases the formulation has been modified to allow the exponents (γ and μ) applied in Equations 1 and 2 to vary with strain level
- Equations 1 and 2 can be re-expressed in terms of total stresses when required. For example, when using a Tresca plasticity formulation (to deal with cases where the shear strength is constant) the shear stiffness- ε_D function given by Equation 1 is normalised by the undrained shear strength S_u

202

- Soil stiffness is often highly anisotropic, with anisotropic very small strain stiffness terms being linked to the magnitudes of the individual effective stress components (see Roesler 1979, Jardine 1995, Tatsuoka et al. 1999, Kuwano & Jardine 2002). Provided the stress increments are similarly oriented within broad areas of the straining soil mass, an approximate 'stress path method' approach can be taken by zoning the area of interest and adopting different sets of parameters for each zone based on appropriate tests. This method can be applied in non-linear analyses made with Equations 1 and 2; see for example the retaining wall and embankment cases described by Jardine et al. (1991) and Jardine (1995)
- Addenbrooke et al. (1997) describe one way of extending the original approach to account more explicitly for anisotropy at small strains, while other models such as MIT-E3 have been used to deal with cases where anisotropy dominates at large strains (see Zdravkovic et al. 2002). It is clear that there is ample scope for improving the general treatment of anisotropy
- The kinematic yielding behaviour of soils at small strains is not accounted for explicitly by Equations 1 and 2, which can cause considerable difficulties with non-monotonic loading cases. This problem has been addressed to some extent by re-zeroing the strain terms in the stiffness expressions when the loading sequence is reversed or its direction is changed substantially
- Well formulated kinematic yield surface approaches (see for example Al Tabbaa & Wood 1989, Simpson 1992 or Stallebrass & Taylor 1997) are more attractive intellectually than the pragmatic approach described above. However, as demonstrated later, Equations 1 and 2 are often satisfactory in practice. They also offer flexibility when selecting failure criteria for materials, while other kinematic surface models have to conform to theoretical frameworks that might not describe other aspects of the behaviour of the particular soil adequately. One example would be the brittle failure behaviour of materials such as London Clay, and their divergence from the classical critical state modelling assumptions. Another difficulty is the present lack of a realistic 'bubble' type model for sands
- When combined with a brittle shear failure model, the simplified non-linear approach can be used to deal with problems involving progressive failure. However, it cannot deal with the more subtle processes by which 'structure' is lost through strains developed before the shear failure criteria are satisfied
- The approach takes no account of the time dependency of soil properties, which has been emphasised by Tatsuoka et al. (2001), Leroueil and Hight

(2002) and others. Coupled consolidation may be considered, along with permeability changing with either effective stress or void ratio. But creep, strain rate and acceleration effects could not be addressed explicitly in any of the FE analyses reported here. Nor could any ageing or creep hardening that may take place under constant effective stress conditions

3 CASE HISTORIES

This paper aims to use field observations both to evaluate how well existing simplified procedures apply in practical geotechnical engineering and identify areas where improvements are necessary. Reference will be made to thirteen high quality case histories, eleven of which concerned one or more of the authors directly. It is not possible to report all the desirable data for every case. However, mention is made wherever possible of the key site characterisation features, including appropriate advanced laboratory test data, before summarising relevant analytical predictions and field measurements. In many cases references to more detailed reports or papers are cited.

Jardine et al. (1991) argued on the basis of six case histories that strong pre-failure non-linearity, the effective stress (p') dependence of stiffness and local contained failure are the key parameters that affect ground movements and soil-structure interaction. Figure 1 shows the spread of results obtained in the six projects, plotting for each case the predictions made for the single most important ground displacement parameter against the corresponding field measurement. The projects included deep excavations

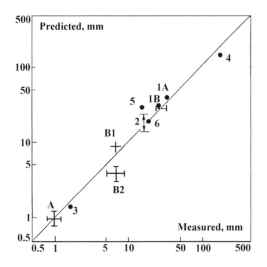

Figure 1. Comparisons between non-linear ground movement predictions and measurements (Jardine 1992).

203

and tunnels in the London area, piled and gravity based North Sea offshore production platforms.

Higgins and Jardine (1998) summarised a further 16 cases, covering a wider range of circumstances. They reported generally good agreement between predictions and measurements. This was highly encouraging, bearing in mind the potential shortcomings of the predictive tools used. It was concluded that advanced laboratory test data could be applied with some confidence, and that non-linearity, p' dependence and local failure are the three most important features of soil behaviour that needed to be modelled explicitly.

Jardine (1995) drew attention to partial drainage effects, the need for fully coupled analyses with clays, and some potential effects of anisotropy. Hight & Higgins (1995) highlighted the influence of construction sequence, showing that predictions could become completely inappropriate if the programme of work differed from that expected. As described later, prediction updating has now become more common with analyses being re-run as new information emerges about the site, the ground water conditions and the construction operations.

Bearing in mind the earlier reviews, the main topics addressed in this paper are:

- Accounting properly for geology and stratification; the potential effects of natural structure on field behaviour
- Dealing with a broader range of geomaterials, ranging from very soft landfill to hard cemented ground
- Using field data and updated analysis as part of an observational approach to overcome uncertainty in construction sequences, in-situ permeability and other parameters to optimise performance and economy
- Considering more complex construction geometries
- Some effects of time, including long-term consolidation, creep, ageing and rate effects
- Cases where much of the ground was approaching failure, or experiencing contained failure; possible brittleness; cracking or shear banding

The case histories are presented under five generic headings, covering four main classes of geomaterial and problems ranging from excavations, spread foundations and piles to tunnels and embankments. The intention is to consider new applications, and draw attention to areas where difficulties have been encountered. Particular emphasis is placed on aspects of behaviour that were not considered fully in earlier evaluations, so the sections dealing with soft clays, sands and cemented jet-grout are more detailed than those covering the more familiar stiff clay sites. In the same way, several sets of long-term field observations are mentioned along with some load tests to failure, while the earlier studies are restricted to relatively short-term measurements of the response under working load conditions.

3.1 Excavations and surface loading cases involving low OCR (soft) clays

3.1.1 The importance of modelling non-linearity and regional groundwater changes in soft materials: medium-term behaviour of Bangkok Metro station excavation, Thailand

It has been argued previously that stiffness behaviour within the classical yield surface (termed Y_3 within the jargon proposed by Jardine 1992, 1995) has less impact on problems involving low OCR (usually soft) clays than when considering other geomaterials. In cases where the clays have highly anisotropic yielding behaviour, anisotropy in undrained shear strength (S_u) can be the feature that most affects ground movements, especially with relatively flexible structures; see for example Whittle & Hashash (1994).

One of the new case histories presented at this symposium by Kovacevic et al. (2003) concerns a four storey 22 m deep and 23 m wide excavation for a new metro station in Bangkok, Thailand. The excavation, which is supported by a very rigid 39 m deep propped reinforced concrete diaphragm wall, has been made in low OCR (soft) clay overlying stiff clay and dense sand deposits, as shown in Figure 2. As a result of groundwater extraction from the sand layers the soft Bangkok clays experience a background rate of compression and regional settlement.

Site investigations were performed by a joint team from AIT (Bangkok) & PHARI (Yokosuka, Japan), who carried out high quality piston sampling, in-situ soil

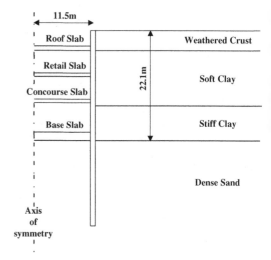

Figure 2. Cross section through Bangkok metro station (Kovacevic et al. 2003).

204

testing and laboratory stress path testing (Shibuya & Tamrakar, 1999). The clays have low OCR, but are not reported as being strongly anisotropic. Analyses of the excavation were undertaken by the research group at Hokkaido University, Japan (Shibuya et al. 2001) using various soil models (Modified Cam Clay, Mohr Coulomb and linear elastic models) to characterise the behaviour of different strata. The respective stiffness parameters were gauged from in-situ shear wave velocity measurements, bender elements and undrained triaxial testing with high resolution small strain stiffness measurements.

The Hokkaido FE analyses show that the patterns of deformation predicted during excavation depended strongly on the constitutive models assumed in the analyses. Tamrakar et al. (2000) reported that a simple model employing a linear elastic pre-yield Young's modulus equal to half of the elastic modulus deduced from a seismic survey ($E' = E'_{max}/2$) predicted lateral wall deflections with reasonable accuracy. However, the predicted surface settlement profiles were far wider and much shallower than those measured. If the stiffness parameters were changed to obtain a better fit to the surface settlement measurements then the lateral wall movements developed during excavation could not be recovered. The combined pattern of field movements could not be matched if the pre-yielding behaviour was assumed to be linear.

Kovacevic et al. (2003) report a recent re-analysis of the same case in which pre Y_3 non-linearity was modelled by incorporating Equations 1 and 2 into the Modified Cam Clay model. The non-dimensional shear and bulk modulus small strain stiffness curves presented in Figure 3 are chosen on the basis of the available in-situ and laboratory tests. No attempt was made to model the background regional subsidence process.

The functions represented by Figure 3 were input into a two dimensional analysis that tried to follow the construction sequence, wall geometries and propping details as closely as possible. Kovacevic et al. give more details of the modelling and note that the construction details could affect predictions considerably.

The spread of predictions is presented in Figures 4 and 5, with the results from the Small Strain Stiffness pre-yield models (Equations 1 and 2) identified as 'SSS'. The non-linear 'small strain' approach provides a better match to the medium-term field behaviour than the optimised linear (pre Y_3 yielding) analyses. Kovacevic et al. note that the final underestimation of settlement across the whole site area may have been due to either: (i) even relatively light site surcharging, (ii) the continuing regional subsidence caused by groundwater extraction being greater than expected or (iii) creep effects acting over the 13 month construction period. The potential long term movements were not considered in this study.

The regional subsidence may well have interacted with the shear distortions caused by the excavation causing settlement rates to accelerate. Stiffness or S_u anisotropy may also have played a role, and the results give the impression that the degree of wall friction assumed in the analysis may have been overestimated. However, the most important conclusion is that it is necessary to simulate the rapid degradation of stiffness with strain, even when dealing with soft clay problems.

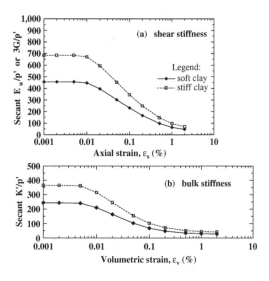

Figure 3. Non-linear stiffness relationships assumed for re-analysis of Bangkok metro station (Kovacevic et al. 2003). Note that shear stiffness curves are defined for undrained test conditions where $\varepsilon_d = \sqrt{3}\varepsilon_s$, and $\varepsilon_s = \varepsilon_a$ the axial strain.

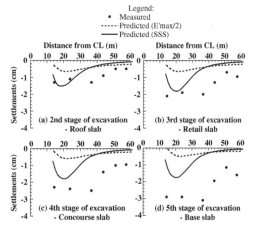

Figure 4. Re-analysis of Bangkok metro station case history: predicted and measured surface settlements (Kovacevic et al. 2003).

205

3.1.2 *Evidence for the effects of long-term loading from tests at Bothkennar, Scotland, UK*

The UK Engineering and Physical Sciences Research Council (EPSRC) developed a national soft clay research test bed site at Bothkennar, Scotland, which operated from the late 1980s until 2001. Intensive high quality sampling, stress path testing and in-situ profiling experiments were performed as described by Hight et al. (1992), (2002), Smith et al. (1992), Albert et al. (2003) and others. Unlike the deposits mentioned in the Bangkok case history, Bothkennar clay is known to have highly anisotropic S_u characteristics. It is also lightly cemented, time/rate dependent and sensitive. Like most soils its stress-strain behaviour is highly non-linear once the clay's small quasi-elastic Y_1 yield surface has been engaged.

Jardine et al. (1995) investigated the influences of the clay's non-linearity, anisotropy, cementing, sensitivity and time dependency on its short and long-term behaviour under foundation loading. Tests were conducted on two instrumented pads (each around 2.4 m square) with one being loaded to failure over four days, while the second was loaded to two thirds of the first's failure load and then left to settle. Paraphrasing some of the conclusions drawn by Jardine et al. (1995):

- Predictions made in parametric non-linear ICFEP analyses by Jardine et al. (1986) were borne out by the field trials. Stress sensors located beneath the foundations also showed that the stress regime is also affected strongly by the soil's pre and post failure non-linearity
- In the same way ground movement patterns are modified considerably by the ground's non-linearity,

with strains and displacements being concentrated towards the loading boundaries and diminishing more rapidly with vertical and lateral distance from the loading boundary than expected by linear elasticity. The latter features are illustrated in Figure 6, where the field measurements and predictions are presented together in a single non-dimensional plot
- The short to medium-term in-situ stress-strain (stiffness) response developed in the field matches that seen in appropriate instrumented laboratory stress path tests. Time dependency (creep) has a very strong influence on long-term behaviour
- The pads' short-term ultimate bearing capacity is affected by cementation, anisotropy and brittleness. The operational global undrained shear strength interpreted at failure drops 20% below the upper limit obtained from CK_0U triaxial compression tests on Sherbrooke 'block quality' samples, but amounted to 2.5 times the lower limit strengths developed in triaxial extension tests
- The Bothkennar pads gave far steeper load-displacement curves and developed greater capacities than similar tests on a footing founded in the non-cemented (and insensitive) Thames Alluvial clays (Hight and Higgins 1995)
- Partial pore pressure dissipation took place in the upper layers during short term loading. It is important to account for such coupled consolidation processes if ground movement predictions are to be accurate

Lehane & Jardine (2003) describe a more recent loading test at Bothkennar. In 2001 they conducted a final load test to failure (denoted Test C) on the second

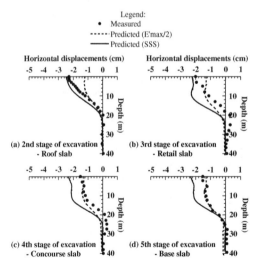

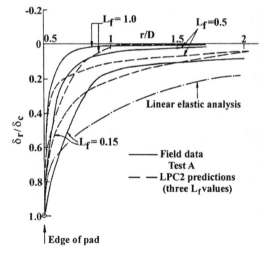

Figure 5. Re-analysis of Bangkok metro station case history: predicted and measured wall deflections (Kovacevic et al. 2003).

Figure 6. Surface settlements δ_r generated at radius r (from centre of footing of diameter D) plotted as δ_r/δ_c against r/D, where δ_c is the centre line settlement (Jardine et al. 1995).

pad, after it had experienced 11 years of maintained loading. Figure 7 shows the footing and instrument layout, along with a simplified borehole log, while Figure 8 shows a photograph of the field trial. Figure 9 shows how the foundation responded to initial loading up to 89 kPa in 1990 (Test B), and then settled under maintained load. Pore pressure observations proved that dissipation was essentially complete within 12 months and at least half of the total settlement resulted from creep.

Test C involved an initial unloading stage, which allowed the load blocks to be re-configured to provide a safer base for the additional weights that were applied incrementally (following the same procedures as in Test A and B in 1990) until failure developed three days layer. The instruments identified in Figure 7 were still functional, although several others that had been installed in 1990 were no longer serviceable. Precise surveying was undertaken continuously during the test, and all the load blocks were carefully pre-weighed.

Figure 10 illustrates the load-displacement responses of the tests, showing the 1990 test to failure (A), the 1990–2001 maintained load test (B) and the response observed in the 2001 test to failure (C). Any linear portions of the pads' loading responses were exceeded during the first increment, but a 10 kPa amplitude cycle applied at the start of Test C developed a nearly closed hysteretic loop, which corresponded to a mean axial strain of around 0.025%. The incremental loading response seen when the first 20 kPa of extra load was applied in Test C was twice as stiff as that at the start of Test A; the footing only developed large deformations when loaded well beyond Test A's ultimate capacity. When Test C failed, at a load around 48% greater than that available in Test A, it involved a

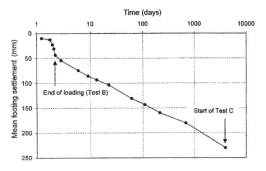

Figure 9. Settlement response of Bothkennar pad in Test B and under maintained load from 1990 to 2001 (Lehane & Jardine 2003).

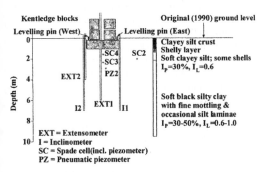

Figure 7. Layout of Bothkennar pad tested in 2001, showing remaining functioning instruments and simplified borehole log (Lehane and Jardine 2003).

Figure 8. Aerial photograph of Bothkennar pad tested in 2001 (Lehane & Jardine 2003).

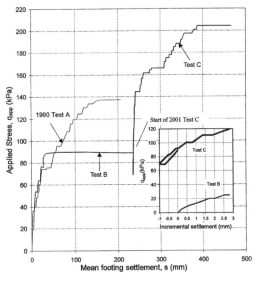

Figure 10. Load- settlement response of Bothkennar pads in Tests A and B (in 1990) and C (in 2001); after Lehane & Jardine (2003).

far larger and deeper mass of soil than that developed in the 1990 test. The latter features are shown in Figure 11.

A site specific ICFEP simulation has yet to be performed for the Bothkennar pad tests. However, Lehane & Jardine compared their results with a parametric study published by Zdravkovic et al. (2003). The latter study employed the isotropic, uncemented, time independent Modified Cam Clay MCC model to compute the effects of pre-loading on the shallow bearing capacity of clays, considering a range of initial conditions. Two cases that bracketed the Bothkennar profile give short term bearing capacity increases of between 15 and 25%. The field tests appear to show gains around 2.4 times larger than the expected value of about 20%. Zdravkovic et al.'s analysis predicts that the failure mechanism will be affected by the consolidation and ageing processes; the strength gains developed in the most heavily loaded area beneath the pads force the mechanism to penetrate to greater depths, where the clay is less affected by the pre-loading. As illustrated in Figure 11, the latter feature is confirmed by the 2001 field test.

Lehane and Jardine noted that some potentially significant important features of ground behaviour cannot be captured by the MCC model including:

• Anisotropy and its evolution under different stress conditions. Jardine et al. (1997) and Zdravkovic & Jardine (2001) report hollow cylinder tests that mimicked consolidation beneath the foundations of stage loaded structures. Consolidation involving re-oriented principal stress directions leads to the evolution of a new anisotropy in each heavily loaded element that tended to be optimally aligned with the pre-loading stress field. This process leads to markedly higher normalised undrained shear strength ratios S_u/p' being available if shearing is renewed while keeping the same σ_1 axis orientation. As an isotropic model the MCC cannot predict these trends

• The volume changes caused by creep under maintained load. The Bothkennar tests show that after 11 years the creep volume changes are at least as significant as those due to consolidation. The large scale Y_3 yield surfaces of the clay are likely to have expanded outwards well beyond the current effective stress point as a result of this 'creep hardening', which contributes to the relatively stiff response to increased loading and additional bearing capacity seen in Figure 10. In contrast, the MCC predicts that the stress states of soil elements that have gained strength significantly are, by definition, on the Y_3 yield surface at the start of the renewed loading stage. Their elastic-plastic response to renewed loading is therefore be expected to be relatively soft

• The organic cements that contribute to Bothkennar clay's undisturbed strength and stiffness are not captured within MCC. The cemented clay structure was probably damaged by the ground movements developed between 1990 and 2001. Any such damage could offset the benefits gained through the two processes described above. However it is also possible that the cemented structure could have reformed as the strain rates slowed down with time

The Bothkennar pad settlement records give direct evidence of the important role played by creep. It is not possible to isolate the influence of evolving anisotropy directly from the results, although this can be investigated by parametric numerical analyses involving more advanced soil models. The role of bonding also remains uncertain. However, it is clear that the unaccounted for features of soil behaviour lead to significantly improved foundation behaviour. Such features offer practical benefits when considering the re-use of foundations in urban redevelopment projects, and when assessing how well existing facilities can withstand unanticipated loads. Potential applications range from extending domestic housing to upgrading offshore structures or raising dams.

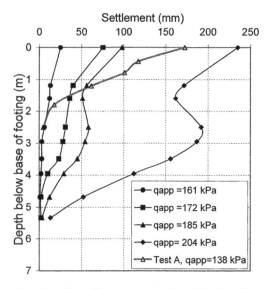

Figure 11. Centre line settlement-depth profiles from 2001 Bothkennar pad (Test C) compared with final profile from 1990 test to failure (Test A); after Lehane & Jardine (2003). Note qapp signifies mean applied vertical bearing stress.

3.2 Excavations and tunnels in stiff clays

Approximately half of the 'small-strain' case histories cited by Jardine et al. (1991) concerned excavations and tunnels in stiff London clay. We return to similar topics in this section and also consider different

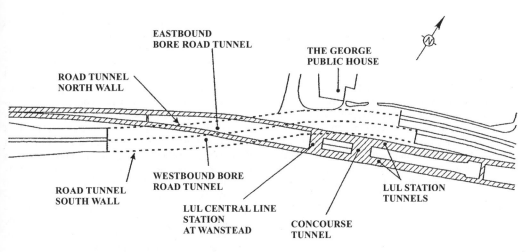

Figure 12. Plan view of new road and existing metro tunnels; Wanstead, London (Higgins et al. 2003).

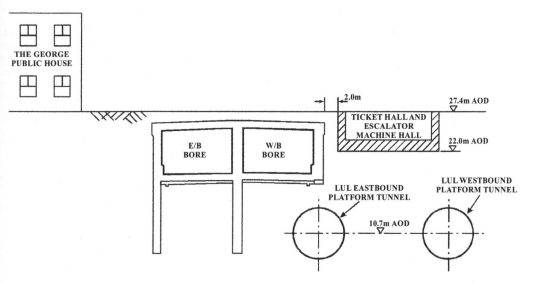

Figure 13. Cross section of new road and existing metro tunnels; Wanstead, London (Higgins et al. 2003).

applications, which illustrate facets of field behaviour and their prediction that may not have been evident earlier. Long term field records are relatively rare and, from necessity, we focus here on medium-term behaviour.

3.2.1 Effects of constructing a cut and cover tunnel on existing bored tunnels and Metro station in Wanstead, East London, UK

Higgins et al. (2003) summarise field data and predictions from a scheme in East London, which involved making a cut and cover tunnel (for the new A12-M11 link road) close to the existing Wanstead

metro station, and its pair of bored tunnels, as shown in Figure 12. The geology is a typical London sequence of 4 to 5 m of made ground (fill) and river terrace gravel (Boyn Hill series in this case) overlying London Clay, with the Woolwich and Reading (now termed Lambeth) and Thanet beds beneath. Figure 13 presents one of the critical cross-sections, showing how the new tunnel passed very close to the metro works. Making predictions of how the new works might affect the existing tunnel and station was one of the key engineering issues.

Data was gathered through a site investigation that included thin walled sampling, in-situ self boring

209

pressuremeter tests and laboratory stress path tests, the latter involving local strain instrumentation. Soil parameters were interpreted for analysis with some uncertainty remaining concerning values of K_0, the degree to which pore water pressures are affected by under-drainage and the detailed construction sequence.

Figure 14 presents the secant shear stiffness-strain relationships that were assumed to apply (via Equation 1) to key strata. The Wanstead parameters for the most important (intact London clay) layer lie within the London clay band summarised by Hight & Higgins (1995), falling slightly below the mid range. Figure 15 shows the corresponding bulk modulus functions obtained from Equation 2. Mohr Coulomb failure parameters were assigned to specify the large-scale

yielding and failure behaviour in each layer, as described by Higgins et al. (2003).

FE single element analyses were run to check the predictions made for various stress paths including K_0 swelling back to very low stresses from an initial vertical effective stress of 350 kPa; the latter can be compared with conventional oedometer tests as shown for the intact London clay in Figure 16. It is encouraging that the triaxial test based model predicts oedometer behaviour well, indicating a slightly softer than average response. Note that the clay undergoes large strains once the effective stresses reduce below 150 to 200 kPa, showing a reverse yielding as the effective stress path meets the passive failure envelope. Two implications are: (i) clay that has experienced unloading to such a level may show untypically low strength and stiffness and (ii) large heave movements can be expected in unloaded excavations.

The non-linear FE analyses were then extended to make Class A predictions for the full field response to tunnel construction at Wanstead involving:

- Construction of the Metro tunnels in the 1940s and the associated stress changes in the ground and tunnel linings
- The subsequent excavation of the 6 m deep station ticket hall
- A 40-year period involving pore water equalisation and stress re-adjustment, with the tunnels acting as drains. At the end of this the strain terms in the stiffness expressions given by Equations 1 and 2 were 're-zeroed' to mimic the kinematic nature of soil stiffness and ageing processes
- The construction of the new cut and cover tunnels. It is known that this would advance in two stages, but it was not clear whether the Eastbound (EB) or the Westbound (WB) section would be the first to be built. Both cases were therefore considered

Monitoring took place inside the Eastbound Metro tunnel, with the tunnel movements being tracked by surveying to within about 1 mm during construction.

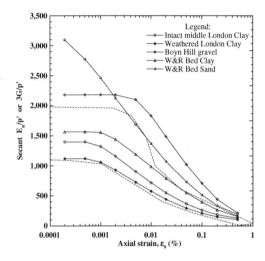

Figure 14. Secant shear stiffness-axial strain functions adopted for key strata in Wanstead analysis.

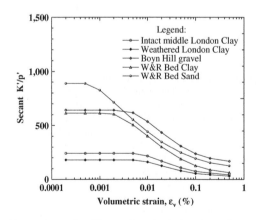

Figure 15. Bulk stiffness-volume strain functions adopted for key strata in Wanstead analysis.

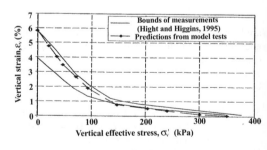

Figure 16. Oedometer swelling behaviour predicted for intact London Clay at Wanstead compared with laboratory test data reported by Hight & Higgins (1995); after Higgins et al. (2003).

Figure 17 displays the results recorded at the end of construction, showing generally good agreement between measurements and predictions. The two-dimensional pattern of movement established by eight targets within the tunnels is essentially as predicted. The tunnel movements, which were around 13 mm at maximum, were generally predicted to within 3 mm without any systematic bias towards over or under prediction.

Surveying scatter of 1 to 2 mm could have contributed a major fraction of the apparent differences between predictions and measurements noted at individual target locations.

Noting the significant factors that could have affected the predictions, including construction timing, temporary works details and uncertainty over ground conditions, Higgins et al. (2003) expressed satisfaction with the predictions made for the medium-term movements. The two dimensional plane strain analyses were fully appropriate to the relatively long Wanstead cut and cover road tunnel and station structures shown in Figure 12. As discussed below, agreement is not always as good when applying the same type of analyses to more complex building geometries or to tunnels formed with TBM boring machines.

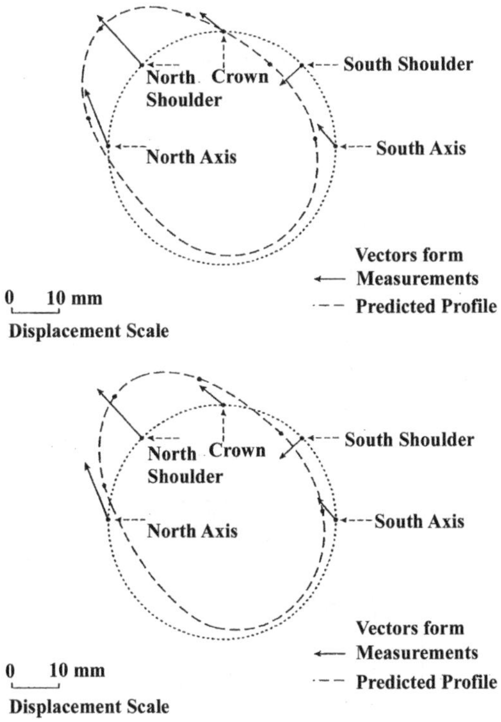

3.2.2 Deep excavations at Victoria Embankment and at proposed Moorgate underground station, City of London, UK: improvements to predictive capabilities offered by 3-D modelling

One of the case histories cited by Jardine et al. (1991), and in more detail by St John et al. (1992), involved the construction of an L-shaped deep basement for a new bank headquarters located near to the Thames at Victoria Embankment in London.

The building's plan shape is shown in Figure 18 and excavation was carried out by first creating an external secant bored pile wall, with top down construction following on, creating stiff bracing floors before mining out each underlying floor space. The ground conditions comprise a typical London profile of made ground (including the remains of a Roman theatre!) and alluvium over London clay, beneath which is the Woolwich and Reading (now termed Lambeth) group of strata, albeit with irregular layering related to former river channels, as summarised in Figure 19.

The site investigations included locally instrumented stress path triaxial tests run at Imperial College. Figures 20 and 21 show the typical secant stiffness functions that were used (along with Mohr-Coulomb parameters) to model the non-linear stiffness behaviour of the stiff clays present; note that

Figure 17. Predicted and measured movements of metro tunnel linings caused by road tunnel construction at Wanstead; after Higgins et al. (2003).

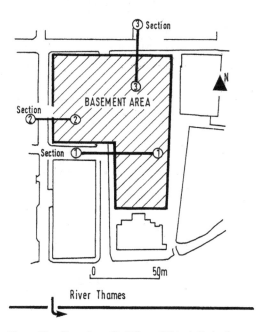

Figure 18. Plan view of building at Victoria Embankment, London, after Jardine et al. (1991).

different curves were applied in the active (compression) and passive (extension) regions, and that marginally softer behaviour was assumed in a 'disturbed' London clay layer that was interpreted as existing beneath the Thames alluvium, as shown in Figure 19. The disturbance was assumed to be due to unloading and weathering.

The tangent versions of Equations 1 and 2 (incorporated into Mohr Coulomb yielding models) were applied in plane strain effective stress analyses to make Class A predictions for the ground movements generated by the new building and the associated stresses developed within the walls. The site is located adjacent to sensitive and important buildings so Jardine et al. (1991) were pleased to report that the medium-term field measurements made at the end of construction fell reasonably close to the predicted values. Figures 22 and 23 compare the recorded

horizontal wall deflections and surface settlement profiles with those predicted by plane strain analyses. It is interesting that the lateral movements agree better than the settlements. Also the lateral movements developed at depth and the settlements observed 20 or more metres away from the excavation were smaller than predicted. Jardine et al. (1991) postulated that

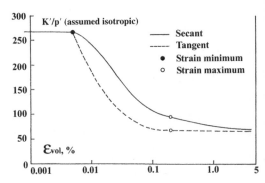

Figure 21. Bulk stiffness-volume strain function adopted for intact London Clay at Victoria Embankment, London, after Jardine et al. (1991).

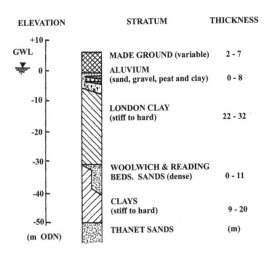

Figure 19. Geology near Victoria Emabankment building, London (Jardine et al. 1991).

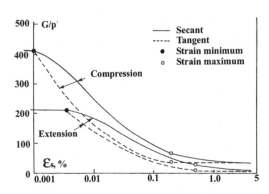

Figure 20. Shear stiffness-axial strain function adopted for intact London Clay at Victoria Embankment, London, after Jardine et al. (1991).

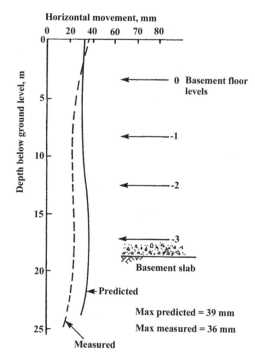

Figure 22. Predicted and measured horizontal movements of propped secant pile basement retaining walls for building at Victoria Embankment, London, after Jardine et al. (1991).

these features might be related to the geometry shown in Figure 18, which did not involve the infinitely long excavation implicitly assumed in the plane strain analyses undertaken.

It is now feasible to perform 3-dimensional analyses of such problems with similar non-linear soil models. Tor-Petersen et al. (2003) report a series of analyses performed with the same code (ICFEP), a similar ground profile, and similar soil parameters to the Victoria Embankment case. They consider a box shaped deep excavation with the plan layout shown in Figure 24. Their 35 m × 35 m box, which will be the deepest yet in London (at 40 m), is being considered as part of a new underground station proposed near to Moorgate and Moor House in the City of London.

Figure 25 shows the 3-D model geometry in comparison with the alternative 2-D simplifications that would more normally be considered for design analyses. A 3-D model was constructed that accounted for the Moorgate ground profile, the proposed seven levels of propping and top-down excavation sequence. Taking account of the box's four fold symmetry the numerical model required about 800 FE elements. The soil models described for the earlier Victoria

Embankment and Wanstead cases were adopted with similar assumed parameters.

Tor-Petersen et al. consider in detail the site's underdrained pore pressure profile and variations of K_0 with depth. They also investigate the effects of the external box wall's construction and consider the out-of-plane bending stiffness, which is generally far smaller than that applying in the vertical plane as a result of wall joints or lack of contact between individual piles in secant or contiguous pile walls.

The most important results are shown in Figures 26 and 27, which illustrate the lateral and vertical ground movements predicted at mid wall and corner locations respectively. Analyses are presented for the 3-D case with various wall construction details being considered; axially symmetric and plane strain 2-D analysis results are also shown, all run with the same soil parameters.

The main trends are:

- Plane strain analysis predicts the largest movements in every case
- Three dimensional analysis gives similar results to the axially symmetric cases at mid wall locations but smaller movements at corners
- The 3-D ground movement patterns resemble those seen at Victoria Embankment (Figures 22 and 23) and also help to explain the imperfections in the original plane strain analyses

Wall construction details affect the ground movements significantly. Assuming an isotropic wall that is not weakened by vertical joints may lead to under-prediction of movements and incorrect wall stresses, particularly at the excavation corners.

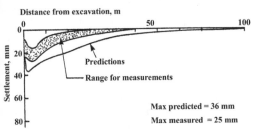

Figure 23. Predicted and measured surface settlements developed around basement excavations for building at Victoria Embankment, London, after Jardine et al. (1991).

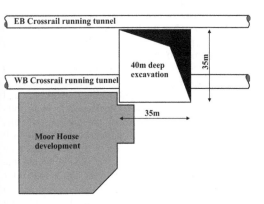

Figure 24. Plan view of development proposed for Moorgate, London, after Tor-Petersen et al. (2003).

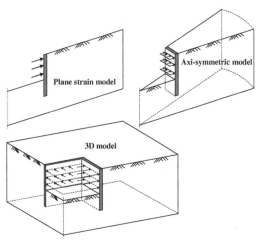

Figure 25. Two and three-dimensional modelling geometries, after Tor-Petersen et al. (2003).

213

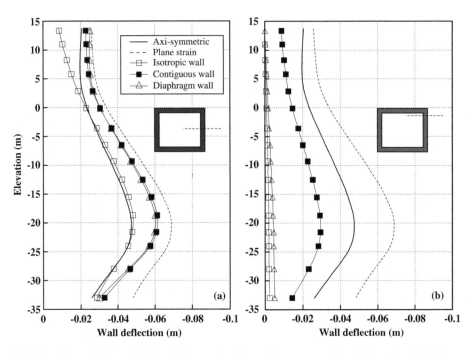

Figure 26. Three-dimensional numerical predictions made with non-linear approach for horizontal wall defections at mid-wall (left) and edge (right) locations for proposed deep basement at Moorgate, London; after Tor-Petersen et al. (2003).

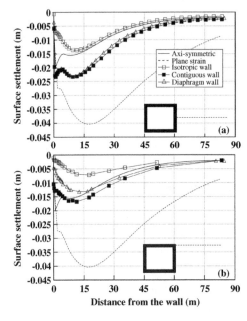

Figure 27. Three-dimensional numerical predictions made with non-linear approach for surface settlements developed at mid-wall (top) and edge (bottom) locations for proposed deep basement at Moorgate, London; after Tor-Petersen et al. (2003).

3.2.3 New Westminster metro station, London UK; analyses including updating of permeability data and construction sequence

The Jubilee Line Extension Project (JLEP) involved building a new metro station at Westminster, London. The station, which was completed in 1997, required a basement to be excavated within 35 m of the national Big Ben landmark and Houses of Parliament, the deepest (38 m) to date in London. The complex work was designed and executed very carefully to the scheme illustrated in Figures 28 and 29.

The J shaped main excavation for the Station Escalator box has an East-West length around 2.5 times its typical width. The ground conditions are broadly similar to those at the Victoria Embankment site (Figures 19 to 21). Crawley & Glass (1998) describe the complex top-down construction sequence, which involved diaphragm walling, strutting, and mini, sheet and bored piles. The deepest level of struts was mined in place between the diaphragm walls before the main excavation started. Non-linear ICFEP (Class A) predictive analyses were undertaken by Geotechnical Consulting Group (GCG) to aid the design work; the ground model was similar to that for Victoria Embankment but further developed to account for site specific borehole data, including locally instrumented triaxial tests run by a commercial laboratory (Higgins & Jardine 1998).

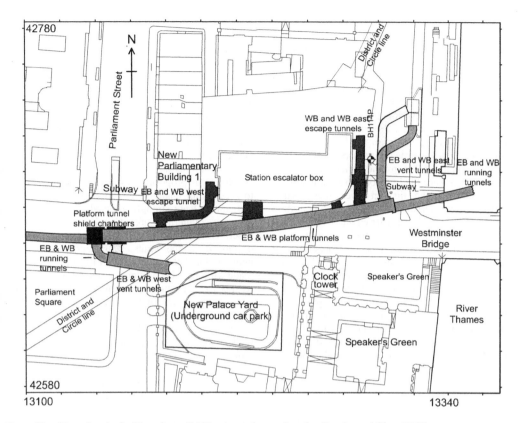

Figure 28. Plan of works for Westminster JLEP metro station works; after Crawley and Glass (1998).

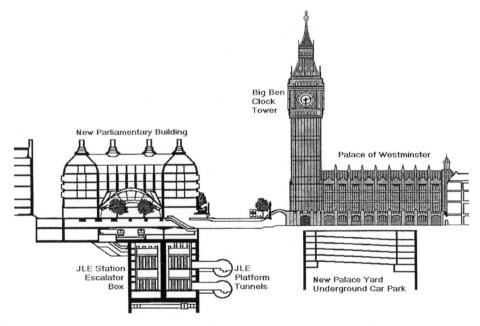

Figure 29. Schematic cross-section of Westminster JLEP metro station strutted excavation; after Harris (2002).

215

One uncertainty was the contractors' construction procedure and the other was the effective rate of drainage in the London clay. The Westminster strata include abundant laminate of silt that had affected considerably the excavation of the nearby House of Commons underground car park. Initial assumptions regarding in-situ bulk permeability and how this might vary with effective stress level were fed into the fully coupled FE analyses. Provision was also made to update the ground model in the light of the site operations and field permeability response.

It was recognised that ground surface settlements (and/or tilting) were likely to become intolerable at key points such as the Big Ben tower unless further steps were taken (Burland 2001). Local compensation grouting (Harris, 2002) was carried out to keep movements within acceptable limits, with generally good results being obtained. These local measures were not modelled in the FE analysis. While they evidently influenced the near surface settlements, it was assumed that they did not affect the lateral movements of the diaphragm walls at greater depths.

Harris (2002) provides the summary given in Figure 30 of the lateral movement observations made as the excavation progressed at the external diaphragm wall's panel 42 location (mid-way along the southern

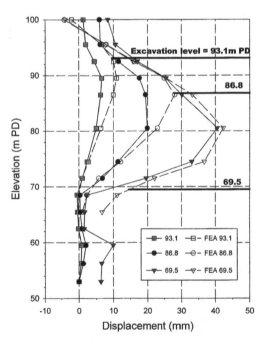

Figure 30. Predictions (FEA) and measurements for lateral movements of wall panel 42, excavation to elevations of 93.1 m, 86.8 m and 69.5 m respectively (taking original ground as 100 mPD); Westminster JLE metro station, London; after Harris (2002).

edge of the station escalator box shown in Figure 28). Predictions from the non-linear ground model are also given for each of the three levels of excavation considered. In general the FE predictions (marked FEA in Figure 30) agree very well with the medium term field data gathered during construction although the plane strain analysis tends to over-predict the field measurements marginally. As explained above, this may be a feature of the geometrical simplification; a 3-D treatment might lead to still better predictions. However, the 2-D ground model provides a valuable tool for evaluating the outcomes of alternative construction proposals, especially when updated to account for changing circumstances. The surface settlements were strongly affected by compensation grouting countermeasures. Crawley & Glass (1998) described these works and also reported the forces developed in the struts, noting good agreement between the latter and the ICFEP FE predictions made with the 2-D non-linear ground model. Overall, it appears that the marginally conservative plane strain analysis was fit for purpose in this sensitive application; the most important updating requirements were to model the ground water conditions realistically and to account for the construction sequence that was actually followed.

3.2.4 Analysing the ground's response to tunnel boring: recent experience at St James's Park, London

The Bangkok, Wanstead, Victoria Embankment and Westminster case histories show that it is now possible to make good analytical predictions for the medium-term ground movements developed around deep excavations if appropriate site investigation, geomaterial characterisation and numerical modelling are carried out, with due account being taken of the construction processes. However, the same stage has not been reached in relation to bored tunnels.

It is usual to adopt a plane strain model when assessing tunnelling movements and not attempt to reproduce the processes that lead to soil squeezing into the tunnel void during boring operations. Tunnelling-induced ground movements are frequently quantified by assuming a volume loss, which is defined as the volume of the soil excavated in excess of the theoretical volume of excavation (per unit advance of tunnel) and is quoted as a percentage of the theoretical volume. Rapid excavation in low permeability clay results in no soil volume change and so the volume of the surface settlement trough is equal to the tunnel's volume loss. Volume loss estimates are usually gauged from earlier case histories involving similar plant and ground conditions, and not derived from any fundamental analysis. The volume loss is often input into an empirical procedure in which ground surface settlements are assumed to follow the shape of a Gaussian probability curve. Volume loss may also be applied as a boundary condition in a 2-D FE analysis, which has the advantage of calculating the ground strains and stresses as well as the surface settlement distributions. However, the FE analyses tend to predict surface settlement troughs that are shallower and much wider than those measured in the field.

The Jubilee Line Extension Project (JLEP) provides a comprehensive set of new case histories, including the metro tunnels that passed alongside the new Westminster station (discussed above) and on towards St James's Park. The tunnel sections under the 100 year old UK government's Treasury Building and under the (literally) green-field St James's Park provide good examples where close observations were made (Viggiani & Standing 2002, Nyren et al. 2002).

Addenbrooke et al. (1997) described 2-D FE analyses of the St James's Park sections that were performed using an excavation boundary condition that simulates the removal of material from the finite element mesh. Prior to excavation the stresses acting on the periphery of what will become the tunnel lining are determined. These stresses are reduced incrementally until the required volume loss is achieved allowing the developing surface settlement trough to be considered in detail. The tunnel's expanded concrete lining is not modelled explicitly. Addenbrooke et al. (1997) investigated the influence of isotropic and anisotropic soil models, with both linear and non-linear assumptions, on the shape of the surface settlement trough, the latter were compared with field data from St James's Park gathered by Standing et al. (1996). Linear (both isotropic and plausible anisotropic) elastic analyses tend to predict the shallowest and broadest troughs; they are also unable to achieve the medium-term volume loss values measured in the field without generating unfeasible tensile radial stresses at the tunnel boundary.

Figure 31 shows some of the predictions made by Addenbrooke et al. by applying a number of models implemented in ICFEP to the first (green-field) tunnel (4.85 m diameter bored at 31 m depth) under St James' Park. Model J4 employed Equations 1 and 2 and a Mohr-Coulomb yield criterion to simulate isotropic non-linear stress–strain behaviour, while Model L4 includes an alternative non-linear formulation; the parameters input into the model to represent the strata are set out by Addenbrooke et al. (1997). Introducing non-linearity allows both the recorded volume losses to be matched and leads to more realistic profiles; but there is little difference between the J4 and L4 predictions. Non-linear analyses in which the J4 model was modified so that the (non-linear but pre- Y_3) stiffness was particularly soft in the G_{vh} mode of shearing led to the best match with field data (see predictions AJ4i and AJ4ii in Figure 31, these being models with different anisotropic characteristics). No laboratory or field test data existed to support the soft G_{vh} assumption of AJ4ii, although intensive research is currently underway at Imperial College to investigate the anisotropy of London Clay.

A second 4.85 m diameter tunnel was bored about 20 m from the first with its axis at 20 m depth, passing through ground that had been influenced by the first tunnel. Figure 32 presents the field measurements for this case, along with 2-D ICFEP predictions made with non-linear models that considered the sequential excavation of both tunnels. Here a distinction is drawn between analyses that either did not (case a) or did (case b) recognise the kinematic effects of soil stiffness by re-invoking the soil's high initial stiffness response when considering the excavation of the second tunnel. In all cases the analyses were run to induce the same overall volume loss. It can be seen that addressing the kinematic features directly leads to better predictions. The anisotropic variant of the 're-zeroed' J4 non-linear model (coded AJ4ii0) give the most realistic overall response, although as mentioned above the assumption of a particularly soft G_{vh} mode has yet to be justified.

It is possible that the three-dimensional nature of real tunnel construction contributes to the discrepancies between field settlements and numerical predictions. This may be particularly relevant at St James's Park where significant over-excavation in front of the shield is known to have taken place. Another potential contributing factor could be assigning accurate values for initial stresses. Improved predictions have also been achieved by artificially varying the ratio of horizontal to vertical effective stress fed into 2-D tunnelling simulations such as those described above, taking substantially lower ratios than the best estimates for K_0. However, the mechanics of such approaches have yet to be explained or justified fully.

A second key point to emerge from the St James's Park case history is the difficulty of estimating in advance the medium-term volume loss. This is the vital parameter fed into 2-D FEM or classical Gaussian curve fitting procedures when making surface settlement assessments. The conventional open face TBM plant employed will normally give rise to a volume loss of around 1.5% in London Clay, and a conservative value of 2% is often adopted for the estimation of ground movements. Earth pressure balance (EPB) machines give significantly less volume loss when well controlled (typical values for different tunnelling methods on the JLEP are given by Burland, 2001). In the case of the open face St James's Park tunnels the values were far higher, amounting to 3.3% and 2.9% for the first and second tunnels respectively in the southern region of the park, reducing to more typical values of less than 2% at the northern end.

In view of the significance of the high volume losses measured at St James's Park and the potential

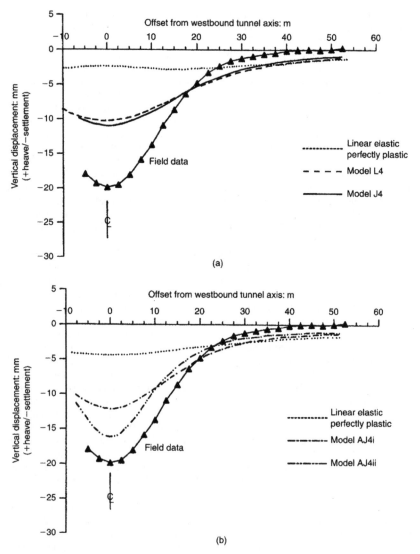

Figure 31. St James Park, London. Surface settlement profiles predicted for first tunnel: (a) isotropic models; (b) anisotropic models, after Addenbrooke et al. (1997); field data after Standing et al. (1996).

detrimental effect these findings might have on future tunnelling proposals, a detailed study was undertaken to establish the reasons for the high volume losses (Standing & Burland 2003). Possible factors included: tunnelling method and workmanship; geology; and geotechnical parameters. Tunnelling method and the practice of over-excavating in front of the shield have already been mentioned; very tight construction control is essential if ground movements are to be kept within acceptable limits and the consequences of particular procedures are to be quantified and understood. Greater control was exercised as the tunnel was

advanced in the northern part of the park and ground movements reduced.

Several new boreholes were drilled across St James's Park to investigate the possible influence of stratigraphy and basic soil parameters. A Thames river terrace boundary is found to straddle the site and divides the southern and northern regions mentioned above. The London Clay profile in which the high volume losses developed had been subjected to about 9 m of additional local erosion during the Quaternary, resulting in stress relief and with time swelling. Referring back to Figure 16, which shows how vertical strains

218

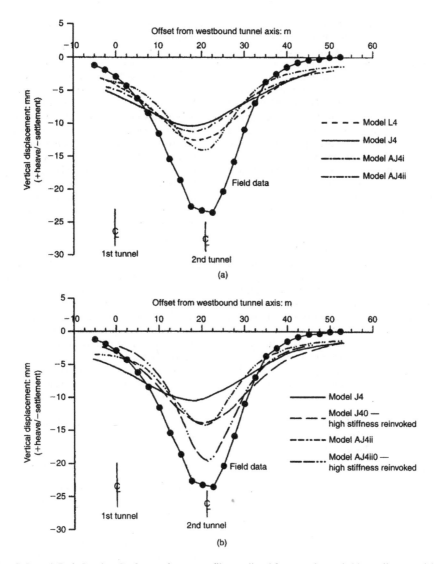

Figure 32. St James's Park, London. Surface settlement profiles predicted for second tunnel: (a) non-linear models; (b) non-linear models with and without re-zeroed strains, after Addenbrooke et al. (1997); field data after Standinget al. (1996).

increase markedly when vertical effective stresses drops to values between about 150 kPa, it is evident that the London Clay above the level of the tunnelling experienced large volumetric strains and softening.

Further variations in engineering characteristics were also sought by identifying the London Clay sub-units defined by King (1981) from their depositional settings and fossilised micro-flora and fauna. King's sub-divisions are shown in Figure 33. Excellent correlations were made between these units and careful geotechnical descriptions for the borehole samples and simple measures such as water content variations

with depth (Standing & Burland, 2003). Between the different clay units there were notable differences in permeability, sand and silt content and the degree of fissuring. The depths of the different units at St James's Park are also shown in the adjoining plot on Figure 33. King's unit A3 has been sub-divided into A3i and A3ii as the upper part of the unit (A3ii) contains numerous sand and silt partings that are significant in terms of engineering behaviour.

The two tunnels pass through different units and some of the units' levels change across the park. The crown of the first tunnel just penetrates the permeable

219

A3ii unit on the southern side of the park. However, the tunnel alignment rises to the north such that most of the northern tunnel section is within the comparatively impermeable overlying B2 unit. Tunnel boring induces the greatest concentrations of shear strains just above the crown, so the apparently minor shift in the stratigraphy along the tunnel line provides a further factor that contributed to the larger volume losses developed along the southern sections. Significant degrees of 'consolidation' would have taken place during the tunnelling period at the locations where material containing sand and silt partings existed above the tunnel crown. The second tunnel (bored at a higher elevation) has its invert in the A3ii division in the southern region but is completely within the B2 unit at the northern end of the park, leading to similar consequences to those noted for the first tunnel.

Although a detailed laboratory investigation was not performed on samples from the units encountered to identify the influences of their different depositional settings, recent stress histories and intrinsic characteristics, these differences would certainly affect the stiffness and possibly anisotropy of London Clay (see Hight & Jardine 1993, Hight et al. 1997, 2002). Ground movements from earlier tunnelling projects have not been assessed in relation to the different sub-units of the London Clay but it is likely that the ground conditions in which most of the earlier

tunnels were bored are similar to those in the northern region of St James's Park (i.e. predominantly constructed in the B2 unit).

The main conclusions to be drawn from the St James's Park case history, and others like it, are:

- It is not yet possible to predict accurately the ground movements caused by real tunnelling from first principles through rigorous analysis, even when relatively advanced soil testing data, constitutive models and computational tools are available. Reliance must still be placed on prior experience
- It is possible to make fair to reasonable predictions for surface ground movements through simplified 2-D FE analyses by: (i) assuming ground losses on the basis of experience, (ii) adopting non-linear soil models. However, it is necessary to apply somewhat arbitrary artifices regarding K_0 or stiffness anisotropy if the predicted settlement troughs are to be as narrow as those seen in the field
- The volume losses developed by open face tunnelling, and the surface settlement troughs generated by it, are sensitive to soil properties, geology and tunnelling procedure. Geotechnical factors may be less important when Earth Pressure Balance TBMs are used
- Tunnelling machines and processes induce volume loss components at different locations along the axis and circumference of the shield that lead to strains developing in the assumed plane strain direction. The boundary conditions of a real TBM are highly three-dimensional
- For open face tunnelling any significant volume loss is likely to be associated with failure developing near to the tunnel face. Fissuring and progressive failure may play an important role, as may anisotropy and time related effects
- Tunnel construction is usually considered to be virtually undrained in clay strata. Fully coupled FE analyses predict additional 'consolidation settlements' as the ground water pressures re-distribute with time as a separate, post-construction, process. Movements can be considerable compared to the short term displacements but they are less damaging to any surface structures as they develop over a much wider flatter trough
- Careful consideration should be given to the geological history and stratigraphy of the ground through which the tunnel is passing. As illustrated in the St James's Park case, there can be very significant variations within one principal stratum such as London Clay. Such sub-divisions are often not taken into account
- The stress conditions applying after a TBM has passed a particular point are unlikely to resemble those in a K_0 consolidated profile that are subjected to simple 2-D plane strain cavity contraction.

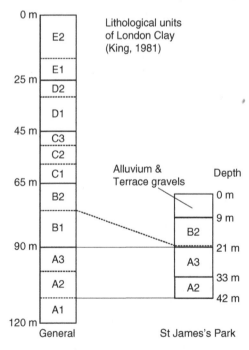

Figure 33. Geological subdivisions of the London Clay as proposed by King (1981) and applied to St James' Park.

The accurate prediction of tunnelling ground movements remains difficult; the shortcomings of most FE approaches have yet to be fully understood or resolved.

3.3 *Embankments, slopes and piles in clay*

In this section we retain our focus on clay materials and turn to case histories involving a wider spread of geotechnical applications.

3.3.1 *Bromborough docks landfill, Merseyside, UK: a waste fill embankment on soft foundations*

A major landfill site is currently being developed at Bromborough, in Merseyside, northwest England, that will provide an interesting ground movement prediction case history. We offer here a brief summary of work to date; a more comprehensive report will be published by the project team after the project is completed (Jardine et al. 2004).

An area of around 20 hectares had been reclaimed over a 30 year period (1935–1965) between the south west bank of the Mersey and an offshore breakwater (termed the lower bund) that enclosed the former Bromborough dock area, as shown in Figure 34. Dredged material was pumped in to form a hydraulic fill composed of clay, with some isolated sand seams, up to the top of the breakwater. A second upper bund was then formed on the reclaimed land (as in tailings dam construction) and filling had continued until the landmass rose 10.5 m above mean sea level. The hydraulic fill impounded by the two bunds shown in Figure 35 is under-drained by the sandstone

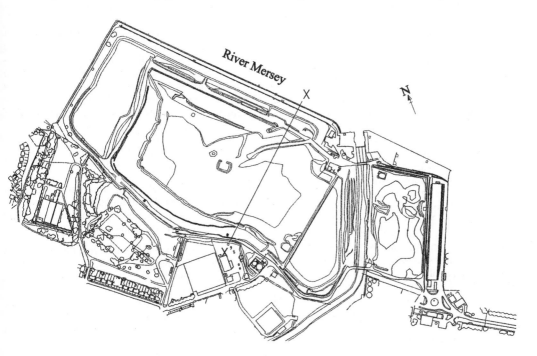

Figure 34.   General plan of landfill facility on reclaimed land, Bromborough, UK.

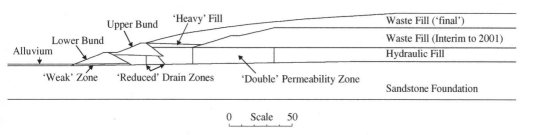

Figure 35.   Typical section through intended landfill facility on reclaimed land, Bromborough, UK.

221

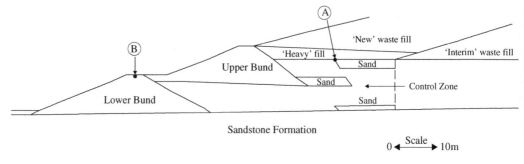

Figure 36. Detailed cross section through toe of landfill facility on reclaimed land, Bromborough, UK.

beneath and is affected at its upper surface by shallow desiccation.

The reclaimed land was developed in the 1990s as an engineered landfill site with the intention of placing compacted domestic waste to a maximum level around 35 m above mean sea level, which would then be sealed and landscaped for future use. An engineered drainage, gas collection and liner system was installed and filling was started along with pore pressure observations in the foundations and lateral ground movement measurements riverward of the fill. It was very important that infrastructure elements installed within the lower bund were not affected and the filling operations were controlled to ensure that no untoward movements developed. However, the pore pressure dissipation observed over the first two years was slower than expected. The project team decided in 1998 to limit filling to the 'Interim' profile shown in Figure 35 and make a more detailed geotechnical analysis of the site, the filling programme and possible ground movements.

A new site investigation was made in 1999, which included high quality (sharp edge) piston sampling, intensive CPTU profiling, and Delft style continuous sampling and careful piezometric monitoring. Multiple CKoU SHANSEP style stress path laboratory tests (with triaxial compression, extension and simple shear modes) were run on samples of the clayey hydraulic fill, with local strain measurements being made in the triaxial tests. The SHANSEP procedure was specified because the landfilling sequence relies on consolidating the soft clayey foundation layers to stresses well above those that act initially. The clayey materials' permeability and compressibility characteristics were also investigated in detail. It was recognised that the laboratory test procedures followed would destroy any 'structural' components of strength, stiffness, or compression resistance that may have developed due to bonding and ageing in-situ, and should provide a ground model that would be marginally conservative when applied to the initial stages of waste filling.

A new filling programme was devised and a dense array of vertical drains was installed in the 20 m wide Control Zone identified in Figure 36, along with a new suite of geotechnical instruments that were specially developed for this application. A comprehensive set of fully non-linear ICFEP analyses was performed by GCG to develop filling control criteria and predictions were made regarding the movements of the sensitive lower bund infrastructure. The analysis developed by Kovacevic and Jardine (2001) proceeded on a marginally conservative basis because of remaining uncertainty in:

- The geotechnical properties and the pore pressures developed within the domestic waste landfill material
- The extent and effect of any permeable layers within the hydraulic fill foundation
- The heterogeneous materials that compose the lower bund and its foundations.

Bearing in mind the above uncertainties, the design was based on a marginally conservative idealisation in which the landfill unit weight was slightly higher than expected. No benefit was taken from thin sand layers identified in the CPTU traces and the closely centred vertical drain system was assumed to allow just 75% pore water pressure dissipation between widely spaced lifts of waste fill; as mentioned above the soft clay foundation parameters were developed through SHANSEP style tests. The latter stress paths tests run on the low OCR hydraulic fill did not show any marked S_u anisotropy, so the Modified Cam Clay (MCC) model was adopted to model these deposits, with the parameters being selected to give:

- A good match to the virgin compression behaviour seen in the laboratory
- A shear strength ratio S_u/σ'_{1c} of 0.25 for K_0 consolidated samples at OCR = 1, which matched the mean of the SHANSEP test results, after making some allowance for potential brittleness

222

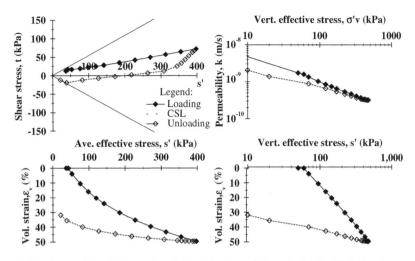

Figure 37. Predictions by non-linear FE ground model for behaviour of clayey hydraulic fill during K_0 compression and swelling, including permeability variations; after Kovacevic & Jardine (2001): note $t = [\sigma'_1 - \sigma'_3]/2$ and $s' = [\sigma'_1 + \sigma'_3]/2$.

As before, the tangent versions of Equations 1 and 2 described the 'non-linear elastic' pre Y_3 response; a non-linear permeability model was also deployed. Figure 37 presents four plots that summarise the one dimensional compression and swelling behaviour predicted for the clayey hydraulic fill. The first (with co-ordinates $t = [\sigma'_1 - \sigma'_3]/2$ and $s' = [\sigma'_1 + \sigma'_3]/2$ details the effective stress path followed during K_0 compression to $\sigma'_v = 500\,\text{kPa}$, and then unloading to $10\,\text{kPa}$. The second shows how volume strain varies with s' on an arithmetic scale over the same stress range while the third and fourth plots show how permeability and volume strain varied with the logarithm of σ'_v. These data matched the oedometer and SHANSEP test data well, although the normally consolidated K_0 value predicted by MCC was (as is often found) too high.

Figure 38 shows the mean shear stiffness-strain curve found from undrained triaxial extension tests at $OCR = 1.0$. The curve adopted for analysis, which is also shown, was based on the extension data as the compression stiffness response (from $OCR = 1$) was considered to be unrepresentatively soft. Figure 39 presents a three-part diagram that shows the stress-strain, excess pore pressure-strain and effective stress paths response computed by the non-linear model for samples that were sheared undrained to failure in compression and extension from lightly overconsolidated states, while Figure 40 presents similar predictions for undrained simple shear tests. These FE single element predictions provided good matches to the available SHANSEP laboratory test data.

The landfill waste was modelled as a Mohr Coulomb material with very low, pressure dependent,

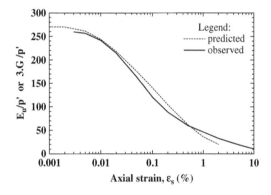

Figure 38. Experimental secant shear stiffness-strain relationship of clayey hydraulic fill during undrained shearing in triaxial extension at OCR = 1, following K_0 compression. Curve fit applying Equation 1, as adopted for FE analysis is also shown; after Kovacevic & Jardine (2001).

pre-Y_3 stiffness. Relatively high pore water pressures were predicted from a non-linear analysis of seepage within the waste material. Further information will be provided in later papers.

A mid-filling re-assessment was carried out in 2003 in which the field measurements and predictions were compared for a two-year period. Figure 41 shows the vertical settlement–time trace recorded at Point A in the Control Zone (identified in Figure 36) on the most critical section, while Figure 42 displays the plots of maximum lateral movements at the lower bund (Point B in Figure 36) as a function of maximum fill level. It can be seen that the settlements to

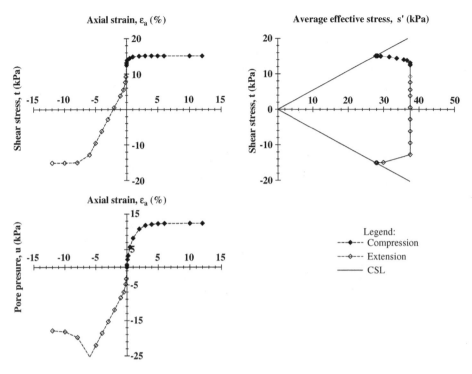

Figure 39. Predictions by non-linear FE ground model for behaviour of clayey hydraulic fill during undrained shearing of low *OCR* samples in triaxial compression and extension following K_0 compression and swelling; after Kovacevic & Jardine (2001): note $t = [\sigma_1' - \sigma_3']/2$ and $s' = [\sigma_1' + \sigma_3']/2$.

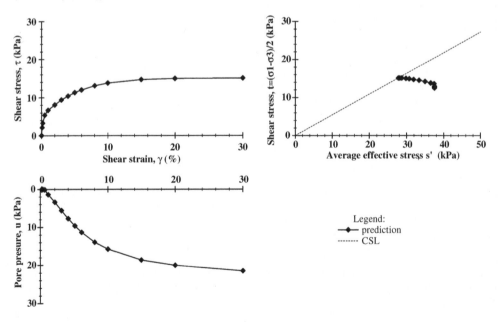

Figure 40. Predictions by non-linear FE ground model for behaviour of clayey hydraulic fill during undrained Simple Shear tests on how OCR samples following K_0 compression and swelling; after Kovacevic & Jardine (2001): note $t = [\sigma_1' - \sigma_3']/2$ and $s' = [\sigma_1' + \sigma_3']/2$.

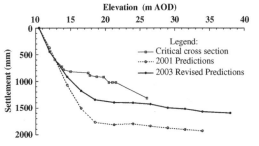

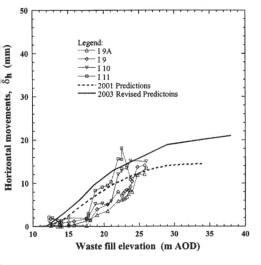

Figure 41. Settlements recorded at Bromborough landfill at location A in the Control Zone (see Figure 36) plotted against maximum fill elevation showing field data for critical cross section and ICFEP predictions made in 2001 and revised in 2003; after Kovacevic & Jardine 2003.

Figure 42. Maximum horizontal movements recorded at Bromborough landfill at location B in the Lower Bund (see Figure 36) plotted against maximum fill elevation showing field data for four tube-way inclinometers near critical cross section and ICFEP predictions made in 2001 and revised in 2003; after Kovacevic & Jardine 2003.

date are large, but around 25% less than originally predicted. The lateral movements of the lower bund showed an interesting initial delay in their response but then picked up to rise towards the predictions. Time dependent straining (creep) is now thought to be influencing the lateral movements and settlements significantly, and this may eventually lead to larger movements developing than had been predicted.

A detailed re-examination of the full set of instrument data (including 35 special piezometers, four remote settlement gauges, multiple inclinometer profiles etc), combined with weighbridge records of the total weight of fill imported, surveys of the volume

raised and logs of the timing of site operations led to a re-appraisal of the geotechnical model. The main features updated in the model were:

- A reduction of 11% in the landfill unit weight (from 13 to $11.5 \, kN/m^3$) to match the long term weighbridge and field survey records made on site
- Explicit modelling of the sand layers in the hydraulic fill identified by the CPTU profiling study
- Revising the pore pressures simulated within the drained hydraulic fill area to match the 90% dissipation that was typically achieved within the Control Zone region between lifts
- Considering the lower bund material to be undrained over the construction period rather than drained, a change made after an analysis of the displacement patterns observed in the field
- Removing a weak layer that had been postulated beneath the lower bund, as the inclinometer records showed no sign of any strain concentration developing towards the bund's base.

The updating exercise corrected the initially cautious assumptions in the light of proven facts, allowing a move away from initially (and necessarily) conservative assumptions. The stress-strain-strength and permeability laws of the soils and fills were not changed.

Figures 41 and 42 show how the revised ground model gives a closer match to the current (2003) ground movements. Some divergence between predictions and measurements remains over the earlier stages that may be related to two factors that were not accounted for fully in either analysis:

- Any initial structure, due to ageing, suction or cementing, in the hydraulic fill that was not identified through the SHANSEP testing procedure. Such features can delay large-scale (Y_3) yielding and stiffen the initial loading response
- Time-dependent material behaviour that becomes more significant as the initial structure is lost, yielding develops and strains increase

Updating the analysis allowed the filling programme to be revised and the site use to be optimised. Substantial benefits were obtained for the landfill operator and the local authority's refuse disposal programme. As mentioned earlier, the case history will be reported in greater detail when the landfilling operations completed.

3.3.2 Other applications in slopes: influence of stiffness response on stability

It is unusual to make detailed analysis of the movements developed by stable slopes. However, predictive analyses involving models that capture behaviour over a wide range of strain can be useful when attempting to set criteria for stability monitoring, as in the

Bromborough case described above. When considering geomaterials that can strain soften, either by swelling, loss of bonding, or residual fabric formation (or all three), progressive failure is possible. The soil pre-peak stiffness response can then have a significant effect on stability, particularly when considering high slopes (Potts et al. 1997).

The authors have been involved in combining the 'small-strain' approaches described above with strain hardening and/or softening models to make ICFEP analyses of both brittle and non-brittle slopes in several practical applications. These have included cut and cover tunnel works in Dublin glacial till (Menkiti et al. 2004), large open cuts in London Clay (Kovacevic et al. 2004) and very large deepwater submarine landslides (Jardine & Kovacevic 2000, 2003). The analyses have accounted for the time-dependent consolidation and swelling phenomena through fully coupled analyses. In some cases the key question is to decide how long a slope can be left open safely, knowing that it would be certain to collapse eventually. Similar progressive failure systems apply to the shafts of long piles, to large spread foundations and to contained failure problems such as pile end bearing. Hopefully, it will be possible in due course to publish summaries made from presently confidential reports on the above projects.

3.3.3 Canons Park, N. London UK: pile ageing in clay and loading rate effects

Before moving on to consider the behaviour of sands and strongly cemented geomaterials, it is instructive to mention field tests that studied the effects of time and strain rate on the behaviour of piles formed in London Clay.

Teams from the UK Building Research Establishment (BRE) and Imperial College London have carried out research at the Canons Park London clay test site in North London for over 25 years; see for example Jardine et al. (1985), Bond & Jardine (1991), Wardle et al. (1992) and Pellew (2002). In each of these studies loading tests showed that pile behaviour is time dependent. A creep yield could be defined during loading tests after which significant extra displacements occurred during loading pause periods and axial capacity became dependent on the time taken for testing. Fast Constant Rate of Penetration (CRP) tests could give capacities 40% greater than those developed in slower Maintained Load (ML) tests in which creep periods were allowed.

The most recent Imperial College study (Pellew 2002) included re-testing steel displacement piles that had been installed and loaded to failure by BRE 17 years earlier. Significant increases in axial capacity were found in three separate cases, as illustrated by the most dramatic example in Figure 43. Pellew's careful research showed that the gains were not related to pore

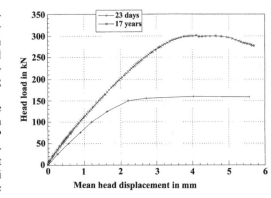

Figure 43. Loading tests to failure on 168 mm diameter, 6.5 m long steel pile jacked into London clay at Canons Park North London, performed 23 days and 17 years after installation by slow jacking; after Pellew (2002).

pressure equalisation (which would have been complete well before the 23 day test; Bond & Jardine 1991) or any purely mechanical effect of creep or ageing. Instead they were due to the disruption of a highly oriented shear surface by Sulphate Reducing Bacteria (SRB). Such additional components of capacity due to ageing (as with those affecting the Bothkennar preloaded foundation presented in Figure 10) can have a significant impact on urban redevelopment engineering, or when considering the impact of changes in the loading applied to existing foundation systems.

3.4 Case histories involving sands

3.4.1 Creep under maintained load: Labenne and Gravelines (France) and at a UK nuclear power plant site

Frank (1992) presented a General report to the Florence ECSMFE on shallow foundation behaviour. One of his themes was the neglected topic of creep in sand. While empirical procedures for estimating field settlements, such as those of Schmertmann (1970) and Burland & Burbridge (1984) recognised that settlement could increase logarithmically with time such a dependency is not commonly addressed in analytical procedures. The evidence for the time effects was to some extent anecdotal; some of Burland & Burbridge's key data points came from an enthusiastic amateur whose hobby was to record building settlements! Frank (1992) referred to systematic long-term loading tests run on 0.5 to 1 m² footings at Labenne, SW France, where the national Central Ponts et Chaussées Laboratory (LCPC) developed a geotechnical test facility on dune sand. Amar et al. (1985) gave further details of the tests and the geotechnical profile, as established by LCPC. One point to note is that the water table fell well below ground level so the soil affected by the footing tests would have been unsaturated and practically dry for most of the year.

Canépa (1998) has presented a more general summary of the LCPC studies into the creep of shallow foundations under constant loads, covering several sites and a range of ground conditions. Figures 44 and 45 show the settlement-time curves developed by two of the Labenne footings when subjected to constant loading. The data shown in the first figure came from an experiment in which a 1 m² footing was loaded to about half of its maximum supportable load and then allowed to creep. As can be seen, the creep settlements tended to follow an approximately semi-logarithmic relationship with time over the first 30 minutes pause period.

Figure 45 shows the longer-term response, as seen in four tests on other footings where deadweight lever loading arrangements were used to apply constant loading over periods of up to 10 years. The semi-logarithmic plots showed initial gradients that increased with the loading level. While the gradients remained relatively steady for the first few months, they all increased very significantly in the longer term, developing settlements after 10 years that could be several times larger than those seen in the short-term experiments.

Canépa (1998) also summarised settlement-time observations made under the much larger scale rafts that supported several French nuclear power stations, covering a range of geotechnical settings. On average, settlements increased by 40 to 50% within 5 years of construction ending. Canépa's data base included the group of six sets of dual (900 MW) nuclear reactors (each with a plan area of 80 m by 130 m) that had been built on a thick sequence of dense marine sands at Gravelines. Beneath the sand is the Argile de Flandres, a heavily overconsolidated plastic clay from the same geological formation as the UK's London Clay; its upper surface is about 20 m below the base of the reactor foundations. Just adjacent to the Gravelines site is the Dunkerque Avant Port Ouest area used by Chow (1997) and Jardine & Standing (2000) for pile research, some of which is reported in later sections.

Measurements were made at 21 locations at Gravelines over a period of more than 10 years. On average the applied surface loads amounted to 370 kPa. The settlements developed at the end of construction ranged from centimetres to more than 10 centimetres, depending on the building considered. The settlements increased by 42% over the average observation period, which extended to 8 years after the end of construction. It is not clear what role consolidation within the deep clay layers played, but it seems probable that creep in the sands provided the main cause for the time-dependent movements. Canépa (1998) analysed the buildings' responses and suggested empirical rules for projecting time-dependent settlements

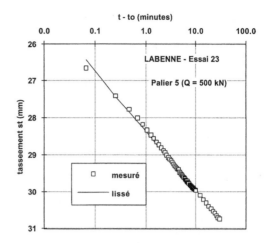

Figure 44. Creep settlements (mm) of 1 m² footing on dry sand at LPC labenne research site South West France. Footing loaded to 900 kPa in five increments over 100 minutes and then held at constant load; after Canépa (1998).

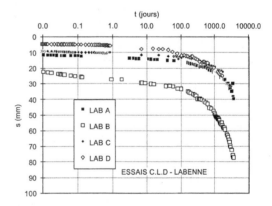

Figure 45. Long-term settlement (mm) versus time (in days) for four footings under constant load at Labenne, South West France. Tests A and B: 0.5 m² footings taken to 480 and 650 kPa respectively; Tests A and B: 1 m² footings taken to 220 and 280 kPa respectively; after Canépa (1998).

on the basis of conventional testing and design methodologies. However, the detailed form of the settlement-time trends was not reported.

Kuwano (1999) performed intensive laboratory stress path testing on the same sands (and on other silica sands) run in support of the Imperial College piling projects performed at Dunkerque. Her work included high resolution studies of their creep behaviour, showing that the strain-time relationships developed under constant load differed considerably from the classical semi-logarithmic functions.

Comparable observations exist for the sands found on the other side of the English Channel. Dun (1973) described short term settlement observations made on a heavily loaded, 3.5 m thick, reinforced concrete raft that supports two nuclear reactors at a UK power station site. A plan of the site is shown in Figure 46, while the soil stratigraphy is depicted in Figure 47. GCG have recently undertaken 3D finite element hindcast analyses of the site's initial excavation, the raft's subsequent construction and the foundations' response to the loads imposed by the reactors and their buildings.

The numerical analyses were made with ICFEP. The plastic behaviours of the sand and gravel strata were modelled with non-associated, Mohr-Coulomb laws, as were the underlying Cretaceous Hastings beds mudstones, siltstones and occasional sandstones. The analysis assumed that the relatively slow rate of construction and commissioning had led to a fully drained response, although the field settlement records suggest that some consolidation settlement developed after the end of construction in the deeper Hastings beds. The pre-Y_3 behaviour of all layers was treated as being non-linear elastic (following Equations 1 and 2), with the soil-stiffness curves being developed from site specific G_{max} profiles obtained from dynamic testing. Standard dynamic shear modulus decay curves were adopted that were modified to allow for the practically static loading conditions involved; bulk modulus curves were assessed from the shear stiffness curves on the basis of experience with similar geomaterials.

The shear and bulk modulus decay curves developed from Equations 1 and 2 are shown in Figure 48; Figure 49 presents the shear stiffness curves normalised by G_{max}.

Analyses were performed in which the curves shown in Figure 48 were normalised by p', with additional runs being made in which stiffness was assumed to be scaled by $[p']^{0.5}$. The latter normalisation led to greater consistency with the available field G_{max}–depth data. However, the choice of power law exponent had surprisingly little impact on the main results from the analyses.

The two reactors were not built simultaneously. However, constraints on computing power and time meant that symmetrical construction loading had to be assumed in the 3-D analysis. Considering this drawback and noting that the soil properties were not adjusted to achieve a good fit, the predicted settlements match the medium-term field observations remarkably well. Figure 50 shows end of construction predictions along an East-West section drawn through the centres of the reactors; also shown are the field settlements for the end of 'primary' consolidation stage interpreted by Hight and Leroueil (2002). Good matching was achieved at other monitoring locations.

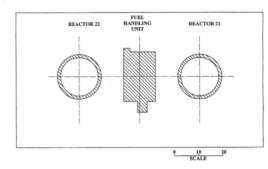

Figure 46. Plan showing raft supporting two nuclear reactors at UK site with dimensions in m; after Dun (1973).

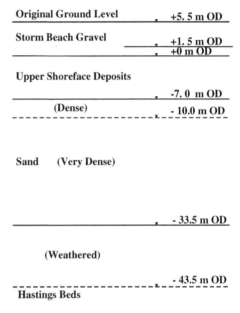

Figure 47. Simplified stratigraphy for UK nuclear power station site; after Dun (1973).

Hight & Leroueil (2002) have recently commented on the foundations' long-term response and Figure 51 reproduces their load-time and settlement-time plots, as recorded over 30 years. The settlements recorded at four typical survey (stud) positions have almost doubled since the end of construction (week 498). Figure 52 presents the post-construction settlements plotted to a logarithmic time base indicating consistently log-linear trends. It appears that the settlement profile scales up by a constant factor that depends on the elapsed time divided by the end of primary period.

The above case history indicates that at moderate loads the medium-term soil-structure interaction can be predicted reasonably accurately with sands by using the analytical tools described in this paper.

228

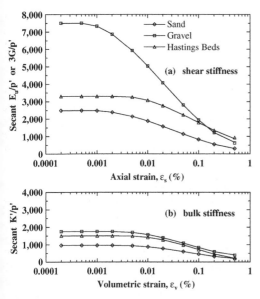

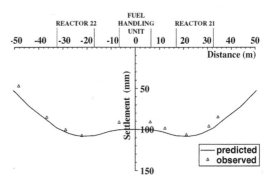

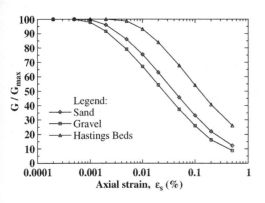

Figure 48. Shear stiffness-invariant shear strain and bulk stiffness-volume strain relationships assumed for three principal layers in GCG re-analysis of the power plant.

Figure 49. Shear stiffness-invariant shear strain relationships assumed for three principal layers in GCG re-analysis of UK nuclear power plant; normalised by G_{max}.

However, it is clear that movements increase very considerably with time and that creep movements can be expected over years or decades that are of comparable magnitude to those developed during the initial loading. It is hard, for practical reasons, to ascertain from the British and French nuclear power station histories whether the conventional assumption of a semi-logarithmic variation of settlement with time holds over generally at full scale, or whether the trend seen at Labenne for the semi-logarithmic gradient $ds/d(\log t)$ to accelerate with time (as in Figure 45) applies. Intensive research, such as that described by Tatsuoka

Figure 50. Observed settlements (shown by triangles) and predictions from non-linear ICFEP analyses for an East-West section through UK nuclear power station shown in Figure 46: end of primary consolidation conditions.

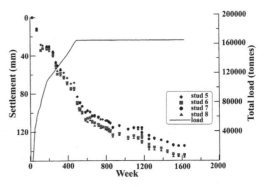

Figure 51. Long-term variation in settlements at four points on main raft for nuclear power station shown in Figure 46: arithmetic axes.

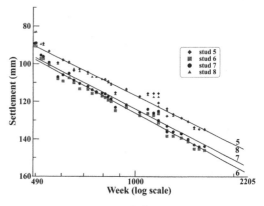

Figure 52. Long-term variation in settlements at four points on main raft for nuclear power station shown in Figure 46: semi-logarithmic axes.

229

et al. (2001) is needed to unravel the underlying mechanical processes.

3.4.2 *Experiments with piles at Dunkerque NW France*

The Port Automone de Dunkerque (PAD) have kindly made available to Imperial College and their collaborators a pile test site which is located adjacent to the Gravelines nuclear power plants. This 'Dunkerque' site was used by Chow (1997) and by the later 'GOPAL' project described by Parker et al. (1999). Chow's research involved a fundamental study of displacement pile behaviour in sands, while the GOPAL project was principally concerned with the behaviour of (i) a novel jet-grouted pile and (ii) plain driven 'control' piles.

The following paragraphs present a brief summary of the test site's geotechnical profile and the Dunkerque sands' properties. This is followed by discussion of the behaviour of the plain driven piles. The experiments and modelling of the jet-grouted pile are considered in the following final section of this paper, offering comments on our experience with hard cemented ground.

Ground conditions at Dunkerque
The PAD test site consists of a largely unoccupied flat expanse formed by placing around 2.5 m of hydraulic

sand fill over Flandrian marine medium sized silica sands. As described by Chow (1997) and Jardine & Standing (1999), the latter extend down to at least 30 m and are thought to be normally consolidated. The hydraulic fill was dredged from the same Flandrian deposits and Figure 53 shows a typical profile from the pile driving area.

The marine sands' cone resistance q_c profile indicates variable, but mainly dense to very dense, conditions. As is typical of this region, sub layers exist where the sands have lower resistances due to higher organic contents (representing the local effects of marine transgressions) and lower unit weights/higher void ratios.

Rotary core samples were taken for laboratory testing and a comprehensive array of CPTU tests was carried out, with practically one test per pile location. Other site investigations included ground water observations with piezometers (showing water at around 4.6 m below ground level), Marchetti Dilatometer DMT and seismic CPT soundings (conducted by the UK Building Research Establishment) and comprehensive laboratory testing at Imperial College. This includes:

- Index, direct shear and interface (soil to steel) shear tests (Chow 1997)
- Comprehensive stress path testing involving multi-axial bender element and high resolution local strain measurements (Kuwano 1999)

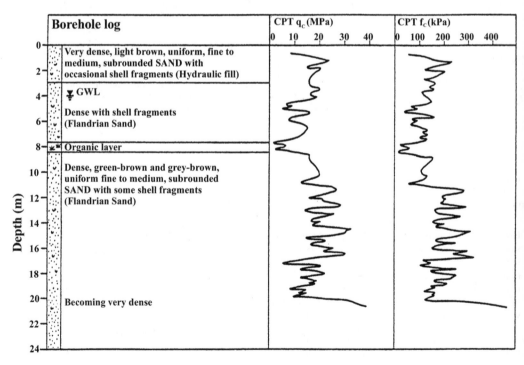

Figure 53. Typical ground profile from Imperial College Dunkerque pile test site (near Gravelines) North West France, including CPT profile; after Chow (1997).

- Torsional shear and resonant column tests on Hollow Cylinder samples (Connolly 1998)

The experiments provided data on the sands' behaviour at small, intermediate and large strains. Kuwano's intensive laboratory measurements allowed the anisotropic behaviour developed within the sands' Y_1 yield surfaces to be explored following the approach described by Kuwano and Jardine (1998, 2002). Kuwano developed expressions that specified how these components vary with void ratio and effective stresses (σ'_v and σ'_h) adopting the forms:

$$E_u = f(e).A_u.\left(p'/p_r\right)^{B_u} \tag{3}$$

$$E'_v = f(e).A_v.\left(\sigma'_v/p_r\right)^{C_v} \tag{4}$$

$$E'_h = f(e).A_h.\left(\sigma'_h/p_r\right)^{D_h} \tag{5}$$

$$G_{vh} = f(e).A_{vh}.\left(\sigma'_v/p_r\right)^{C_{vh}}.\left(\sigma'_h/p_r\right)^{D_{vh}} \tag{6}$$

$$G_{hh} = f(e).A_{hh}.\left(\sigma'_v/p_r\right)^{C_{hh}}.\left(\sigma'_h/p_r\right)^{D_{hh}} \tag{7}$$

$$f(e) = (2.17-e)^2/(1+e) \tag{8}$$

Equation 8 is the well known Hardin & Richart (1963) void ratio function, while in Equations 3 to 7 A_{ij}, B_{ij}, C_{ij} and D_{ij} are non-dimensional material constants and p_r is a reference pressure taken here as standard atmospheric pressure. With Dunkerque sand the values of B_u and the sum $[C_{ij} + D_{ij}]$ of the exponents applying to Equations 3 to 7 fell between 0.5 and 0.6.

The equations are evaluated and plotted against depth in Figure 54 by adopting Kuwano's sets of material coefficients (A_{ij}, B_{ij}, C_{ij} and D_{ij}) combined with Chow's unit weight profile and K_0 values. A single void ratio (0.61) has been adopted for this illustration that matches the expected mean, although the CPT q_c profiles point to substantial fluctuations in the sands' states (and void ratios) with depth.

The main feature to note from Figure 54 is the strong anisotropy, with the stiffest response being found in axial compression. E'_v /E'_h is predicted to be ~1.7 and E'_v/G_{vh} ~3.9 under the expected stress regime. Also shown is the in-situ G_{vh} profile measured with seismic CPT tests; on average this falls marginally (12%) above the curve projected from the laboratory Bender Element tests.

The laboratory experiments showed a wide range of stiffness degradation with strain. Figures 55 and 56 plot the variations of shear stiffness, normalised by mean effective stress p' (with an exponent of unity) for K_0 consolidated samples tested from p' = 200 kPa with OCR = 1, or at OCR = 2 with p' = 121 kPa. In

each case suitable ageing periods were imposed before conducting an overconsolidation (swelling) stage, or undrained shearing to failure. The ground is thought to be normally consolidated with mean effective stresses around the driven piles varying from local extremes over 500 kPa (at deeper points near the shaft) to zero. Bearing in mind that pile loading imposes vertical shearing on the shaft and axial loading at the base, emphasis was placed on the higher p', OCR = 1, torsional shear test data in selecting the composite

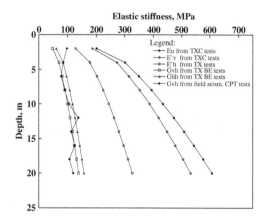

Figure 54. Profiles of quasi-elastic stiffness components with depth at Dunkerque, North West France, evaluating expressions derived by Kuwano (1999) for case of constant void ratio of 0.61. Seismic CPT G_{vh} profile is also shown.

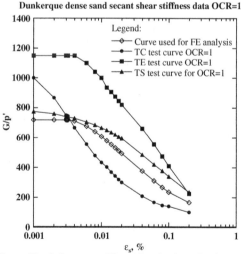

Figure 55. Laboratory stiffness data for dense Dunkerque sand at OCR = 1; TC = Triaxial Compression; TE = Triaxial Extension; TS = Hollow Cylinder Torsional Shear; Curve (Equation 1) adopted for analysis is also shown.

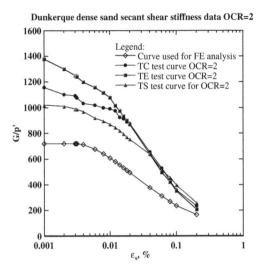

Dunkerque dense sand secant shear stiffness data OCR=2

Legend:
—◇— Curve used for FE analysis
—●— TC test curve OCR=2
—■— TE test curve OCR=2
—▲— TS test curve for OCR=2

Figure 56. Laboratory stiffness data for dense Dunkerque sand at OCR = 2; TC = Triaxial Compression; TE = Triaxial Extension; TS = Hollow Cylinder Torsional Shear; Curve adopted for analysis is also shown; note $\varepsilon_D = \sqrt{3}\varepsilon_S$.

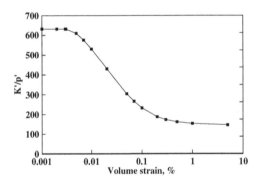

Figure 57. Bulk stiffness-volume strain curve (Equation 2) adopted for analysis of Dunkerque sands.

curve input via Equation 1 into the predictive FE analyses; the latter is also shown in Figures 55 and 56.

The bulk stiffness-volume strain curve that was chosen to represent (through Equation 2) the laboratory volumetric swelling/re-compression data is presented in Figure 57. This gave reasonable predictions for the K_0 swelling effective stress paths, an important constraint in many numerical analyses (see Jardine et al. 1991).

Recalling the nuclear power station case history described earlier, it was decided to retain the linear p′ stiffness functions given as Equations 1 and 2 when attempting to model the pile experiments conducted at Dunkerque. Softer curves (factored by 0.8) were assumed to apply within the (low CPT qc) organic

layers found at each pile location. Other test data allowed Mohr Coulomb failure criteria and dilation angles to be selected. Interface shear tests identified the limiting ratios of τ_{rz}/σ'_r that could apply over the pile shafts.

Experiments on plain driven piles
Comprehensive tension testing was performed on the well spaced out reaction piles that supported the GOPAL test frame. Six steel tubes, each with 457 mm outside diameter (D), were driven to penetrations of 19.3 m. They were tested individually to research:

• Time effects, as part of the GOPAL project, and
• Cyclic loading, under a separate project funded by the UK Health and Safety Executive (HSE) which included supporting numerical studies and some two-way testing on the GOPAL 'control' pile

More than 25 load tests were conducted on the six reaction piles between August 1998 and April 1999. The main focus was on pile capacity, which was strongly affected by time, testing history and cyclic loading (see Jardine and Standing 2000, and Jardine 2000). Jardine et al. (2001) described a predictions competition that focused on the main two 'virgin' compression tests conducted for the GOPAL project. The results emphasise the poor reliability of most design approaches for piles driven in sand. The predictions made for the plain driven piles yielded a strong conservative bias and a global standard deviation amounting to 60% of the value proven in the field.

The two most difficult features to predict are:

• Radial effective stress profile acting on the driven pile's shaft as a result of its installation and loading to failure
• The end bearing component available at the defined failure condition (either peak load or at a settlement of D/10)

Jardine & Chow (1996) proposed simplified design procedures that predict both features relatively well at most sites, including Dunkerque, giving little bias and standard deviations of 30% (or less) with sands when tested against large independent data bases of pile tests (with the latter typically being conducted within a few days or weeks of driving).

One of the reasons for the variations between predictions and measurements is the strong trend for shaft capacity to grow with time, as illustrated in Figure 58 by the failure loads measured in four Dunkerque 'virgin' tension tests conducted at different times and an 'end of installation' capacity estimate based on driving records. The oldest 'CLAROM' pile test was carried out by Chow (1997) on a 22 m long steel 327 mm diameter steel tubular pile that had been driven at Dunkerque as part of the earlier research programme

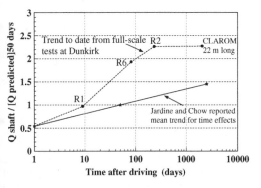

Figure 58. Variations in tensile (shaft) pile capacity with time at Dunkerque, North West France for 'virgin', previously untested piles. Piles R1, R6 and R2 were 19.3 m long, 457 mm diameter. CLAROM pile was 22 m long and 327 mm diameter. Results are normalised by Jardine & Chow (1996) predictions to allow for local variations; after Jardine & Standing (2000).

described by Brucy et al. (1991). Set up in sands is rarely mentioned in text books or addressed in design, but the shaft capacities grew by more than 400% in six months. In this case the individual capacities are normalised by the Jardine & Chow predictions made for each pile on the basis of the nearest CPT profile in order to allow for pile sizes and local geotechnical variations.

When discussing a different set of data Chow et al. (1998) argued that the primary cause for capacity growing with time is creep in the sand allowing relaxation of high circumferential stresses that shield the shaft from higher radial effective stresses by an arching action. While the Jardine & Chow procedure give good predictions for the ten-day capacities, corrections need to be applied when considering piles at different elapsed times.

Figure 59 presents the load-displacement plot for a 'virgin' tension test to failure on reaction pile R6 which was conducted around 80 days after driving. The test procedure involved applying load increments and then pausing under constant load until creep rates fell below specified thresholds. The tests were designed to take about 12 hours each and it is clear that creep (movement under constant load) became significant at loads exceeding about half of the ultimate capacity, growing progressively more important as loads increased. Testing was terminated when (i) creep rates became large, (ii) the load-displacement curve was tending to a plateau and (iii) pile head movement had reached about 7% of the pile diameter. Over half of the total recorded movement developed during constant load pause periods and approximately 70% of the movement was found to be irrecoverable on unloading.

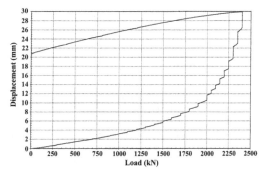

Figure 59. Field load test to failure on 19.3 m long, 456 mm diameter steel driven pile R6 at Dunkerque, North West France, 80 days after driving.

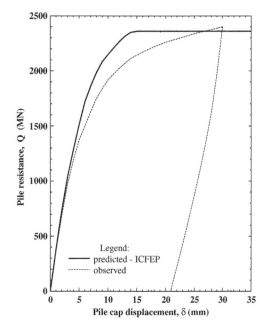

Figure 60. Comparison between non-linear FE prediction and simplified field data for test on driven pile R6 at Dunkerque, North West France.

An interesting feature was that the movements recorded during pause periods were often non-monotonic, with the movements suddenly accelerating and then slowing down during constant load periods; these features are thought to be related to stick-slip phenomena and associated load re-distribution along the pile shafts.

Figure 60 presents a simplified version of the R6 pile load test, along with predictions from a non-linear ICFEP finite element analysis based on simplified layering established at the nearest CPT location and the soil characteristics described above. The main features of the analysis were:

- A suitably fine mesh with eight main strata including three 1 m thick looser 'organic' layers with bulk unit weights between 17.1 and 20 kN/m^3
- Effective stress regimes that were simplified to give constant stress ratios σ_r/σ'_{v0}, $\sigma'_\theta/\sigma'_{v0}$ and σ'_v/σ'_{v0} at the pile face within each block, where σ'_{v0} is the 'far field' undisturbed vertical effective stress. The stress ratios were derived following the Jardine & Chow (1996) procedures, adjusted to account for the pile's 80 day age
- An undisturbed K_0 stress regime acting at radial distances more than 3.5 pile diameters (D) from the pile axis, with a semi-logarithmic variation between the near and far-field cases
- Equations 1 and 2 specifying non-dimensional pre-failure non-linear shear and bulk stiffness functions as illustrated in Figures 55 to 57, with parameters reduced by 20% in the 'organic' layers
- Interface shear effective stress failure parameters with $\delta' = \tan^{-1}[\tau_{rz}/\sigma'_r]$ taken as 28°, and φ' values of 35° and 32° for the dense and loose sands; these were established from interface shear tests and triaxial test to failure
- The sands' dilation angles ψ were taken as $\varphi'/2$
- The pile was represented as a solid body with stiffness values chosen to give the same compliance as the steel tubular pile

All of the above parameters can impact significantly on the predicted load displacement curves. Figure 60 shows that the predictions made with the best estimate parameter set were accurate while loads remained less than half the capacity, Q_{ult}. The pile's overall capacity was well predicted, but the movements developed over the later test stages were underestimated, with the discrepancy reaching 13 mm just before failure. Although adequate for most engineering purposes, the analysis did not account for the large creep movements that developed, following a stick-slip pattern, as failure was approached. Parallel analyses run with the higher normalised stiffnesses indicated by overconsolidated samples (see Figure 56) led to more marked under-predictions for field settlements.

An extended suite of numerical analyses was made to help understand the 25 load tests performed on the Dunkerque piles. In all cases the piles were predicted to fail progressively from the top downwards, with local slip starting at a relatively early stage. The predictive inaccuracy appears to stem from not considering the time dependency of the local interface shear failure; the latter is controlled by the radial effective stresses applied by the highly stressed and partially arching surrounding soil mass. The same processes

cause pile capacities to be dependent on test procedures and timing: slow Maintained Loading (ML) tests give lower capacities and settlements that are more representative of service conditions than rapid Constant Rate of Penetration (CRP) tests.

The main conclusions to flow from the plain pile tests at Dunkerque are:

1. As proven by an open competition, it is difficult to predict the load-displacement behaviour and capacity of plain piles driven in sand
2. The field tests identify both the strong positive effects of time on capacity and the importance of creep movements when piles are subject to high load factors
3. Although not discussed here, high level cyclic loading can lead to large displacements and reduced capacity. Low level cycling can accelerate the ageing processes (Jardine & Standing 2000)
4. Nevertheless, it is possible to develop a realistic effective stress analysis of load-displacement behaviour up to working loads provided account is taken in parameter selection of the detailed stress-strain behaviour of the sands (anisotropy, stress dependency, non-linearity), the stress regime set up by installation and the effects of time
5. The time-dependent behaviour seen at high loading levels is more difficult to model

Another key feature that has not been discussed is the brittleness of shaft capacity in sand: capacity can be halved in the short term by load testing to failure. Some recovery takes place with time but pre-testing leads to a capacity-time relationship that never manages to regain that applying to 'virgin' piles (Jardine & Standing 2000).

The effects of scale on pile end resistance were also investigated at Dunkerque. A driven cone-ended micro-pile with the same diameter as the standard CPT (36.5 mm) is expected to develop an end bearing pressure q_b similar to the CPT resistance q_c, giving q_b/q_c approaching unity. The larger (101 mm diameter) cone-ended Imperial College model pile employed by Chow (1997) developed an average $q_b/q_c = 0.72$ at a displacement of D/10 in compression tests run at Dunkerque. Full-scale piles driven in sand tend to develop lower base bearing pressures; the open ended (but fully plugged) CLAROM piles driven at Dunkerque (with D = 327 mm) could only mobilise an average $q_b/q_c = 0.245$. Chow (1997) argued that still lower q_b/q_c ratios apply to larger piles and Hight et al. (1997) reported a similar scale effect from an independent field and laboratory study. While the apparent trends require more research and explanation, Jardine & Chow (1996) proposed a provisional, empirical, scale relationship that both matched a significant database of field tests better than existing approaches and avoided non-conservative errors when considering large diameter piles.

3.5 Experience with hard cemented ground

In this final section we consider the behaviour of a composite system involving a strong steel pile and an artificial hard cemented geomaterial embedded within a soil mass.

3.5.1 Jet-grouted pile installation at Dunkerque: properties of jet-grouted sand, field performance and numerical analyses

The primary aim of the EU funded GOPAL project was to study the behaviour of steel driven piles enhanced by jet grouting through field, laboratory and analytical research (Parker et al. 1999). The project team consisted of D'Appolonia from Italy (project design, management and reporting), Bachy-Soletanche from France (jet grouting technology) and Imperial College (laboratory research, pile testing and numerical analysis). The main results have been reported by Jardine & Standing (2000), Standing & Jardine (1999) and D'Appolonia (2000).

Two large compression pile tests were undertaken at Dunkerque for GOPAL, with reaction being provided by the six driven piles (R1 to R6) described in the last section. Ancillary field research included:

- Forming a series of trial jet grout bulbs which were both sampled by rotary coring and exhumed for visual inspection
- A further rotary core taken through the jet-grouted pile after load testing to failure
- Proving the diameter and depth of the jet-grouted pile bulb by CPT soundings conducted after testing to failure

The two GOPAL compression piles (C1 and JP1) were formed by driving piles similar to R1 to R6 (456 mm open steel tubes) to penetrations of 10 m. While the Control pile C1 was left as a plain pile, an enlarged jet grouted base was formed under the twin pile JP1 thirty days after it was driven. An air-shrouded injection monitor was installed to a depth 15 m below ground level by drilling down through the pile's axis. The jet grout was made with a special slag cement that could be mixed with fresh or salt water. After a series of trial bulbs had been formed to optimise the grouting parameters, the desired 2.8 m diameter and 5 m high column was formed centrally under the driven pile as shown in Figure 61.

Grouting continued within the overlying pile shaft but with a wetter and less sandy grout mix. Degassing, grout seepage and bleeding led to the grout level falling within the steel pile in the following days so topping up was carried out with concrete and backfill.

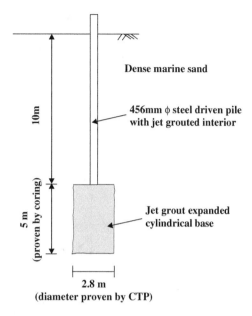

Figure 61. General arrangement of jet grouted pile JP1 tested at Dunkerque, North West France.

A curing time of 38 days was allowed between jet grout installation and pile load testing to failure.

Jet grout properties

A large programme of tests was carried out at Imperial College before the field testing. Mortar samples were formed by mixing Dunkerque sand, fresh water and the jet grout cement at a wide range of cement-to-water and cement-to-sand ratios. The samples were cured under water for up to two months prior to testing. Over 50 cylindrical samples were subjected to strain controlled Unconfined Compression (UC) tests; suites of direct shear interface (mortar to rough steel) tests were also performed. The UC samples had low permeabilities and were tested with relatively high degrees of saturation. However, their tendency to dilate would have led to rapid cavitation as they approached failure and so the tests may be considered to have been virtually drained. The samples underwent highly brittle failures with vertical and oblique cracks forming post peak. Curing time and cement-to-water ratio influence the UC results strongly, as illustrated in the summary plot given in Figure 62; the sand-cement ratio is less important.

Interface shear tests showed a brittle adhesion component (ranging from 0.1 to 1.5 MPa depending on mix etc). Once this had been overcome a frictional resistance was developed with a 'critical state' $\delta' = 42°$, respective of mix type and curing time. Rotary coring was carried out through the trial bulbs

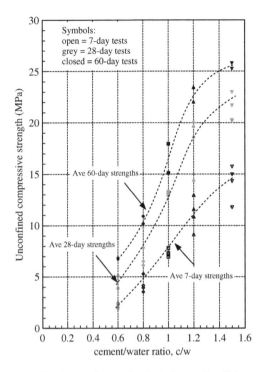

Figure 62. Scatter diagram showing UC strengths of laboratory made Dunkerque sand-slag cement grout: influence of cement/water ratios and curing duration (after Standing & Jardine 1998).

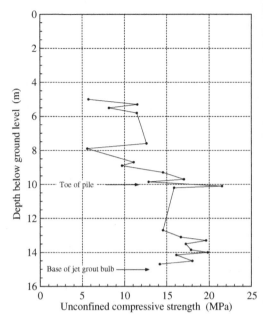

Figure 63. Profile of UC strengths from cores taken through Dunkerque JP1 pile axis 240 days after jet grouting.

30 days after initial grouting. The samples resembled weak, green-grey sandstones and displayed a heterogeneous macro-structure. Multiple UC tests were performed on 50 mm diameter, 100 mm high samples, along with index testing, petrographic and chemical analyses. The main results were:

- Laboratory mortar samples provided a good analogue for the field jet grout UC data provided the comparisons were made at the same overall cement-to-water ratio and curing times
- The initial state of the sand has little influence on the field UC strengths
- Grout quality did not vary significantly across the grout bulbs, with the bulbs developing a distinct but rough interface with the surrounding sand

The profile of UC strengths established from cores taken on the pile axis 240 days after it was jet grouted is shown in Figure 63. The jet-grouted bulb has UC strengths that are perhaps 50% above those measured on samples taken from within the steel pile section, but there is no clear trend with depth within the bulb. Material within the pile was lighter, giving saturated $\gamma = 15 \, kN/m^3$, while values around $20 \, kN/m^3$ were found within the sand-grout bulb.

P and S body wave velocity measurements made in the Imperial College Rock Mechanics laboratory indicated maximum shear moduli of between 5 and 10 GPa and Poissons ratios of around 0.45 under unconfined conditions. High pressure drained stress path triaxial tests were undertaken in the same laboratory on saturated field jet-grout samples. The samples were machined to give very flat ends, but they did not carry local strain sensors. The stress path tests were conducted slowly. While an elevated back pressure was applied at one end, pore water pressure measurements were made at the other to reveal any pore pressure gradients; the latter gave estimates of the grout's effective coefficients of consolidation. The stress-strain curves were non-linear but were probably subject to considerable bedding errors at small strains, making it impossible to characterise the pre Y_3 stiffness-strain relationships accurately.

Figure 64 shows a family of effective stress paths followed in experiments where mean effective stresses p' of up to 65 MPa were applied. Standing and Jardine's (1999) analysis of the tests identified the large scale Y_3 yield surface shown in Figure 64. Note that the maximum deviator stress q developed in each test increased sharply with the level of applied effective cell pressure rose and the q/p' ratio fell. The jet-grout samples hardened and gave q values exceeding the UC strengths under any test conditions that led to $p' > 4$ MPa at failure. The post-peak behaviour also

236

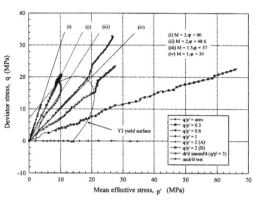

Figure 64. Effective stress paths followed in drained and undrained stress path tests run on jet-grout cores from Dunkerque site; after Standing & Jardine (1999).

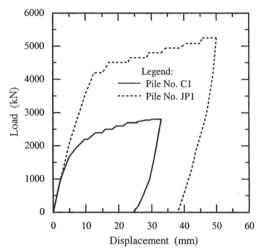

Figure 65. Load tests to failure on Jet Grouted pile (JP1) and Control plain driven pile (C1) at Dunkerque, North West France; after Jardine & Standing (2000).

becomes progressively more ductile as p' and σ'_3 increased. The maximum shear strengths available at elevated effective stresses appear to be controlled by φ' value of around 48°, an angle approximately 6° above the typical δ' value of 42°.

The key result from the study was that the grout samples behaved like a weak sandstone and could carry far greater shear strengths under triaxial conditions (where $q/p' < 3$) than in UC tests. For example, tests performed under undrained conditions with a large confining pressure could take twice the deviator stress sustained by unconfined samples. However, they require a radial effective stress of least 3 MPa to achieve this condition. The UC samples probably failed with σ'_r equal to the cavitation suction, which could be far less than 1 MPa.

3.5.2 Jet-grouted pile tests and their numerical simulation

Tests to failure were conducted on the jet-grouted (JP1) and plain control pile (C1) within two days of each other in November 1998, 68 days after driving and 38 days after jet grouting. The two piles were spaced just 10 m apart and a maintained load test procedure was followed, as with the R1 to R6 reaction piles. The load-displacement curves were measured by an independent pile head load cell and four displacement transducers. Rod extensometer systems were attached to both piles with anchor points attached at four points along the steel shafts to help gauge the load transfer with depth; these worked better with JP1 than when recovered and used for a second time in C1.

The test on JP1 was complete within 24 hours, with C1 being terminated after a shorter time and at a slightly smaller displacement than intended because of reaction frame problems. The overall compression load-settlement curves are shown in Figure 65; the two tests followed common, practically creep free,

shaft resistance controlled curves up to about 1 MN. They then started to diverge with JP1 ultimately developing about twice the capacity of the control pile; while C1 exhibited a relatively soft and time dependent response at loads greater than 1.7 MN. JP1 sustained a load of around 4.2 MN before it appeared to both yield and develop large creep settlements.

The nearest CPT profiles and the driving records showed that, if anything, C1 should have been able to develop a higher capacity than JP1 up to the time of the jet-grouting intervention. Furthermore, interpretation of the extensometer data indicated that jet-grouting probably reduced the shaft capacity of JP1 by undermining the driven steel section – leaving a soft slurry pocket that would have taken some days to achieve the same stiffness as the in-situ sand. This would certainly have eased the radial stresses acting on the lower part of the driven pile shaft. The greater field resistance of JP1 can only have been due to the extra base capacity offered by the jet grout bulb.

Use was made of the extensometer interpretation to project the load-displacement response of the jet grout bulb on its own in Figure 66. It is clear that local failure of the grout bulb started when the load reached about 3.2 MN, after which the bulb's response was both soft and time dependent. Cores taken through the pile axis showed that cracking had taken place in the lower part of the grout column contained within the steel pile, and over the upper third of the grout column depth. The failure involved the steel pile and grouted core punching into the grout bulb, causing the latter to fracture and expand out radically into the surrounding sand mass.

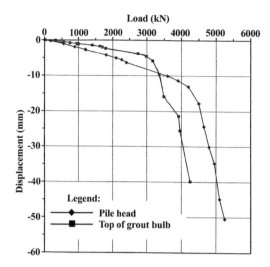

Figure 66. Jet-grout bulb contribution to load carrying capacity of pile JP1 in test and Dunkerque, North West France.

Table 1. Properties assumed for jet-grout in ICFEP analyses of pile JP1.

Case	Unit weight (kN/m^3)	S_u (MPa)	Friction angle, δ'	$G = G_{max}/2$ (GPa)
Sand-free grout in steel pile	15.0	4.0	42	2.5
Sand-grout in main jet grouted bulb	20.0	7.0	42	5.0

A numerical analysis was undertaken by the Authors after the pile test had been completed. The approach was generally similar to that described above for the plain reaction pile R6, with the same parameters and non-linear elastic plastic models being adopted for the sand layers and pile-sand interface. Naturally the different geometry (shorter steel pile and bulb) was accounted for in the new FE mesh; other modifications made to reflect the presence of the jet grout bulb included:

- The stresses developed over the lower part of the pile were reduced to allow for the undermining effects of jet grouting. Stresses were reduced linearly over the final 3.75 m of shaft from the 'best estimate' driven values to free-field K_0 values.
- The jet-grout was modelled as an elastic-plastic Tresca material with a no-tension cut off. It was argued that the test duration was too short to allow much local drainage, but that the test boundary

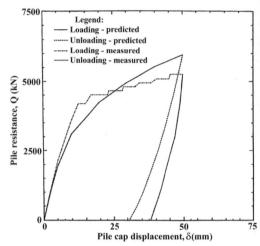

Figure 67. Load test to failure on Dunkerque jet-grouted pile: predictions and field measurements.

conditions would only provide partial confinement. Initial static pore water pressures of just 50 to 100 kPa were expected along with relatively low mean effective stresses (100 to 300 kPa) in the grout bulb vicinity. Under these conditions the Tresca 'undrained shear strengths' could be similar to half the UC test values, although the post peak response might be more ductile. In the absence of any other data, the operational shear stiffnesses were taken as half the G_{max} values deduced from the laboratory shear wave velocity measurements.

The parameters adopted for analysis are summarised below.

Only one set of parameters was adopted: no attempt was made to fine-tune or optimise the input parameters by comparing predictions with the field data. Figure 67 shows how the single main ICFEP analysis compared with the test results.

The main points to note are:

- Predictions and measurements are in good general agreement. The load predicted at a settlement of D/10 is within 10% of that measured, as is the load-settlement curve developed up to half of the failure load
- The numerical analysis predicted the observed failure mode, including the pattern of cracking that was associated with low radial total stresses within the grout mass
- However, yielding was predicted in the grout plug at a lower load than was measured; the peak shear strength of the grout may have been marginally under-estimated by neglecting all effects of confinement

- The final slope of the predicted load-displacement curve was twice as steep as the field measurement. Lack of knowledge concerning the grout's non-linear pre-failure behaviour and neglecting the brittle post peak response and time dependency were probably the main factors contributing to this error

- One further possible shortcoming in the ground model concerns the stiffness parameters assumed for the sand around the grout bulb. As indicated in Figure 54, the sand's small strain response to purely radial loading is 40% softer than that due to vertical loading. Neglecting this anisotropy would have led to an over-prediction of the radial stresses generated by the expanding grout bulb; the assumption of a linear relationship between stiffness and p' may have had a similar effect

Despite the discrepancies developed over the later stages of the simulation, the ground model was considered sufficiently realistic to be employed in parametric studies of how design variables such as grout strength and bulb dimensions would affect the overall load-displacement response. Improved predictions could be obtained by using more advanced models to consider (i) non-linearity, brittleness, confining pressure and time dependency in the grout, and (ii) stiffness dependence on pressure and anisotropy in the sand.

The main conclusions to be drawn from the jet grouted pile research are:

1 Jet grouting can considerably enhance the capacity of piles driven in sand
2 Non-linear ground models developed from advanced laboratory testing and realistic pile capacity prediction tools can be used to analyse and design such interventions
3 Jet grout properties can be characterised in the laboratory through careful testing on either artificial mixtures or field cored samples. The cement-to-water ratios and curing times are the key variables when considering sands; local strain instrumentation is necessary if the stress-strain behaviour of such hard cemented materials are to be characterised accurately
4 Predictions made for the settlements developed under working loads are likely to be more reliable than those for states near to failure
5 Factors such as non-linearity, brittleness, time dependency and anisotropy can be expected to play a more important role as failure is approached
6 Considering the predictive errors made by practitioners in relation to routine problems such as the capacity of plain driven piles, field tests play a vital role in building our experience and checking our analytical approaches.

4 SUMMARY

We offer below brief summaries of the lessons learned from the 13 main sites covered in this keynote paper, drawing conclusions from the above described geotechnical investigations, field measurements, non-linear numerical analyses and associated studies.

4.1 Bangkok Metro station strutted excavation (Case 1)

The key conclusions from the recent re-analysis of the Bangkok Metro station deep excavation depicted in Figure 1 are: (i) the importance of modelling the full range of pre-yield non-linearity, even with soft ground, when attempting to predict the ground movements associated with stiff structures (such as the described strutted diaphragm walls), and (ii) the need to establish and model any global settlements that might be related to site surcharge, wall seepage or regional subsidence (or uplift) due to continuing groundwater variations. While it was shown that the non-linear approach advocated by the Authors led to a better set of predictions for the medium-term response than simpler models, it is believed that unaccounted factors such as possible site surcharging and interactions with regional subsidence may have reduced the accuracy of the predictions presented in Figures 4 and 5.

4.2 Bothkennar pad tests, Scotland (Case 2)

The first (1990) tests on instrumented pads founded on soft clay at Bothkennar provided field evidence that confirmed the analytical prediction that soil non-linearity concentrates ground strains far more closely to the loading boundaries than had been appreciated from classical elastic solutions (see Figure 6). Observations made between 1990 and 2001 have shown the crucial role that creep plays in soft clay problems, with settlements more than doubling after the end of full primary consolidation (Figure 9). The 2001 test to failure on the pre-loaded pad proved that pre-loading increased the pad's capacity by 48%, a larger margin than could be expected from a time independent, isotropic critical state modelling approach. Factors such as creep volume straining, evolving anisotropy and the clay's cemented structure may be adding strength components that augment those relating to classical primary consolidation. The same processes may also explain the pad's initially stiff loading response; see Figure 10.

4.3 Metro tunnel movements at Wanstead, London (Case 3)

The study of the interactions between a new cut and cover tunnel and an existing metro station and its tunnels (see Figures 12 and 13) emphasised the importance of considering the full history of such a site, including the long-term coupled consolidation processes. Combining this with a careful characterisation of the non-linear properties of the geological layers present led to accurate (Class A) predictions being made for the medium-term effects of the cut and cover tunnels on the existing deeper bored tunnels, as shown by the field data and predictions presented in Figure 17. The study also pointed to the potentially large swelling strains that can develop due to natural erosion, or excavation by man, in materials such as London clay. The swelling strains can lead to both large heave movements and to the loss of strength and stiffness through destructuring, as mentioned also in the St James' park case history.

4.4 Deep basement construction at Victoria Embankment and Moorgate London (Cases 4 and 5)

Readers were reminded of an earlier deep basement case history from London that had been reported by Jardine et al. (1991) and St John et al. (1992). Here, the Authors' non-linear approach had led to generally successful medium-term Class A predictions, but with some interesting differences between predictions and measurements. Use was made of more recent 3-D analyses of the Moorgate development described by Tor-Petersen et al. (2003) to show that one probable reason for the over-prediction of far-field ground settlement and deep lateral wall movements (shown in Figures 22 and 23) was the 2-D simplification that had been made in the earlier analyses of the L-shaped excavation illustrated in Figure 18. Predictive accuracy can be improved by fully non-linear 3-D analysis for projects where the structure cannot be idealised as being infinitely long, or circular. However, great care is needed when considering the assumptions made concerning out-of-plane wall stiffnesses and other structural details; the latter have a considerable impact on ground movements, strut loads and wall bending stresses.

4.5 Westminster Metro station, JLEP, London (Case 6)

The Jubilee Line Extension Project (JLEP) in London offered many opportunities to check numerical predictions at full scale. Among these was the new box for the Westminster station (see Figures 28 and 29) where special measures were taken to protect nationally important buildings. The advocated non-linear, fully coupled, ground modelling technique was applied to predict the lateral ground movements of the deep retained excavation, with the ground model being updated to account for uncertainty regarding

the in-situ permeability of the strata present. Excellent agreement was found between predictions and measurements made during construction (see Figure 30). The soil modelling and analytical procedures, combined with an observational approach, provided the project engineers with a valuable tool in their optimisation of the station's construction.

4.6 St James' Park JLEP tunnels, London (Case 7)

The second JLEP case history concerned two new tunnels that were bored under St James' park, in central London. It was shown that simplified 2-D numerical predictions of ground surface settlements tended to distribute the volume losses generated by tunnelling over a wider trough than was observed in the field. Predictions can be improved by adopting the advocated fully non-linear approach and by rezeroing the strain terms in Equations 1 and 2 at appropriate stages to model the kinematic nature of soil stiffness (see Figures 31 and 32). Other artifices have been proposed, including invoking strong anisotropy and adopting unlikely K_0 values in the soil layers, but these have yet to be justified rationally. It remains difficult to predict accurately the ground movements caused by real tunnelling procedures without relying on experience in similar ground, with comparable machines. The case history fully highlights the difficulty of predicting ground loss from first principles, showing that subtle changes in geology can be highly influential, as can the details of the Tunnel Boring Machine.

4.7 Bromborough docks landfill, Merseyside, UK (Case 8)

The Bromborough landfill case illustrated the application of the same ground modelling approach to a large landfill embankment built on very soft foundations. Despite careful investigation of the properties of the governing soft hydraulic fill layer (Figures 37 to 40), uncertainties existed regarding the distribution of sand layers within the fill and the properties of the landfill material (see Figure 35). Cautious Class A predictions had been made regarding the movements developed over two years beyond the landfill's toe, and the settlements developed under the fill. Predictions and measurements for points A and B on a critical section, as identified in Figure 36, showed generally reasonable agreement (see Figures 41 and 42). The initial ground model was updated to account for further ground investigations and field records regarding pore pressure responses, landfill unit weight and other factors. The revised ground model led to improved predictions and had been very helpful in optimising the use of the space available for landfilling, while assuring the landfill's overall stability and ensuring that ground movements did not affect sensitive infrastructure elements installed at point B. The SHANSEP testing procedure adopted for the hydraulic fill layers may have led to some over-prediction of ground movements in the early stages of land filling.

4.8   Canons Park, North London (Case 9)

All of the remaining case histories drew attention to some unexpected effects of time on geotechnical problems. Pile tests at the UK Building Research Establishment (BRE)'s Canons Park test bed site illustrated one such case. Research into pile ageing processes revealed that the axial capacity of a steel displacement pile had practically doubled after being left to rest for 17 years (see Figure 43). In this instance Sulphate Reducing Bacteria (SRB) are believed to have affected behaviour by modifying the properties of the shear zone created near to the shaft during pile installation and prior testing.

4.9   Labenne, South West France (Case 10)

Data were presented from a long-term research programme into the settlement of surface foundations conducted by the French Laboratoire des Ponts et Chaussées (LCPC). As shown in Figures 44 and 45, small scale surface footing tests conducted on dry sand at Labenne indicated very significant creep settlements. A semi-logarithmic relationship appeared to apply over short time intervals, but the long-term plots showed gradients that increased with time, giving settlements after 10 years that could be several times those developed shortly after loading. It was noted that the much larger raft foundations of French nuclear power stations also developed large long-term settlements, including those located on marine sands at Gravelines, near to Dunkerque, North West France.

4.10   UK nuclear power station foundations on sand (Case 11)

More details were given of the behaviour of a UK nuclear power station built on sand underlain by mixed strata, giving data on the layout, soil profile, geotechnical properties, numerical analyses and field measurements (Figures 46 to 52). The non-linear approach advocated by the Authors was applied in a 3-D analysis of load-displacement response of the main raft. Good hindcast predictions were made of the raft's end of primary settlements and deflected shape. However, settlements have continued to grow in the 30 years following construction, now amounting to double those seen at the end of loading. The settlements measured over the last 20 or so years tend to follow a log-linear trend with time, showing little or no change in gradient.

4.11   Research with plain driven piles at Dunkerque, North West France (Case 12)

A significant section of the paper was devoted to describing the ground profile of the piling test site used by Imperial College at Dunkerque (see Figure 53) and research into the properties of the marine sand layers present (Figures 54 to 57). Emphasis was placed on the sands' non-linearity, anisotropy and pressure dependency.

Attention was then turned to predictions and measurements for the behaviour of plain tubular steel piles driven at the site. The difficulties of predicting the effective stress conditions around such piles were discussed and it was shown that applying recently developed effective stress methods led to good predictions for medium-term axial capacities. However, axial capacity increased rapidly with the time elapsed after driving (Figure 58) following an unexpected trend which is thought to be due to creep processes breaking down an arching mechanism developed around the lower shaft.

Numerical analyses that took account, to some degree, of all of the above features gave good predictions for field test data until the loads exceeded around half the failure values. However, the displacements developed over the later stages of the tests were larger than predicted (see Figure 60). The field records show that the differences between the predictions and measurements were principally associated with creep movements, which became increasingly important as the loading level increased (see Figure 59).

4.12   Research with jet grouted piles at Dunkerque, North West France (Case 13)

The final case discussed concerns the jet grouted pile installed at Dunkerque as part of the GOPAL project. A jet grout column was installed under a plain driven tubular steel pile as shown in Figure 61. Considerable research was carried out into the properties of the jet grout formed in Dunkerque sand (see Figures 62 to 64), showing that its strength and stiffness properties were mainly affected by cement to water ratio, and curing time. Samples collected from the field resembled weak sandstones and stress path tests on such samples proved that effective confining pressure increased its shear strength and reduced its degree of post-peak brittleness. Analysis of the laboratory test data reinforced the importance of making local strain measurements, even when testing carefully machined rock or hard grout samples. Laboratory body wave velocity measurements gave useful data guidance on the maximum stiffness values.

A field loading test to failure was performed on the jet grouted pile, and the data obtained were compared with non-linear numerical predictions (Figures 65 to

Table 2. Summary of ICFEP predictions made incorporating Equations 1 and 2 compared with measurements from nine of the case histories.

No	Movement parameter considered	Predicted movement (mm)	Measured movement (mm)
1	Max settlement &	18	30
	lateral movement:	24	16
	end of construction		
3	Largest tunnel	13	14
	movement (vector		
	length)		
4	Max settlement &	36	25
	lateral movement:	39	36
	end of construction		
6	Max lateral movement	40	42
	end of construction		
7	Max settlement at end		
	of construction:		
	First tunnel	10	20
	Second tunnel	14	34
8	Settlement at point A:		
	Original analysis	1850	1300
	Revised analysis	1450	1300
	Lateral movement at		
	point B:		
	Original analysis	11	14
	Revised analysis	15	14
11	Max raft settlement at	108	108
	end of consolidation		
	Max settlement after	108	148
	30 years		
12	Settlement on loading	3.7	4.1
	to half capacity		
13	Settlement on loading	7.5	6
	to half capacity		

67). The pile failed by the steel shaft punching into the jet grout column, which then expanded out into the confining soil mass. As with the plain pile, the FE predictions provided a very good match up to around half of the failure load and offered a reasonable prediction for the ultimate capacity.

However, the final parts of the load-displacement curve were softer than had been predicted. This could reflect: (i) the effects of a progressive brittle failure within the jet grout (which was not modelled), or (ii) the anisotropy of the sand (which had a far softer response to radial stress changes than to vertical or horizontal shear stresses – see Figure 54), or (iii) an overestimation of the operational stiffness of the jet grout.

4.13   Overall predictive reliability

A summary is made above of the predictions made (using the approach described by the Authors) of the most important ground movements relating to nine of the case histories described and the recorded field data. The ratio between prediction and measurement ranges between around 0.5 and 2, with a mean value just below unity (0.98) and a standard deviation of 0.33. Excluding the tunnelling cases reduces the standard deviation to around 0.25.

While this degree of agreement is encouraging, the paper has attempted to show how further improvements might be made by improving site investigation and geological analysis, consideration of three dimensional geometry and construction details, taking account of soil anisotropy and brittleness, addressing time dependent properties and processes, and updating analyses during the course of a project to account for aspects or parameters that were either unknown or uncertain at the design stage.

5   GENERAL CONCLUSIONS

The material reviewed in this paper leads to six main conclusions:

1 Field observations and carefully recorded case histories provide the vital means of assessing the applicability of predictive tools, experimental procedures and analytical approaches
2 The case histories discussed here, along with others published previously, confirm the value of high quality site investigations that include careful sampling and laboratory stress path testing involving local strain measurements. Field and laboratory geophysical techniques are also of great value, as are other in-situ tests, especially in sands
3 Given adequate geological and geotechnical input data, the non-linear FE predictive approach advocated by the Authors can provide relatively accurate and useful predictions for medium-term ground movements and soil-structure interaction. It has become possible to expect the most important end of construction ground movements to be predictable to within about 30%, even when the displacements are only of the order of 10 mm
4 Problem areas that were highlighted include: the analysis of ground movements caused by tunnel boring machines (particularly those with open shields) and other highly 3-dimensional problems; modelling anisotropy over a wide range of strain; and the effects of time. The latter are particularly significant when loading levels are high and when considering the installation's long-term response
5 There are significant practical benefits in adopting an 'observational approach' where uncertainties exist before construction and field measurements are made to allow analysis updating as construction proceeds, providing a valuable tool for decision making

6 Research and development are likely to offer further improvements in geotechnical modelling, but comparisons with full-scale field performance will be vital to check the new approaches' predictive reliability.

ACKNOWLEDGEMENTS

The Authors acknowledge the contributions made by Prof. David Potts in developing the computer code ICFEP and taking part in several of the projects described. They also thank the LCPC and EDF (France) for permission to publish data from Labenne and Gravelines. They also thank Mr David Harris, Dr David Hight and Dr Lidija Zdravkovic for providing other data, and acknowledge the help provided by Mr Aleksandar Djordjevic, Mr NguyenAnh Minh and Dr Akihiro Takahashi with the figures and manuscript. The field work at Bothkennar was funded by the UK Engineering and Physical Sciences Research Council (EPSRC).

REFERENCES

Addenbrooke, T.I., Potts, D.M. & Puzrin, A.M. 1997. The influence of pre-failure stiffness on the numerical analysis of tunnel construction. *Geotechnique* 47(3): 693–712.

Albert, C., Zdravkovic, L. & Jardine, R.J. 2003. Behaviour of Bothkennar clay under rotation of principal stresses. In Vermeer, Schweiger, Karstunen & Cudney (eds), *Int workshop on Geotechnics of soft soils – theory and practice*.

Al-Tabbaa, A. & Wood, D.M. 1989. An experimentally based bubble model for clay. *Numerical methods in Geomechanics NUMOG III*: 91–99. Elsevier Applied Science.

Amar, S., Baguelin, F., Canépa, Y. & Frank, R. 1985. Comportement à long terme des fondations superficielles, Prévisions et observations, Proceedings *XI ICSMFE*: 2155–2158. San Francisco: Balkema.

Bond, A.J. & Jardine, R.J. 1991. The effects of installing displacement piles in a high OCR clay. *Geotechnique* 41(3): 341–363.

Bond, A.J. & Jardine R.J. 1995. Shaft capacity of displacement piles a high OCR clay. *Geotechnique*, 45(1): 3–23.

Brucy, F., Meunier, J. & Nauroy, J.F. 1991. Behaviour of pile plug in sandy soils during and after driving. In *Proc 23rd Offshore Technology Conference*, OTC 6514: 145–154. Houston.

Burland, J.B. & Burbridge, M.C. 1985. Settlement of foundations on sand and gravel. *Proc Inst. Civil Engrs*. 78: 1325–1381. London: Thomas Telford.

Burland, J.B. 2001. Mechanisms of behaviour in foundation-structure interaction – some case histories. In Jamiolkowski, Lancellotta & Lo-Presti (eds), *Pre-failure Deformation Characteristics of Geomaterial* 2: 1143–1160. Torino: Swets & Zeitlinger.

Canépa, Y. 1998. Le comportement des foundations superficielles sous charge constante – Analyse des données expérimentales; Application aux calculs des ouvrages. *LCPC CT24 – 2.24.19.4 - Dossier LREP 1.4.0.9889*, 104 p.

Chow, F.C. 1997. "Investigations into displacement pile behaviour for offshore foundations". *Ph.D Thesis*, Imperial College, University of London.

Chow, F.C., Jardine, R.J., Brucy, F. & Nauroy, J.F. 1998. The effects of time on the capacity of pipe piles in dense marine sand. *ASCE, JGE* 124(3): 254–264.

Connolly, T. 1998. Report on Hollow Cylinder and Resonant Column testing on Dunkerque sand. *Internal Report*, Imperial College, University of London.

Crawley, J.D. & Glass, P. 1998. Westminster Station London – Deep foundations and limiting movements in the deepest excavation in London. *Proc. Conf on Deep Foundations, DFI*, Paper 5.18.

D'Appolonia, 2000. Report for European Commission on GOPAL project. Doc No 96-322-H11, *D'Appolonia Consultants, Genova, Italy*.

Dun, C.S. 1974. Settlement of a large raft foundation on sand. *Proc BGS Symposium on Settlement of structures*: 14–21. Cambridge: Pentech Press.

Frank, R. 1992. Some recent developments on the behaviour of shallow foundations. General Report. *10th ECSMFE* 4: 115–1146. Florence: Balkema.

Hardin, B.O. & Richart, F.E. 1963. Elastic wave velocities in in granular sands: measurements and parameter effects. *ASCE Journal* 89(1): 33–65.

Harris, D.I. 2001. The Big Ben Clock Tower and the Palace of Westminster. In Burland Standing & Jardine (eds), *Building Response to Tunnelling. Case studies from construction of the Jubilee Line Extension, London CIRIA SP199*: 453–508. London: Thomas Telford.

Higgins, K.G. & Jardine, R.J. 1998. Experience of the use of non-linear prefailure constitutive soil models. *Prefailure behaviour of geomaterials*: 409–412. London: Thomas Telford.

Higgins, K.G. & Jardine, R.J. 1999. Case histories relating to the use of laboratory small strain testing in practical applications. *Report to ISSMGE Committee TC-29*.

Higgins, K.G., Paterson, J., Moriarty, J., Potts, D.M. & Jardine, R.J. 2001. The effect of an excavation in a stiff fissured clay on existing tunnels in close proximity. *Conference on Response of buildings to excavation induced ground movements. CIRIA SP199*: 313–324. London: Thomas Telford.

Hight, D.W., Bond, A.J. & Legge, J.D. 1992. Characterisation of Bothkennar clay: an overview. *Geotechnique*, 42(2): 303–348.

Hight, D.W. & Jardine, R.J. 1993. Small strain stiffness and strength characteristics of hard London Tertiary clays. *Proc. Int. Symposium on Hard Soils – Soft Rocks* 1: 533–522. Athens: Balkema.

Hight, D.W. & Higgins, K.G. 1995. An approach to the prediction of ground movements in engineering practice: background and application. Keynote Lecture, In Shibuya, S., Mitachi, K. & Miura, N. (eds). 1995. *Prefailure Deformation of Geomaterials. Proceedings of TC-29 International Symposium* (2):909–946. Hokkaido: Balkema.

Hight, D.W., Bennell, J.D., Chana, B., Davis, P.D., Jardine, R.J. & Porovic, E. 1997. Wave velocity and stiffness measurements of the Crag and Lower London Tertiaries at Sizewell. *Geotechnique*: 47(3) 451–474.

Hight, D.W., Lawrence, D.M., Farquhar, G.N. & Potts, D.M. 1996. Evidence for scale effects in the end bearing capacity of open-ended piles in sand. *Proc 28th OTC, Houston, OTC 7975*: 181–192.

Hight, D.W. 1998. Soil characterisation: the importance of structure, anisotropy and natural variability. *38th Rankine Lecture*.

Hight, D.W. & Leroueil, S. 2002. Characterisation of soils for engineering purposes. In Tan et al. (eds), *Characterisation and Engineering Properties of Natural soils*: 255–362. Singapore: Swets & Zeitlinger.

Hight, D.W., Paul, M.A., Barras, B.F., Powell, J.J.M., Nash, D.F.T., Smith, P.R., Jardine, R.J. & Edwards, D.H. 2002a. The characterisation of Bothkennar clay. In Tan et al. (eds), *Characterisation and Engineering Properties of Natural soils*: 543–598. Singapore: Swets & Zeitlinger.

Hight, D.W., McMillan, F., Powell, J.J.M., Jardine, R.J. & Allenou, C.P. 2002b. Some characteristics of London clay. In Tan et al. (eds), *Characterisation and Engineering Properties of Natural soils*: 851–908. Singapore: Swets & Zeitlinger.

Hight, D.W. 2003. Personal communication.

Jamiolkowski, M., Lancellotta, R. & Lo-Presti, D. (eds) 2001. *Pre-failure Deformation Characteristics of Geomaterials* Volumes 1 & 2. Torino: Balkema.

Jardine, R.J., Symes, M.J. & Burland, J.B. 1984. The measurement of soil stiffness in the triaxial apparatus. *Geotechnique* 34(3): 323–340.

Jardine, R.J. 1985. Investigations of pile-soil behaviour, with special reference to the foundations of offshore structures. *PhD Thesis*, Imperial College, University of London.

Jardine, R.J., Fourie, A.B., Maswoswse, J. & Burland, J.B. 1985. Field and laboratory measurements of soil stiffness. *11th ICSMFE*: (2)511–514. San Francisco: Balkema.

Jardine, R.J., Potts, D.M., Fourie, A.B., & Burland, J.B. 1986. Studies of the influence of non-linear stress-strain characteristics in soil-structure interaction. *Géotechnique*, 36(3): 377–396.

Jardine, R.J. & Potts, D.M. 1988. Hutton Tension Leg Platform Foundations: an approach to the prediction of pile behaviour. *Géotechnique* 38(2): 231–252.

Jardine, R.J., St John, H.D., Hight, D.W. & Potts, D.M. 1991. Some practical applications of a non-linear ground model. *Proc. Xth ECSMFE*, 1: 223–228. Florence: Balkema.

Jardine, R.J. 1992. Some observations on the kinematic nature of soil stiffness. *Soils and Foundations* 32 (2): 111–124.

Jardine, R.J. 1992. Panel contribution on laboratory determination of soil properties. Session IA. *Proc. 10th ECSMFE*: (4)1209–1212. Florence: Balkema.

Jardine, R.J. & Potts, D.M. 1993. Magnus foundations: Soil properties and predictions of field behaviour. *Large scale pile test in clay*: 69–83. London: Thamas Telford.

Jardine, R.J. 1995. One perspective on the pre-failure deformation characteristics of some geomaterials. Keynote lecture. In Shibuya, S., Mitachi, K. & Miura, N. (eds). 1995. *Pre-failure Deformation of Geomaterials. Proceedings of TC-29 International Symposium* (2): 855–866. Hokkaido: Balkema.

Jardine, R.J., Lehane, B.M., Smith, P.R. & Gildea, P.A. 1995. Vertical loading experiments on rigid pad foundations at Bothkennar. *Geotechnique*, 45(4): 573–599.

Jardine, R.J. & Chow, F.C. 1996. *New design methods for offshore piles. MTD Publication 96/103*. London: MTD (now CMPT).

Jardine, R.J., Zdravkovic, L., & Porovic, E. 1997. Anisotropic consolidation including principal stress axis rotation: experiments, results and practical implications. *Proc 14th ICSMFE* :(4)2165–2172. Hamburg: Balkema.

Jardine, R.J., Davies, M., Hight, D., Smith, A. & Stallebrass, S. 1998. Pre-failure deformation behaviour of geomaterials. *Symposium in Print*. London: Thomas Telford.

Jardine, R.J. & Standing, R.J. 2000. Pile load testing performed for HSE cyclic loading study at Dunkirk, France. Volume 1 & 2. *Offshore Technology Report OTO 2000-007*. London: Health and Safety Executive.

Jardine, R.J. 2000. Some surprising results from research into the behaviour of piles driven in sand. Keynote paper for AGI 1999. *Revista Italiana Geotecnica*. 43(3): 5–17. Parma: AGI.

Jardine R.J. & Kovacevic, N. 2003. Confidential Reports on Ormen Lange Slope Stability studies, *Imperial College Consultants*.

Jardine R.J., Standing, J.R., Jardine, F.M., Bond, A.J. & Parker, E. 2001a. A competition to assess the reliability of pile prediction methods. *Proc. 15th ICSMGE*: (2) 911–914. Istanbul: Balkema.

Jardine, R.J., Kuwano, R., Zdravkovic, L. & Thornton, C. 2001b. Some fundamental aspects of the pre-failure behaviour of granular soils. In Jamiolkowski, Lancellotta & Lo-Presti (eds), *Pre-failure Deformation Characteristics of Geomaterial*: 1077–1112. Torino: Swets & Zeitlinger.

Jardine, R.J., Kovacevic, N., Collins, R., Devine, J. & Ellis, R. 2004. Geotechnical Aspects of the Bromborough Docks landfill project. In preparation.

Kovacevic, N. & Jardine, R.J. 2003. Confidential Reports on Bromborough Docks landfill project, *Geotechnical Consulting Group*.

Kovacevic, N., Hight, D.W. & Potts, D.M. 2003. A comparison between observed & predicted behaviour of a deep excavation in soft Bangkok clay. *Proc. 3rd Int. Conf. Deformation Characteristics of Geomaterials*: (1) 983–990, Lyon, Balkema.

Kovacevic, N., Hight, D.W. & Potts, D.M. 2004. Temporary slope stability in London Clay – back analyses of two case histories. Submitted for *Skempton Memorial Conference London 2004*.

Kuwano, R. & Jardine, R.J. 1998. Stiffness measurements in a stress path cell. In Jardine et al. (eds), *Pre-failure behaviour of geomaterials*: 391–395. London: Thomas Telford.

Kuwano, R. 1999. The stiffness and yielding anisotropy of sand. *PhD Thesis*, Imperial College, University of London.

Kuwano, R. & Jardine, R.J. 2002. On measuring creep behaviour in granular materials through triaxial testing. *Canadian Geotechnical Journal* 39(5): 1061–1074.

Kuwano, R. & Jardine R.J. 2002. On the applicability of cross anisotropic elasticity to granular materials at very small strains. *Geotechnique* 52(10): 727–750.

Lehane, B.M., Jardine, R.J., Bond, A.J. & Frank, R. 1993. Mechanisms of shaft friction in sand from instrumented pile tests. *ASCE Geot. Journal* 119(1): 19–35.

Lehane, B.M. & Jardine, R.J. 2003. Effects of long term preloading on the performance of a footing on clay. *Geotechnique* 53(8): 689–697.

244

Leroueil, S., & Hight, D.W. 2002. Behaviour and properties of natural soils and soft rocks In Tan et al. (eds), *Characterisation and Engineering Properties of Natural soils*: 29–524. Singapore: Swets & Zeitlinger.

Menkiti, C.O., Long, M., Kovacevic, N., Edmonds, H.E., Milligan, G.W.E. & Potts, D.M. 2004. Trial excavation for cut and cover tunnel construction in glacial till – a case study from Dublin. Submitted for *Skempton Memorial Conference London 2004*.

Nyren R.J., Standing J.R. & Burland J.B. 2002. Surface displacements at St James's Park greenfield reference site above twin tunnels through the London Clay. In Burland et al. (eds), *Building response to tunnelling. Case studies from the Jubilee Line Extension, London*: (2)387–400. London: CIRIA & Thomas Telford.

Parker, E.J., Jardine, R.J., Standing, J.R. & Xavier, J. 1999. Jet grouting to improve offshore pile capacity. *Offshore Technology Conference, Houston, OTC 10828*.

Pellew, A. 2002. Field investigations into the behaviour of piles in clay. *PhD Thesis*, Imperial College, University of London.

Potts, D.M., Kovacevic, N. & Vaughan, P.R. 1997. Delayed collapse of cut slopes in stiff clay. *Geotechnique* 47(5): 953–982.

Potts, D.M. & Zdravkovic, L. 1999. *Finite element analysis in geotechnical engineering: theory*. London: Thomas Telford.

Potts, D.M. & Zdravkovic, L. 2001. *Finite element analysis in geotechnical engineering: application*. London: Thomas Telford.

Potts, D.M. 2003. Numerical Analysis: a virtual dream or practical reality? The 42nd Rankine Lecture. *Geotechnique* 53(6): 535–573.

Roesler, S.K. 1979. Anisotropic shear modulus due to stress anisotropy. *Journ Geot Eng, ASCE* 105(7): 871–880

St John, H.D., Potts, D.M., Jardine, R.J. & Higgins, K.G. 1992. Prediction and performance of ground response due to construction of a deep basement at 60 Victoria Embankment. *Wroth Memorial Symposium on Predictive Soil Mechanics* 581–608. Oxford. Thomas Telford.

Schmertmann, J.H. 1970. Static cone to compute settlement over sand. *ASCE JSMFE Div.*: (96) 1011–1043.

Shibuya, S., Mitachi, T. & Miura, N. (eds). 1995. *Pre-failure Deformation of Geomaterials. Proceedings of TC-29 International Symposium*. Hokkaido: Balkema.

Shibuya, S. & Tamrakar, S.B. 1999. In-situ and laboratory investigations into engineering properties of Bangkok clay. *Proc. Int. Symp. Characterisation of Soft Marine Clays – Bothkennar, Drammen, Quebec and Ariake Clays*.: (1) 107–132. Yokosuka: Balkema.

Shibuya, S., Tamrakar, S.B. & Mitachi, T. 2001. Deep excavation in Bangkok – characterisation, measurement and prediction. In Addachi et al. (eds), *Proc. Conf. Modern Tunneling Scienec and Technology*: 243–248.

Simpson, B. 1992. 32nd Rankine Lecture. Retaining structures – displacement and design. *Geotechnique* 42(4): 539–576.

Simpson, B. 2002. Engineering needs. In Jamiolkowski, Lancellotta & Lo-Presti (eds), *Pre-failure Deformation Characteristics of Geomaterial*: 1011–1026. Torino: Swets & Zeitlinger.

Smith P.R., Jardine, R.J. & Hight, D.W. 1992. On the yielding of Bothkennar clay. *Geotechnique*, 42(2): 257–274.

Stallebrass, S.E. & Taylor, R.N. 1997. The devlopment and evaluation of a constitutive model for the prediction of ground movements in overconsolidated clay. *Geotechnique* 47(2): 235–253.

Standing J.R., Nyren, R.J., Burland J.B. & Longworth T.I. 1996. The measurement of ground movements due to tunnelling at two control sites along the Jubilee Line Extension. In Mair R.J. & Taylor R.N. (eds), *Proc. Int. Symp. Geotechnical Aspects of Underground Construction in Soft Ground*: 751–756. London: Balkema.

Standing, J.R. & Jardine, R.J. 1999. Grout testing for the GOPAL Project. *Internal report*, Imperial College, University of London.

Standing, J.R. & Burland, J.B. 1999. Ground characterisation to explain JLEP tunnelling volume losses in the Westminster area. *Report*, Imperial College, University of London.

Standing, J.R. & Burland, J.B. 2003. Explanations for variations in the JLE tunnelling volume losses in the Westminster area. Submitted to *Geotechnique*.

Tan et al. (eds) 2002. Characterisation and Engineering Properties of Natural Soils. *International Symposium*. Volumes 1 & 2. Singapore: Balkema.

Tatsuoka, Jardine, Lo Presti, Di Benedetto & Kodaka. 1999. Characterising the Pre-Failure Behaviour of Geomaterials. Theme Lecture, *Proc 14th ICSMGE* (4): 2129–2164. Hamburg: Balkema.

Tatsuoka, F., Uchimura, T., Hayano, K., Di Benedetto, H., Koseki, J. & Siddiquee, M.S.A. 2001. Time dependent deformation characteristics of stiff geomaterials in engineering practice. In Jamiolkowski, Lancellotta & Lo-Presti (eds), *Pre-failure Deformation Characteristics of Geomaterial*: 1161–1250. Torino: Swets & Zeitlinger.

Tor-Petersen, G., Zdravkovic, L., Potts, D.M. & St. John, H.D. 2003. The prediction of ground movements associated with the construction of deep station boxes. In *Claiming the Underground space*: 1051–1058. Swets & Zeitlinger.

Viggiani G. & Standing J.R. 2002. The Treasury. Building response to tunnelling. In Burland et al. (eds), *Case studies from the Jubilee Line Extension, London*, Burland J.B., Standing J.R. & Jardine F.M.: (2) 401–432. London: CIRIA & Thomas Telford.

Wardle, I.F., Price, G. & Freeman, T.J. 1992. Effect of time and maintained load on the ultimate capacity of piles in stiff clay. In Sand M.J. (ed), *Piling European practice and worldwide trend*: 92–99.

Zdravkovic L. & Jardine, R.J. 2001. The effects on anisotropy of principal stress rotation during consolidation, *Geotechnique* 51(1): 69–83.

Zdravkovic, L., Potts, D.M. & Jackson, C. 2003. A numerical study of the effect of preloading on undrained bearing capacity. *Int. J. Numerical Methods in Geomechanics, ASCE*, September: 1–10.

Whittle, A.J. & Hashash, Y.M.A. 1994. Soil modelling and prediction of deep excavation behaviour. In Shibuya, S., Mitachi, K. & Miura, N. (eds). 1995. *Pre-failure Deformation of Geomaterials. Proceedings of TC-29 International Symposium*: (1)589–594. Hokkaido: Balkema.

Author index